Multimodality Imaging

Qifa Zhou · Zhongping Chen
Editors

Multimodality Imaging

For Intravascular Application

Springer

Editors
Qifa Zhou
Department of Ophthalmology
and Biomedical Engineering
University of Southern California
Los Angeles, CA, USA

Zhongping Chen
Department of Biomedical Engineering
University of California, Irvine
Irvine, CA, USA

ISBN 978-981-10-6306-0 ISBN 978-981-10-6307-7 (eBook)
https://doi.org/10.1007/978-981-10-6307-7

This Springer imprint is published by the registered company Springer Nature Singapore Pte Ltd.
The registered company address is: 152 Beach Road, #21-01/04 Gateway East, Singapore 189721, Singapore

Preface

Atherosclerosis is a progressive disease that is characterized by the accumulation of lipids, cholesterol, fibrous constituents, monocytes, and various other inflammatory cells in the arterial wall. Atherosclerosis is one of the major causes of morbidity and mortality in developed countries. Mortality from heart attack (86%) is mainly due to "vulnerable plaques" which rupture suddenly and trigger a blood clot or thrombus that blocks blood flow. Early detection of plaque lesions is the first and necessary step in preventing the lethal consequences of atherosclerosis. Diagnosis of the latent vulnerability of a plaque lesion relies on both tissue structural and chemical compositions. Multimodality intravascular imaging that can provide both structural and molecular information will provide clinicians with a critically important tool for diagnosing vulnerable plaques, monitoring the progression of disease, and evaluating the efficacy of intervention.

We have selected the top experts in the field as chapter authors, many of whom have worked continuously on multimodality intravascular imaging since their Ph.D. work. Therefore, this book will cover recent research progress on integrated multimodal intravascular imaging systems which combine intravascular ultrasound (IVUS), optical coherence tomography (OCT), intravascular photoacoustic imaging (IVPA), fluorescence life imaging (both system and contrast), and therapeutic IVUS.

In this book, we will first introduce the multimodality intravascular imaging (Chap. 1) and the integration of multi-frequency intravascular ultrasound (Chap. 2), and then, we will introduce the integration of IVUS and OCT (Chap. 3). In Chap. 4, intravascular photoacoustic imaging of lipid-laden plaques will be introduced. Following in Chap. 5, we will introduce contrast-enhanced dual-frequency superharmonic intravascular ultrasound (IVUS) imaging. We will introduce dual-modality fluorescence lifetime and intravascular ultrasound for label-free intravascular coronary imaging and intravascular dual-modality imaging (NIRF/IVUS, NIRS/IVUS, IVOCT/NIRF, and IVOCT/NIRS) in Chaps. 6 and 7, respectively. In Chaps. 8 and 9, tri-modality intravascular imaging systems and acoustic radiation force optical coherence elastography will be reported, respectively. Finally, therapeutic IVUS and contrast imaging as well as high-resolution intravascular ultrasound imaging systems will be presented in Chaps. 10 and 11, respectively.

We greatly appreciate all the authors and laboratory members who gave their time and contributed significant research work for this compilation. Without their help, this book would not have reached fruition. We also wish to acknowledge the work of Dr. Ruimin Chen who contributed the editorial help for this book. We hope that this book will help biomedical engineers as well as clinicians.

Los Angeles, USA Qifa Zhou
Irvine, USA Zhongping Chen

Contents

Contributors

Julien Bec Department of Biomedical Engineering, University of California, Davis, Davis, CA, USA

Zhongping Chen Department of Biomedical Engineering, Beckman Laser Institute, University of California, Irvine, Irvine, CA, USA

Ji-Xin Cheng Department of Electrical and Computer Engineering, Department of Biomedical Engineering, Photonics Center, Boston University, Boston, MA, USA

Youmin He Department of Biomedical Engineering, Beckman Laser Institute, University of California, Irvine, Irvine, CA, USA

John A. Hossack Department of Biomedical Engineering, University of Virginia, Charlottesville, VA, USA

Jie Hui Department of Electrical and Computer Engineering, Photonics Center, Boston University, Boston, MA, USA

Xiaoning Jiang Department of Mechanical and Aerospace Engineering, North Carolina State University, Raleigh, NC, USA

Jiawen Li Adelaide Medical School, Australian Research Council Centre of Excellence for Nanoscale Biophotonics, Institute for Photonics and Advanced Sensing, The University of Adelaide, Adelaide, SA, Australia

Yan Li Department of Biomedical Engineering, Beckman Laser Institute, University of California, Irvine, Irvine, CA, USA

Jianguo Ma School of Instrumentation and Optoelectronic Engineering, Beijing Advanced Innovation Center for Big Data-Based Precision Medicine, Beihang University, Beijing, China

Teng Ma Paul C. Lauterbur Research Center for Biomedical Imaging, Institute of Biomedical and Health Engineering, Shenzhen Institutes of Advanced Technology, Chinese Academy of Sciences, Shenzhen, China

Laura Marcu Department of Biomedical Engineering, University of California, Davis, Davis, CA, USA

Jennifer E. Phipps Department of Biomedical Engineering, University of California, Davis, Davis, CA, USA

Weibao Qiu Paul C. Lauterbur Research Center for Biomedical Imaging, Institute of Biomedical and Health Engineering, Shenzhen Institutes of Advanced Technology, Chinese Academy of Sciences, Shenzhen, China

Yueqiao Qu Department of Biomedical Engineering, Beckman Laser Institute, University of California, Irvine, Irvine, CA, USA

Hairong Zheng Paul C. Lauterbur Research Center for Biomedical Imaging, Institute of Biomedical and Health Engineering, Shenzhen Institutes of Advanced Technology, Chinese Academy of Sciences, Shenzhen, China

Qifa Zhou Department of Biomedical Engineering, Roski Eye Institute, University of Southern California, Los Angeles, CA, USA

Chapter 1
Introduction to Multimodality Intravascular Imaging

Zhongping Chen and Qifa Zhou

Atherosclerosis is a progressive disease that is characterized by the accumulation of lipids, cholesterol, fibrous constituents, monocytes, and various other inflammatory cells in the arterial wall. These deposits form vascular lesions known as atheromatous plaques, which contain necrotic cores and are separated from the arterial intima by a fibrous cap composed of collagen and smooth muscle cells (Narula and Strauss 2005; Virmani et al. 2005b). Upon plaque maturation, the fibrous caps become thin and increasingly susceptible to tearing, which increases the vulnerability to plaque rupture. Rupture of these vulnerable plaques releases the inflammatory elements of the necrotic core into the artery, causing thrombosis. This leakage may lead to obstruction of arterial blood flow and angina and/or myocardial infarction, which can be lethal (Marcu et al. 2005). Atherosclerosis is one of the major causes of morbidity and mortality in developed countries. The major cause of deaths from heart attacks (86%) and brain aneurysms (45%) is due to "vulnerable plaques" that rupture suddenly and trigger a blood clot or thrombus that blocks blood flow (Narula and Strauss 2007; Weber and Noels 2011; Virmani et al. 2005a; Narula and Strauss 2005).

Intravascular imaging techniques that enable early detection and classification of vulnerable plaque segments are essential to understand, diagnose, and manage vascular diseases. Although the understanding of vulnerable plaques is still at an early stage, previous research based on pathological studies has demonstrated that a plaque's stability is strongly affected by the plaque's morphology and tissue chemical composition (Virmani et al. 2005a; Narula and Strauss 2005; Puri et al. 2011; Moreno

Z. Chen
Beckman Laser Institute, University of California, Irvine, Irvine, CA 92697, USA
e-mail: z2chen@uci.edu

Q. Zhou (✉)
Roski Eye Institute, University of Southern California, Los Angeles, CA 90033, USA
e-mail: qifazhou@usc.edu

Department of Biomedical Engineering, University of Southern California, Los Angeles, CA 90089, USA

Q. Zhou and Z. Chen (eds.), *Multimodality Imaging*,
https://doi.org/10.1007/978-981-10-6307-7_1

et al. 2002; Kolodgie et al. 2001). Structurally, the thickness of the fibrous cap is a reliable indicator of plaque vulnerability (Sawada et al. 2008; Puri et al. 2011). Chemically, the intra-lesion lipid density and the cholesterol content are important parameters that correlate with the vulnerability of the lesion (Waxman et al. 2006; Puri et al. 2011; Yoo et al. 2011). Therefore, an optimal imaging modality for diagnosis and characterization of plaques should combine high spatial resolution capable of resolving fibrous cap thickness, deep imaging depth capable of assessing plaque burden and vessel remodeling, and molecular sensitivity capable of determining tissue composition (Puri et al. 2011; Guo et al. 2018).

Many biomedical imaging techniques aimed at imaging and assessing vulnerable plaques have been reported in the literature (Puri et al. 2011; Li and Chen 2018; Abran et al. 2015; Piao et al. 2015; Cao et al. 2016). Intravascular ultrasound (IVUS) and optical coherence tomography (OCT) are currently the two most commonly used modalities in the clinic for diagnosing cardiovascular diseases which allow direct tomographic visualization of cross-sectional images from inside the vessel lumen (Potkin et al. 1990; Landini and Verrazzani 1990; Huang et al. 1991; Tearney et al. 2006; Puri et al. 2011). IVUS is a catheter-based technique that provides high-resolution, cross-sectional images of the coronary vessel in vivo. In daily clinical practice, IVUS is increasingly being used for the visualization of coronary lumen, vessel wall, and atherosclerotic plaque formation (Nissen and Yock 2001; Iida and Mano 2019; Gomez-Lara et al. 2016). Although current IVUS has limited resolution and sensitivity to assess the thickness of the thin fibrous cap and for plaque classifications (Sawada et al. 2008; Puri et al. 2011), recent work in IVUS backscattering analysis demonstrates the feasibility and limitation of using IVUS to characterize specific lesions and identify plaques that lead to various clinical syndromes (Mintz and Weissman 2006; Bermejo et al. 1998; Hanekamp et al. 1999).

In recent years, significant progress has been made in the development of optical diagnostics for cardiovascular diseases. In particular, intravascular OCT (IVOCT), a technique sensitive to structural density variations in the arterial wall, was clinically proven to be a sensitive method for determining the thickness of the fibrous cap (Cilingiroglu et al. 2006; Tearney et al. 2006). Intravascular OCT has been demonstrated by several groups for imaging and evaluation of vulnerable plaques (Fujimoto 2003; Yun et al. 2006; Brezinski et al. 1996; Jang et al. 2002, 2005; Fujimoto et al. 1995; Brezinski 2007, 2006; Raffel et al. 2008; Li et al. 2017a). Although OCT has limited imaging depth and cannot image the full depth of a large lipid pool in plaques (Puri et al. 2011; Sawada et al. 2008), it has been used for vulnerable plaque evaluation and is capable of measuring microscopic features with high spatial resolution.

Both IVUS and IVOCT provide structural information regarding the arterial wall but lack molecular specificity for identification of plaque composition. Near-infrared reflectance spectroscopy (NIRS) has been used to characterize the intra-lesion lipid content and is currently under investigation in large-scale clinical studies (Moreno et al. 2002; Wang et al. 2002; Negi et al. 2015). In addition, near-infrared fluorescence (NIRF) imaging utilizes molecular probes or autofluorescence to provide complementary information with regard to plaque activity and inflammation (Giovanni et al. 2016; Lee et al. 2014; Abran et al. 2015). Although both NIRS and NIRF lack the depth resolution to generate cross-sectional image mapping of tissue composition, they provide molecular contrast to characterize plaque lesions.

Photoacoustic tomography (PAT) is an emerging biomedical imaging modality that has the advantage of providing optical absorption contrast at ultrasound resolution (Wang et al. 2003, 2010, 2011, 2012b; Brecht et al. 2009; Yang et al. 2009; Sethuraman et al. 2007a; Jansen et al. 2011; Wei et al. 2011; Hui et al. 2017; Li and Chen 2018; Cao et al. 2016; Li et al. 2015b; Jansen et al. 2014). PAT detects acoustic waves generated by the absorption of pulsed light in tissue (Wang 2009). Several groups have shown that PAT can be used to image and identify intima, media, and adventitia of a vascular wall based on different absorption coefficients of these tissues (Sethuraman et al. 2007b; Wei et al. 2011). In addition, PAT can identify different constituents of fibro-cellular inflammatory plaque. Because lipid has a distinct absorption spectrum in the NIR wavelength range, several groups have investigated spectroscopic intravascular imaging to detect the presence of lipid in atherosclerotic plaque (Wang et al. 2010, 2011, 2012b Sethuraman et al. 2007a; Jansen et al. 2011; Li et al. 2015b; Jansen et al. 2014). The enhanced absorption peak of lipids due to the first overtone of CH vibration near 1730 nm and the second overtone of the CH bond stretch near 1200 nm has been identified by several groups for imaging and mapping of lipids in an atherosclerotic lesion (Wang et al. 2010, 2011, 2012b; Sethuraman et al. 2007a; Jansen et al. 2011; Piao et al. 2015; Wu et al. 2016; Hui et al. 2017). Although miniature probes have been developed and intravascular imaging of atherosclerotic specimens from cadaver and animal models has been demonstrated, clinical translation of this technology is still in the early stage.

In addition, label-free optical techniques, such as second harmonic generation (SHG) imaging of collagen, two-photon excited fluorescence (TPEF) imaging of elastin, CARS imaging of lipids, and optical coherence elastography (OCE) imaging of tissue elasticity, have not yet reached the stage of clinical studies but have shown great potential for atherosclerotic research (Campagnola et al. 2002; Lilledahl et al. 2007; Wang et al. 2008, 2009, 2012a; Zoumi et al. 2004; Jansen et al. 2011; Wei et al. 2011; Qu et al. 2017).

Unfortunately, atherosclerosis exhibits an asymptomatic nature, as vulnerable plaques grow without causing any detrimental side effects until rupturing (Narula and Strauss 2005). Due to this complication, the information provided by a single clinical arterial imaging technique is often insufficient to diagnose vulnerable plaque formation at an early stage. Integration of several modalities is necessary to gather the information required to establish a robust method for early detection of plaque vulnerability. Several multimodality imaging techniques that provide complementary information have been developed. We have developed an integrated OCT/US system for intravascular imaging applications (Yin et al. 2010, 2011; Li et al. 2010, 2014, 2015a). Furthermore, integration of intravascular OCT and fluorescence imaging as well as integrated PAT and US has been reported by a number of groups (Yoo et al. 2011; Liang et al. 2012; Wang et al. 2010, 2011, 2012a, b; Sethuraman et al. 2007a; Jansen et al. 2011; Wei et al. 2011; Piao et al. 2015). Furthermore, integrated NIRS/IVUS, NIRF/IVUS, IVOCT/NIRS, and IVOCT/NIRF imaging systems have also been demonstrated and translated to clinical imaging (Roleder et al. 2014; Fard et al. 2013; Lee et al. 2014; Abran et al. 2015). Finally, multimodality imaging that integrated three or more imaging systems has also been reported (Yang et al. 2011; Liang et al. 2014; Abran et al. 2014; Li et al. 2017b).

This book will cover recent research progress on the integrated multimodal intravascular imaging systems that combine IVUS, OCT, PAT, NIRF, NIRS OCE, and fluorescence lifetime imaging, etc., as well as their clinical applications for imaging and characterizing atherosclerosis. In addition, therapeutic IVUS and contrast imaging are also included.

The clinical need and value for an imaging system that can identify patients with vulnerable plaques with a high risk of rupture have been discussed extensively in the literature (Braunwald 2006; Kusters et al. 2012; Puri et al. 2011; Suri et al. 2011) and also highlighted in the NIH/NHLBI Working Group Report on Detection of High-Risk Atherosclerotic Plaque (Narula and Dilsizian 2008). Currently, there is no single imaging modality that can reliably identify vulnerable plaque or predict late occlusion after drug-eluting stent placement (Brezinski 2012; Kusters et al. 2012; Puri et al. 2011). Several interventional procedures to treat vulnerable plaques at high risk of rupture are under clinical trials (Meier 2004; Wykrzykowska et al. 2012; Kereiakes et al. 2003). The widespread clinical application of these measures requires improved risk stratification of vulnerable plaque with a better predictive power (Narula and Dilsizian 2008; Oberhoff and Karsch 2003). The ability to detect these vulnerable plaques noninvasively is likely to serve as a powerful stimulus for increased effort in the development of such therapies (Meier 2004; Wykrzykowska et al. 2012; Kereiakes et al. 2003). An integrated intravascular imaging modality that can detect and characterize vulnerable plaques will provide a critically important tool for monitoring the progression of disease and evaluating the efficacy of intervention.

Acknowledgments We would like to thank many of our colleagues who have contributed to the multimodality intravascular projects at the Beckman Laser Institute and the Department of Biomedical Engineering at UCI, and the Department of Biomedical Engineering at USC. We would like to acknowledge the research grants awarded from the National Institutes of Health (R01EB-10090, R01HL125084, and R01HL127271). Please address all correspondence to Dr. Z. Chen (z2chen@uci.edu), who first proposed and initiated the research project on integrated OCT/US for intravascular imaging and wrote this introduction chapter. Dr. Z. Chen has a financial interest in OCT Medical Imaging Inc., which, however, did not support this work.

References

Abran M, Cloutier G, Cardinal MH, Chayer B, Tardif JC, Lesage F (2014) Development of a photoacoustic, ultrasound and fluorescence imaging catheter for the study of atherosclerotic plaque. IEEE Trans Biomed Circuits Syst 8(5):696–703. https://doi.org/10.1109/TBCAS.2014.2360560

Abran M, Stahli BE, Merlet N, Mihalache-Avram T, Mecteau M, Rheaume E, Busseuil D, Tardif JC, Lesage F (2015) Validating a bimodal intravascular ultrasound (IVUS) and near-infrared fluorescence (NIRF) catheter for atherosclerotic plaque detection in rabbits. Biomed Opt Express 6(10):3989–3999. https://doi.org/10.1364/BOE.6.003989

Bermejo J, Botas J, Garcia E, Elizaga J, Osende J, Soriano J, Abeytua M, Delcan JL (1998) Mechanisms of residual lumen stenosis after high-pressure stent implantation: a quantitative coronary angiography and intravascular ultrasound study. Circulation 98:112–118

Braunwald E (2006) Epilogue: what do clinicians expect from imagers? J Am Coll Cardiol 47(8 Suppl):C101–C103. https://doi.org/10.1016/j.jacc.2005.10.072. S0735-1097(06)00163-X [pii]

Brecht HP, Su R, Fronheiser M, Ermilov SA, Conjusteau A, Oraevsky AA (2009) Whole-body three-dimensional optoacoustic tomography system for small animals. J Biomed Opt 14(6):064007. https://doi.org/10.1117/1.3259361

Brezinski ME (2006) Optical coherence tomography for identifying unstable coronary plaque. Int J Cardiol 107(2):154–165. https://doi.org/10.1016/j.ijcard.2005.07.066. S0167-5273(05)01071-5 [pii]

Brezinski ME (2007) Applications of optical coherence tomography to cardiac and musculoskeletal diseases: bench to bedside? J Biomed Opt 12(5):051705. https://doi.org/10.1117/1.2795689

Brezinski ME (2012) Current capabilities and challenges for optical coherence tomography as a high-impact cardiovascular imaging modality. Circulation 123(25):2913–2915. https://doi.org/10.1161/CIRCULATIONAHA.111.034272. 123/25/2913 [pii]

Brezinski ME, Tearney GJ, Bouma BE, Izatt JA, Hee MR, Swanson EA, Southern JF, Fujimoto JG (1996) Optical coherence tomography for optical biopsy. Properties and demonstration of vascular pathology. Circulation 93(6):1206–1213

Campagnola PJ, Millard AC, Terasaki M, Hoppe PE, Malone CJ, Mohler WA (2002) Three-dimensional high-resolution second-harmonic generation imaging of endogenous structural proteins in biological tissues. Biophys J 82(1):493–508

Cao Y, Hui J, Kole A, Wang P, Yu Q, Chen W, Sturek M, Cheng JX (2016) High-sensitivity intravascular photoacoustic imaging of lipid-laden plaque with a collinear catheter design. Sci Rep 6:25236. https://doi.org/10.1038/srep25236

Cilingiroglu M, Oh JH, Sugunan B, Kemp NJ, Kim J, Lee S, Zaatari HN, Escobedo D, Thompson S, Milner TE, Feldman MD (2006) Detection of vulnerable plaque in a murine model of athereosclerosis with optical coherence tomography. Catheter Cardiovasc Interv 67(6):915–923

Fard AM, Vacas-Jacques P, Hamidi E, Wang H, Carruth RW, Gardecki JA, Tearney GJ (2013) Optical coherence tomography—near infrared spectroscopy system and catheter for intravascular imaging. Opt Express 21(25):30849–30858. https://doi.org/10.1364/Oe.21.030849

Fujimoto JG (2003) Optical coherence tomography for ultrahigh resolution in vivo imaging. Nat Biotechnology 21:1361–1367

Fujimoto JG, Brezinski ME, Tearney GJ, Boppart SA, Bouma B, Hee MR, Southern JF, Swanson EA (1995) Optical biopsy and imaging using optical coherence tomography. Nature Med 1:970–972

Giovanni J, Ughi P, Wang H, Gerbaud E, Gardecki JA, Fard AM, Hamidi E, Vacas-Jacques P, Rosenberg M, Jaffer FA, Tearney GJ (2016) First-in-human dual-modality OCT and near-infrared autofluorescence imaging of coronary artery disease. JACC Cardiovasc Imaging 9(11):1304–1314

Gomez-Lara J, Salvatella N, Gonzalo N, Hernández-Hernández F, Fernandez-Nofrerias E, Sánchez-Recalde A, Bastante T, Marcano A, Romaguera R, Ferreiro JL, Roura G, Teruel L, Ariza-Solé A, Miranda-Guardiola F, Rodríguez García-Abad V, Gomez-Hospital JA, Alfonso F, Cequier A (2016) IVUS-guided treatment strategies for definite late and very late stent thrombosis. EuroIntervention 12(11):e1355–e1365. https://doi.org/10.4244/EIJY15M12_08

Guo X, Giddens DP, Molony D, Yang C, Samady H, Zheng J, Mintz GS, Maehara A, Wang L, Pei X, Li ZY, Tang D (2018) Combining IVUS and optical coherence tomography for more accurate coronary cap thickness quantification and stress/strain calculations: a patient-specific three-dimensional fluid-structure interaction modeling approach. J Biomech Eng 140(4). https://doi.org/10.1115/1.4038263

Hanekamp C, Koolen J, Pijls J, Michels H, Bonnier H (1999) Comparison of quantitative coronary angiography, intravascular ultrasound, and coronary pressure measurement to assess optimum stent deployment. Circulation 99:1015–1021

Huang D, Swanson EA, Lin CP, Schuman JS, Stinson WG, Chang W, Hee MR, Flotte T, Gregory K, Puliafito CA et al (1991) Optical coherence tomography. Science 254(5035):1178–1181

Hui J, Cao Y, Zhang Y, Kole A, Wang P, Yu G, Eakins G, Sturek M, Chen W, Cheng JX (2017) Real-time intravascular photoacoustic-ultrasound imaging of lipid-laden plaque in human coronary artery at 16 frames per second. Sci Rep 7(1):1417. https://doi.org/10.1038/s41598-017-01649-9

Iida O, Mano T (2019) Role of IVUS in the endovascular treatment of calcified femoropopliteal lesions. J Endovasc Ther. https://doi.org/10.1177/1526602819838991

Jang IK, Bouma BE, Kang DH, Park SJ, Park SW, Seung KB, Choi KB, Shishkov M, Schlendorf K, Pomerantsev E, Houser SL, Aretz HT, Tearney GJ (2002) Visualization of coronary atherosclerotic plaques in patients using optical coherence tomography: comparison with intravascular ultrasound. J Am Coll Cardiol 39:604–609

Jang IK, Tearney GJ, MacNeill B, Takano M, Moselewski F, Ftima N, Shishkov M, Houser S, Aretz HT, Halpern EF, Bouma BE (2005) In vivo characterization of coronary atherosclerotic plaque by use of optical coherence tomography. Circulation 111:1551–1555

Jansen K, van der Steen AF, van Beusekom HM, Oosterhuis JW, van Soest G (2011) Intravascular photoacoustic imaging of human coronary atherosclerosis. Opt Lett 36(5):597–599. 210116 [pii]

Jansen K, Wu M, van der Steen AF, van Soest G (2014) Photoacoustic imaging of human coronary atherosclerosis in two spectral bands. Photoacoustics 2(1):12–20. https://doi.org/10.1016/j.pacs.2013.11.003

Kereiakes DJ, Szyniszewski AM, Wahr D, Herrmann HC, Simon DI, Rogers C, Kramer P, Shear W, Yeung AC, Shunk KA, Chou TM, Popma J, Fitzgerald P, Carroll TE, Forer D, Adelman DC (2003) Phase I drug and light dose-escalation trial of motexafin lutetium and far red light activation (phototherapy) in subjects with coronary artery disease undergoing percutaneous coronary intervention and stent deployment: procedural and long-term results. Circulation 108(11):1310–1315. https://doi.org/10.1161/01.CIR.0000087602.91755.19. 01.CIR.0000087602.91755.19 [pii]

Kolodgie FD, Burke AP, Farb A, Gold HK, Yuan J, Narula J, Finn AV, Virmani R (2001) The thin-cap fibroatheroma: a type of vulnerable plaque: the major precursor lesion to acute coronary syndromes. Curr Opin Cardiol 16(5):285–292

Kusters DH, Tegtmeier J, Schurgers LJ, Reutelingsperger CP (2012) Molecular imaging to identify the vulnerable plaque–from basic research to clinical practice. Mol Imaging Biol 14(5):523–533. https://doi.org/10.1007/s11307-012-0586-7

Landini L, Verrazzani L (1990) Spectral characterization of tissues microstructure by ultrasounds: a stochastic approach. IEEE Trans Ultrason Ferroelectr Freq Control 37(5):448–456. https://doi.org/10.1109/58.105251

Lee S, Lee MW, Cho HS, Song JW, Nam HS, Oh DJ, Park K, Oh WY, Yoo H, Kim JW (2014) Fully integrated high-speed intravascular optical coherence tomography/near-infrared fluorescence structural/molecular imaging in vivo using a clinically available near-infrared fluorescence-emitting indocyanine green to detect inflamed lipid-rich atheromata in coronary-sized vessels. Circ Cardiovasc Interv 7(4):560–569. https://doi.org/10.1161/CIRCINTERVENTIONS.114.001498

Li Y, Chen Z (2018) Multimodal intravascular photoacoustic and ultrasound imaging. Biomed Eng Lett 8(2):193–201. https://doi.org/10.1007/s13534-018-0061-8

Li X, Yin J, Hu C, Zhou Q, Shung KK, Chen Z (2010) High-resolution coregistered intravascular imaging with integrated ultrasound and optical coherence tomography probe. Appl Phys Lett 97(13):133702. https://doi.org/10.1063/1.3493659

Li JW, Li X, Mohar D, Raney A, Jing J, Zhang J, Johnston A, Liang SS, Ma T, Shung KK, Mahon S, Brenner M, Narula J, Zhou QF, Patel PM, Chen ZP (2014) Integrated IVUS-OCT for real-time imaging of coronary atherosclerosis. JACC Cardiovasc Imaging 7(1):101–103. https://doi.org/10.1016/J.Jcmg.2013.07.012

Li J, Ma T, Mohar D, Steward E, Yu M, Piao Z, He Y, Shung KK, Zhou Q, Patel PM, Chen Z (2015a) Ultrafast optical-ultrasonic system and miniaturized catheter for imaging and characterizing atherosclerotic plaques in vivo. Sci Rep 5:18406. https://doi.org/10.1038/srep18406

Li Y, Gong X, Liu C, Lin R, Hau W, Bai X, Song L (2015b) High-speed intravascular spectroscopic photoacoustic imaging at 1000 A-lines per second with a 0.9-mm diameter catheter. J Biomed Opt 20(6):065006. https://doi.org/10.1117/1.jbo.20.6.065006

Li Y, Jing J, Heidari E, Zhu J, Qu Y, Chen Z (2017a) Intravascular optical coherence tomography for characterization of atherosclerosis with a 1.7 micron swept-source laser. Sci Rep 7(1):14525. https://doi.org/10.1038/s41598-017-15326-4

Li Y, Jing J, Qu Y, Miao Y, Zhang B, Ma T, Yu M, Zhou Q, Chen Z (2017b) Fully integrated optical coherence tomography, ultrasound, and indocyanine green-based fluorescence tri-modality system for intravascular imaging. Biomed Opt Express 8(2):1036–1044. https://doi.org/10.1364/BOE.8.001036

Liang S, Saidi A, Jing J, Liu G, Li J, Zhang J, Sun C, Narula J, Chen Z (2012) Intravascular atherosclerotic imaging with combined fluorescence and optical coherence tomography probe based on a double-clad fiber combiner. J Biomed Opt 17(7):070501. https://doi.org/10.1117/1.JBO.17.7.070501

Liang S, Ma T, Jing J, Li X, Li J, Shung KK, Zhou Q, Zhang J, Chen Z (2014) Trimodality imaging system and intravascular endoscopic probe: combined optical coherence tomography, fluorescence imaging and ultrasound imaging. Opt Lett 39(23):6652–6655. https://doi.org/10.1364/OL.39.006652

Lilledahl MB, Haugen OA, Lange-Davies Cd, Svaasand LO (2007) Characterization of vulnerable plaques by multiphoton microscopy. J Biomed Opt 12(4):044005

Marcu L, Fang Q, Jo J, Papaioannou T, Dorafshar A, Reil T, Qiao J, Baker J, Freischlag J, Fishbein M (2005) In vivo detection of macrophages in a rabbit atherosclerotic model by time-resolved laser-induced fluorescence spectroscopy. Atherosclerosis 181(2):295–303

Meier B (2004) Plaque sealing by coronary angioplasty. Heart 90(12):1395–1398. https://doi.org/10.1136/hrt.2004.034983. 90/12/1395 [pii]

Mintz GS, Weissman NJ (2006) Intravascular ultrasound in the drug-eluting Stent Era. J Am Coll Cardiol 48:421

Moreno PR, Lodder RA, Purushothaman KR, Charash WE, O'Connor WN, Muller JE (2002) Detection of lipid pool, thin fibrous cap, and inflammatory cells in human aortic atherosclerotic plaques by near-infrared spectroscopy. Circulation 105(8):923–927

Narula J, Dilsizian V (2008) From better understood pathogenesis to superior molecular imaging, and back. JACC Cardiovasc Imaging 1(3):406–409. https://doi.org/10.1016/j.jcmg.2008.02.002. S1936-878X(08)00028-4 [pii]

Narula J, Strauss HW (2005) Imaging of unstable atherosclerotic lesions. Eur J Nucl Med Mol Imaging 32(1):1–5. https://doi.org/10.1007/s00259-004-1580-3

Narula J, Strauss WH (2007) The popcorn plaques. Nat Med 13:532–534

Negi SI, Didier R, Ota H, Magalhaes MA, Popma CJ, Kollmer MR, Spad MA, Torguson R, Suddath W, Satler LF, Pichard A, Waksman R (2015) Role of near-infrared spectroscopy in intravascular coronary imaging. Cardiovasc Revasc Med 16(5):299–305. https://doi.org/10.1016/j.carrev.2015.06.001

Nissen SE, Yock P (2001) Intravascular ultrasound: novel pathophysiological insights and current clinical applications. Circulation 103:604–616

Oberhoff M, Karsch KR (2003) Who wants his plaque sealed? Eur Heart J 24(6):494–495. S0195668X02008187 [pii]

Piao Z, Ma T, Li J, Wiedmann MT, Huang S, Yu M, Kirk Shung K, Zhou Q, Kim CS, Chen Z (2015) High speed intravascular photoacoustic imaging with fast optical parametric oscillator laser at 1.7 m. Appl Phys Lett 107(8):083701. https://doi.org/10.1063/1.4929584

Potkin BN, Bartorelli AL, Gessert JM, Neville RF, Almagor Y, Roberts WC, Leon MB (1990) Coronary artery imaging with intravascular high-frequency ultrasound. Circulation 81(5):1575–1585

Puri R, Worthley MI, Nicholls SJ (2011) Intravascular imaging of vulnerable coronary plaque: current and future concepts. Nat Rev Cardiol 8(3):131–139. https://doi.org/10.1038/nrcardio.2010.210. nrcardio.2010.210 [pii]

Qu Y, Ma T, He Y, Yu M, Zhu J, Miao Y, Dai C, Patel P, Shung KK, Zhou Q, Chen Z (2017) Miniature probe for mapping mechanical properties of vascular lesions using acoustic radiation force optical coherence elastography. Sci Rep 7(1):4731. https://doi.org/10.1038/s41598-017-05077-7

Raffel OC, Merchant FM, Tearney GJ, Chia S, Gauthier DD, Pomerantsev E, Mizuno K, Bouma BE, Jang IK (2008) In vivo association between positive coronary artery remodelling and coronary plaque characteristics assessed by intravascular optical coherence tomography. Eur Heart J 29(14):1721–1728. https://doi.org/10.1093/eurheartj/ehn286. ehn286 [pii]

Roleder T, Kovacic JC, Ali Z, Sharma R, Cristea E, Moreno P, Sharma SK, Narula J, Kini AS (2014) Combined NIRS and IVUS imaging detects vulnerable plaque using a single catheter system: a head-to-head comparison with OCT. Eurointervention 10(3):303–311. https://doi.org/10.4244/Eijv10i3a53

Sawada T, Shite J, Garcia-Garcia HM, Shinke T, Watanabe S, Otake H, Matsumoto D, Tanino Y, Ogasawara D, Kawamori H, Kato H, Miyoshi N, Yokoyama M, Serruys PW, Hirata KI (2008) Feasibility of combined use of intravascular ultrasound radiofrequency data analysis and optical coherence tomography for detecting thin-cap fibroatheroma. Eur Heart J 29:1136–1146

Sethuraman S, Aglyamov SR, Amirian JH, Smalling RW, Emelianov SY (2007a) Intravascular photoacoustic imaging using an IVUS imaging catheter. IEEE Trans Ultrason Ferroelectr Freq Control 54(5):978–986

Sethuraman S, Amirian JH, Litovsky SH, Smalling RW, Emelianov SY (2007b) Ex vivo characterization of atherosclerosis using intravascular photoacoustic imaging. Opt Express 15(25):16657–16666. 148272 [pii]

Suri JS, Kathuria C, Molinari F (eds) (2011) Atherosclerosis disease management, 1st edn. Springer

Tearney GJ, Jang IK, Bouma BE (2006) Optical coherence tomography for imaging the vulnerable plaque. J Biomed Opt 11(2):021002. https://doi.org/10.1117/1.2192697

Virmani R, Kolodgie FD, Burke AP, Finn AV, Gold HK, Tulenko TN, Wrenn SP, Narula J (2005a) Atherosclerotic plaque progression and vulnerability to rupture: angiogenesis as a source of intraplaque hemorrhage. Arterioscler Thromb Vasc Biol 25(10):2054–2061. https://doi.org/10.1161/01.atv.0000178991.71605.18. 01.ATV.0000178991.71605.18 [pii]

Virmani R, Kolodgie FF, Burke AP, Finn AV, Gold HK, Tulenko TN, Wrenn SP, Narula J (2005b) Atherosclerotic plaque progression and vulnerability to rupture: angiogenesis as a source of intraplaque hemorrhage. Artheriosler Thromb Vasc Biol 25(10):2054–2061

Wang LV (ed) (2009) Photoacoustic imaging and spectroscopy. Taylor & Francis/CRC Press Boca Raton, Florida

Wang HW, Le TT, Cheng JX (2008) Label-free imaging of arterial cells and extracellular matrix using a multimodal CARS microscope. Opt Commun 281:1813–1822

Wang J, Geng YJ, Guo B, Klima T, Lal BN, Willerson JT, Casscells W (2002) Near-infrared spectroscopic characterization of human advanced atherosclerotic plaques. J Am Coll Cardiol 39:1305–1313

Wang X, Pang Y, Ku G, Xie X, Stoica G, Wang LV (2003) Noninvasive laser-induced photoacoustic tomography for structural and functional in vivo imaging of the brain. Nat Biotechnol 21(7):803–806. https://doi.org/10.1038/nbt839. nbt839 [pii]

Wang HW, Langohr IM, Sturek M, Cheng JX (2009) Imaging and quantitative analysis of atherosclerotic lesions by CARS-based multimoddal nonlinear optical microscopy. Artherioscler Thromb Vasc Biol 29:1342–1348

Wang B, Su JL, Amirian J, Litovsky SH, Smalling R, Emelianov S (2010) Detection of lipid in atherosclerotic vessels using ultrasound-guided spectroscopic intravascular photoacoustic imaging. Opt Express 18(5):4889–4897. 196110 [pii]

Wang HW, Chai N, Wang P, Hu S, Dou W, Umulis D, Wang LV, Sturek M, Lucht R, Cheng JX (2011) Label-free bond-selective imaging by listening to vibrationally excited molecules. Phys Rev Lett 106(23):238106

Wang B, Karpiouk A, Yeager D, Amirian J, Litovsky S, Smalling R, Emelianov S (2012a) In vivo intravascular ultrasound-guided photoacoustic imaging of lipid in plaques using an animal model of atherosclerosis. Ultrasound Med Biol. https://doi.org/10.1016/j.ultrasmedbio.2012.08.006. S0301-5629(12)00470-X [pii]

Wang P, Wang HW, Sturek M, Cheng JX (2012b) Bond-selective imaging of deep tissue through the optical window between 1600 and 1850 nm. J Biophotonics 5(1):25–32. https://doi.org/10.1002/jbio.201100102

Waxman S, Ishibashi F, Muller JE (2006) Detection and treatment of vulnerable plaques and vulnerable patients: novel approaches to prevention of coronary events. Circulation 114(22):2390–2411. https://doi.org/10.1161/circulationaha.105.540013. 114/22/2390 [pii]

Weber C, Noels H (2011) Atherosclerosis: current pathogenesis and therapeutic options. Nat Med 17(11):1410–1422. https://doi.org/10.1038/nm.2538. nm.2538 [pii]

Wei W, Li X, Zhou Q, Shung KK, Chen Z (2011) Integrated ultrasound and photoacoustic probe for co-registered intravascular imaging. J Biomed Opt 16(10):106001. https://doi.org/10.1117/1.3631798

Wu M, Fw van der Steen A, Regar E, van Soest G (2016) Emerging technology update intravascular photoacoustic imaging of vulnerable atherosclerotic plaque. Interv Cardiol 11(2):120–123. https://doi.org/10.15420/icr.2016:13:3

Wykrzykowska JJ, Diletti R, Gutierrez-Chico JL, van Geuns RJ, van der Giessen WJ, Ramcharitar S, Duckers HE, Schultz C, de Feyter P, van der Ent M, Regar E, de Jaegere P, Garcia-Garcia HM, Pawar R, Gonzalo N, Ligthart J, de Schepper J, van den Berg N, Milewski K, Granada JF, Serruys PW (2012) Plaque sealing and passivation with a mechanical self-expanding low outward force nitinol vShield device for the treatment of IVUS and OCT-derived thin cap fibroatheromas (TCFAs) in native coronary arteries: report of the pilot study vShield evaluated at cardiac hospital in rotterdam for investigation and treatment of TCFA (SECRITT). EuroIntervention. 20101220-01 [pii]

Yang JM, Maslov K, Yang HC, Zhou Q, Shung KK, Wang LV (2009) Photoacoustic endoscopy. Opt Lett 34(10):1591–1593. 179902 [pii]

Yang Y, Li X, Wang T, Kumavor PD, Aguirre A, Shung KK, Zhou Q, Sanders M, Brewer M, Zhu Q (2011) Integrated optical coherence tomography, ultrasound and photoacoustic imaging for ovarian tissue characterization. Biomed Opt Express 2(9):2551–2561. https://doi.org/10.1364/BOE.2.002551. 150674 [pii]

Yin J, Yang HC, Li X, Zhang J, Zhou Q, Hu C, Shung KK, Chen Z (2010) Integrated intravascular optical coherence tomography ultrasound imaging system. J Biomed Opt 15(1):010512. https://doi.org/10.1117/1.3308642

Yin J, Li X, Jing J, Li J, Mukai D, Mahon S, Edris A, Hoang K, Shung KK, Brenner M, Narula J, Zhou Q, Chen Z (2011) Novel combined miniature optical coherence tomography ultrasound probe for in vivo intravascular imaging. J Biomed Opt 16(6):060505. https://doi.org/10.1117/1.3589097

Yoo H, Kim JW, Shishkov M, Namati E, Morse T, Shubochkin R, McCarthy JR, Ntziachristos V, Bouma BE, Jaffer FA, Tearney GJ (2011) Intra-arterial catheter for simultaneous microstructural and molecular imaging in vivo. Nat Med 17(12):1680–1684. https://doi.org/10.1038/nm.2555. nm.2555 [pii]

Yun SH, Tearney GJ, Vakoc BJ, Shishkov M, Oh WY, Desjardins AE, Suter MJ, Chan RC, Evans JA, Jang IK, Nishioka NS, de Boer JF, Bouma BE (2006) Comprehensive volumetric optical microscopy in vivo. Nat Med 12:1429–1433

Zoumi A, Lu X, Ghassan S, Tromberg BJ (2004) Imaging coronary artery microstructure using second-harmonic and two-photon fluorescence microscopy. Biophys J 87:2778–2786

Chapter 2
Advances in Multi-frequency Intravascular Ultrasound (IVUS)

Teng Ma and Qifa Zhou

Background

Coronary heart disease (CHD) remains the leading cause of death in developed countries. Acute coronary syndromes (ACS) are the clinical manifestations of a sudden reduction in perfusion and oxygenation to the myocardium, typically resulting in heart attacks. Each year, more than 20 million patients worldwide with CHD experience ACS, and one-third of these individuals die from complications of CAD (Go et al. 2014). Atherosclerosis, a chronic disease typically asymptomatic at early stages, is characterized by the thickening of the arterial vessel wall due to the buildup of athermanous plaque in the inner lining of arteries (Ross 1993, 1999). Vulnerable atherosclerotic plaque, a particularly risk-laden plaque vulnerable to sudden rupture, is widely recognized to be the main "troublemaker" underlying ACS (Moreno 2010; Finn et al. 2010). Although the understanding of vulnerable plaques remains to be elucidated, histological studies have demonstrated that thin-cap fibroatheroma (TCFA) is the most common phenotype of vulnerable plaques (shown in Fig. 2.1). TCFA is composed of a lipid-rich necrotic core with an overlying thin-cap-rich in macrophages (white blood cells that attack foreign substances) (Libby 1995). Quantitatively, TCFA is further defined as an atherosclerotic plaque with a fibrous cap <65 μm in thickness associated with macrophage infiltration (>25 cells per 0.3-mm-diameter field) and a large lipid-rich necrotic core occupying nearly 35% of plaque

T. Ma
Paul C. Lauterbur Research Center for Biomedical Imaging, Institute of Biomedical and Health Engineering, Shenzhen Institutes of Advanced Technology, Chinese Academy of Sciences, Shenzhen 518055, China
e-mail: teng.ma@siat.ac.cn

Q. Zhou (✉)
Roski Eye Institute, University of Southern California, Los Angeles, CA 90033, USA
e-mail: qifazhou@usc.edu

Department of Biomedical Engineering, University of Southern California, Los Angeles, CA 90089, USA

Q. Zhou and Z. Chen (eds.), *Multimodality Imaging*,
https://doi.org/10.1007/978-981-10-6307-7_2

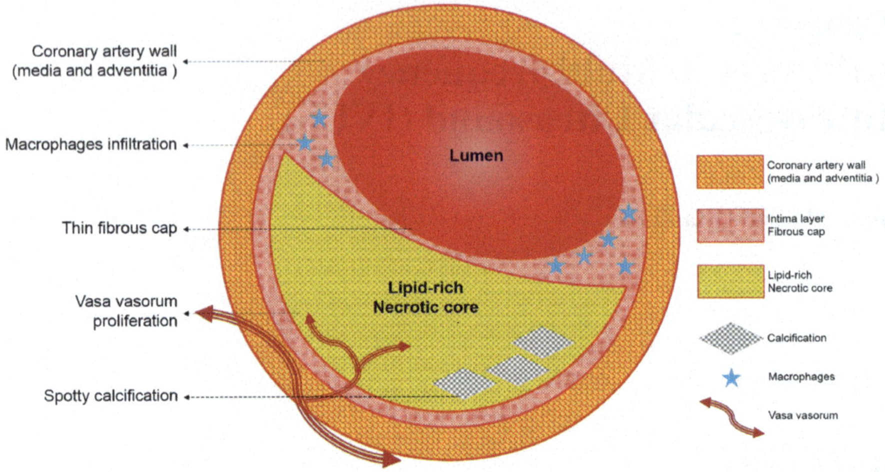

Fig. 2.1 Scheme of the phenotype of vulnerable atherosclerotic plaque—thin-cap fibroatheroma (TCFA). Morphologic and biologic function markers distinctive to TCFA are illustrated, including thin fibrous cap, large lipid-rich necrotic core, macrophages infiltration, vasa vasorum proliferation and spotty calcifications (Virmani et al. 2003)

volume (Virmani et al. 2003). Therefore, both the thickness of TCFA and the size of the lipid-rich necrotic core are considered to be the major predictors of ACS. As a corollary, the presence of the inflammatory molecules and cells, such as increased macrophages, is useful in both identifying and staging the vulnerable plaques. Additional markers of TCFA are micro-calcifications and proliferation of the vasa vasorum (vessels that supply the walls of large arteries). To precisely identify intravascular TCFA in vivo, the imaging techniques employed must recognize key morphological structures as well as biological features of the TCFA. Thus, early detection and staging of TCFA will not only guide the interventional or pharmacological strategy to prevent plaque rupture, but also contribute to the study of epidemiology of vulnerable plaques.

Therefore, an optimal intravascular imaging technology for plaque characterization, especially for the identification of vulnerable plaque with TCFA, should meet the following requirements: (1) visualizing the endoluminal structure in detail and scaling the degree of stenosis; (2) quantifying the entire plaque volume and plaque burden; (3) identifying plaque components such as calcification, lipid-rich necrotic core, fibrous tissue, and inflammatory markers; (4) providing adequate spatial resolution to measure the thickness of thin fibrous cap; (5) monitoring plaque rupture and thrombus formation. Each intravascular imaging technology possesses unique features that yield valuable information while exhibiting inherent limitations that can be difficult to overcome; therefore, an integration of multiple imaging modalities seems a synergistic solution (Bourantas et al. 2013; Bourantas and Serruys 2014; Garcia-Garcia et al. 2008; Honda and Fitzgerald 2008; Maehara et al. 2009; Puri et al. 2011).

Catheter-based intravascular ultrasound (IVUS) has been used clinically for over the last two decades to image coronary arteries for atherosclerotic lesions, to evaluate the lumen and plaque dimensions, and to guide intervention and stent deployment. The mechanically scanning IVUS transducer (20–40-MHz) or the radial array transducer (10–20-MHz), transmitting and receiving the high-frequency ultrasonic waves, are capable of delineating the cross-sectional anatomy of coronary artery wall in real time with 70–200 μm axial resolution, 200–400 μm lateral resolution, and 5–10 mm imaging depth (Elliott and Thrush 1996; Brezinski et al. 1997). In the late 1990s, Bernard Sigel et al. first demonstrated the feasibility of using ultrasonic spectrum analysis to characterize vulnerable plaques of carotid arteries (Noritomi et al. 1997a, 1997b; Lee et al. 1998). Later on, the newly developed radiofrequency backscatter spectrum analysis algorithm, quantitatively analyzing the back-reflected ultrasonic signal in the frequency domain and determining the tissue composition, was implemented in two commercial intracoronary artery imaging systems: Virtual Histology (VH-IVUS, Volcano Therapeutics, CA, USA) and iMap (Boston Scientific, CA, USA) (Nair et al. 2002, 2007; Nasu et al. 2006; Shin et al. 2011). The IVUS-based elastography technique, intravascular palpography, is able to assess local mechanical properties during arterial deformation caused by the intraluminal pressure, which can be used to perform high-risk plaque assessment (Schaar et al. 2003, 2006; Deleaval et al. 2013). However, based on clinical studies in patients with ACS, the reliability of using ultrasonic spectrum analysis and intravascular palpography to detect vulnerable plaque was subpar (Brugaletta et al. 2012). This was caused by the insufficient resolution of IVUS to reliably characterize different tissue types and to precisely detect TCFA at such small scales. Nevertheless, IVUS remains an important tool for assessing plaque burden and monitoring artery remodeling (Maresca et al. 2014).

Optical coherence tomography (OCT), considered as the optical analog of ultrasound, utilizes back-scattered infrared light to achieve high spatial resolution (10–30 μm) and high-speed microstructural coronary artery images (>100 frames per second, 20–40 mm/s pull back speed) (Brezinski et al. 1996; Regar et al. 2003). The optical pulse, or broad bandwidth infrared light, is irradiated into the tissue at different angular positions. 2D cross-sectional image can then be reconstructed based on the echo time delay and the intensity of the detected optical echo from tissue. Under rapid development in scientific research and proliferation in medical device industry, intravascular OCT has gained wide recognition in clinical practice and has become the top contender to challenge the status of IVUS in the intravascular imaging field. However, the major disadvantages of OCT are the limited penetration depth (1–2 mm) and lacking the reliability of tissue characterization as compared to IVUS. Moreover, similar to other optical intravascular imaging techniques, another important drawback of OCT is that it requires the temporal clearance of high-scattering luminal blood by using flushing agents such as iohexal and iodixanol, which may cause life-threatening reactions during or after the imaging procedures (Li et al. 2015; Dawson 1989).

Recently, using chemical composition for tissue characterization has further increased the feasibility of assessing the metabolic state of vulnerable plaques in molecular level. Different tissue compositions have different optical absorption and scattering effect on near-infrared (NIR) light (400–2400 nm). Near-infrared spectroscopic (NIRS) is the first intravascular imaging technique to achieve lipid content characterization within plaques by analyzing the absorbance of emitted NIR light at different wavelengths (Moreno et al. 2002; Waxman et al. 2009). However, this technology lacks the quantitative information about the size and location of the lipid core, which potentially limits its clinical utility. Intravascular photoacoustic (IVPA) detects the acoustic waves generated by thermal expansion induced by pulsed light to provide unique optical absorption contrast at ultrasound resolution. Because lipid has a distinct absorption spectrum in the NIR wavelength range, IVPA imaging is a promising technique for detecting and quantifying the amount of lipid in atherosclerotic plaques (Wang et al. 2010; Jansen et al. 2014). Other emerging optical imaging techniques such as near-infrared fluorescence (NIRF) imaging (Yoo et al. 2011), Raman spectroscopy (Buschman et al. 2000; van de Poll et al. 2002) and fluorescence spectroscopic imaging (Stephens et al. 2009; Sun et al. 2011) are advancing the field of catheter-based technology by providing the contrast that involves chemical specificity, which can be used to identify tissue composition. However, these optical image techniques lack the capability to perform cross-sectional mapping for tissue structure; thus, IVUS and OCT remain essential.

Historically, IVUS has served as the de facto catheter-based intravascular imaging modality as the most established device in the clinical setting; however, it is by no means the "gold standard" for diagnosing coronary atherosclerosis and assessing plaque vulnerability. New intravascular imaging techniques are emerging to supplement deficiencies in IVUS, each with its own strengths and limitations. This review aims to expound upon the several advances in IVUS-based intravascular imaging systems and address their innovations, challenges, and strategies for improvement.

Ultrahigh-Frequency IVUS Imaging at 80-MHz

Introduction

A typical IVUS probe consists of a rotating shaft with a side-looking unfocused single-element transducer, which leads to a radial imaging geometry of a cross section of a vessel. The size of an IVUS probe in clinical applications ranges from 2.6 to 3.5 French gauge (0.90–1.17 mm), limiting the aperture of the transducer within the catheter to be less than 0.8 mm. The center frequencies of commonly used IVUS transducers are between 20 and 50-MHz, implying that the axial/lateral resolutions are on the order of 60/200 μm. This value is inferior to intravascular OCT imaging, of which the resolution is on the order of 10–30 μm, which could provide much more detailed information about the microstructures of vessel and plaque compositions. Its drawback is that it has a limited penetration depth of approximately 1 mm. Increasing the IVUS center frequency to 80-MHz or higher is a compromise between resolution and penetration. To date, little IVUS work has been done at such a high-frequency.

The major concern is the strong tissue attenuation at high frequencies (Foster et al. 2000b). At 80-MHz, an attenuation coefficient of 10 dB/mm is expected for coronary artery, which means that a penetration depth of 3 mm can only be achieved for a system with a dynamic range of 60 dB. Yet such a system requires a highly sensitive miniaturized IVUS transducer, which is a great challenge.

The difficulty of such a high-frequency transducer comes from the preparation of a very thin layer of piezoelectric material with piezoelectric properties similar to the bulk material. Traditional machining of bulk materials down to the thickness of 20–30 μm is extremely difficult and time-consuming. Additionally, quality deterioration and the brittle nature of the lapped bulk materials may severely downgrade the sensitivity of a high-frequency transducer. A more feasible solution is to pursue piezoelectric thin-film technology. PMN-PT thin films have been extensively studied (Calzada et al. 2007; Kuscer et al. 2009; Luo et al. 2007; Shih et al. 2006; Ursic et al. 2008). Due to the high k_t and $\varepsilon_r/\varepsilon_0$ values, it could be a good candidate for IVUS transducer fabrication. Here we present the fabrication of novel PMN-PT freestanding thin films with enhanced piezoelectric properties. A highly sensitive miniaturized 80-MHz IVUS transducer was built from these films. In vitro imaging of a rabbit aorta has been carried out to verify the feasibility of the transducer for intravascular imaging.

Design and Fabrication of 80-MHz IVUS Probe

One piece of 7×10 mm^2 PNM-PT film was used as the active piezoelectric material to fabricate a side-looking miniature transducer. The PMN-PT film was first sputtered with Cr/Au (500 Å/1000 Å) layers as electrodes on top and bottom. A matching layer made from Insulcast 501, Insulcure 9 (American Safety Technologies, Roseland, NJ) and 2–3 μm silver particles (Sigma-Aldrich Inc., St. Louis, MO) was then cured over the top of the film and lapped to 5 μm. A conductive backing material, E-solder 3022 (VonRoll Isola, New Haven, CT) was applied to the bottom of the film and lapped to 0.6 mm. The active stack was diced along the thickness direction into small posts with the aperture of 0.4×0.4 mm^2. The post was housed within a 0.57-mm-ID polyimide tube (MedSource Technologies, Trenton, GA), on the side of which a window was opened to allow the transducer to slant toward side face. A 0.1-mm-OD electrical wire was connected to the conductive backing using E-solder 3022 inside the polyimide tube. The polyimide tube provided the electrical isolation from the outer stainless steel needle housing. The outer needle housing with an ID of 0.66 mm and OD of 0.92 mm had a window on the side for acoustic wave to go through. The epoxy was filled into the gap between piezoelectric post and needle housing to insulate the inner electrode. Another Cr/Au electrode was sputtered over the silver matching layer and stainless steel needle housing to form the ground connection. A 3-μm-thick parylene layer was vapor-deposited onto the aperture and needle housing to serve as second matching and protecting layer. The transducer was finally connected to a brass holder and SMA connector for mechanical holding and electrical connection.

To enhance the piezoelectric activity of the PMN-PT film, the finished transducer, shown in Fig. 2.2, was poled in a DC electric field of 300 kV/cm for five minutes at room temperature.

The side-looking miniature transducer's performance was measured in a deionized water bath at room temperature. Pulse-echo test (Cannata et al. 2003) was conducted with an X-cut quartz as signal-reflecting target. A single sinusoidal wave centered at 85-MHz with of approximately 100 V_{pp} and 200 Hz repetition rate emitted from a monocycle function generator (Avtech Electrosystems Ltd., Ontario, Canada) was used to excite the transducer. Echo signal was received and digitized by a 1 GHz oscilloscope (LC534, LeCroy Corp., Chestnut Ridge, NY). The frequency response of the transducer was analyzed from the echo waveform, shown in Fig. 2.3. The peak-to-peak amplitude was 601 mV. The measured center frequency was 84-MHz and −6 dB fractional bandwidth was 35%. Two-way insertion loss was measured to be 25 dB, which indicated the transducer's sensitivity is comparable with that of an 80-MHz large aperture lithium niobate ($LiNbO_3$) single-crystal transducer (10–25 dB). Six-μm-diameter tungsten wire targets were imaged to determine axial and lateral resolutions of the transducer, as shown in Fig. 2.4a. The envelopes of echo signals from the wire located at 1.2 mm away from the transducer surface, or point spread function (PSF), were displayed in Fig. 2.4b, c. The axial (R_{axial}) and lateral ($R_{lateral}$)

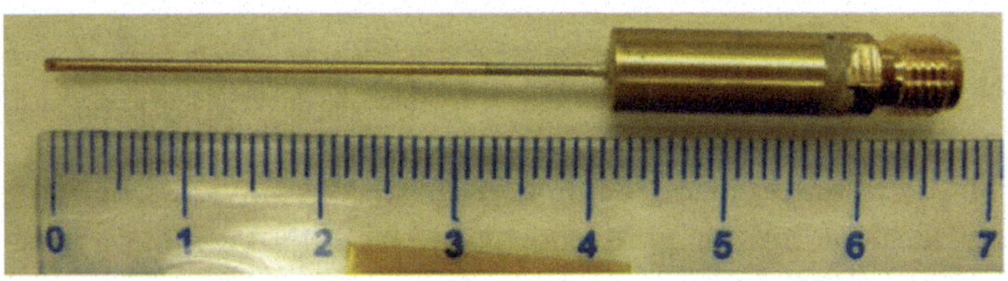

Fig. 2.2 Side-looking IVUS transducer at 80-MHz (Li et al. 2011)

Fig. 2.3 Pulse-echo measurement of 80-MHz IVUS transducer (Li et al. 2011)

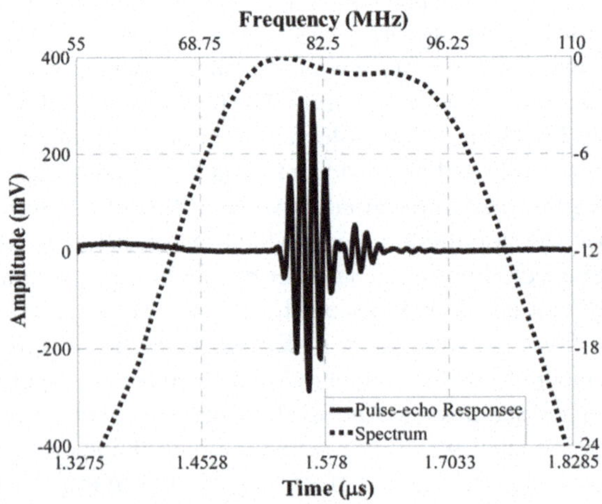

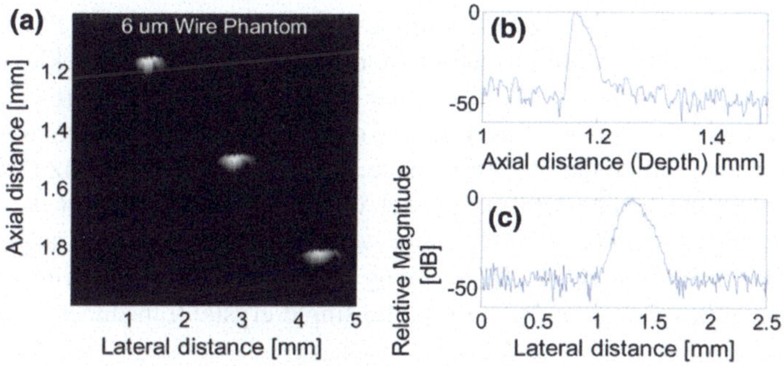

Fig. 2.4 Ultrasound wire phantom (**a**), displayed with a dynamic range of 45 dB; axial (**b**) and lateral (**c**) envelopes of echo signals from the wire located at 1.2 mm away from the transducer surface (Li et al. 2011)

resolutions were determined from the −6 dB envelope width, which were 35 and 176 μm, respectively.

Ex Vivo 80-MHz IVUS Imaging

In vitro imaging of a normal rabbit aorta was performed to test the transducer's ability for intravascular application. The side-looking PMN-PT freestanding film transducer was used to image the cross section of an aorta, shown in Fig. 2.5a. For comparison purpose, another image of the same aorta, but with a 35-MHz PMN-PT single-crystal needle transducer, which had the same aperture size (0.4×0.4 mm^2),

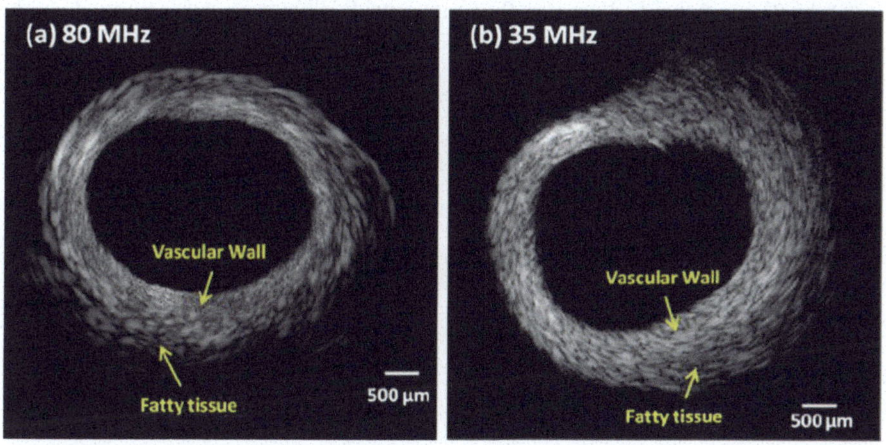

Fig. 2.5 **a** Rabbit aorta image from 80-MHz PMN-PT freestanding film transducer; **b** The same aorta image from 35-MHz PMN-PT single-crystal transducer (Li et al. 2011)

shown in Fig. 2.5b, was obtained. From these images, the 80-MHz PMN-PT film transducer appeared to exhibit a much better resolution (denser speckles) than the 35-MHz PMN-PT single-crystal transducer. Due to the improved resolution, the vascular wall and the surrounding fatty tissues in the 80-MHz IVUS image were better differentiated than the 35-MHz image. Meanwhile, the 80-MHz IVUS could easily visualize the whole depth of vessel wall, though not the full depth of the hypoechoic fatty tissue. The 80-MHz image was depicted at a dynamic range of 51 dB and 35-MHz image was at 54 dB, implying 80-MHz freestanding film transducer had a comparable signal-to-noise ratio to the single-crystal transducer.

80-MHz Intravascular Photoacoustic Imaging (IVPA)

The catheter-based intravascular photoacoustic (IVPA) imaging for diagnosing atherosclerosis, which can provide optical absorption contrast of the arterial wall besides acoustic scattering contrast from the conventional intravascular ultrasound (IVUS) imaging, has been intensively researched recently. The resolution of IVPA is determined by the frequency bandwidth of an ultrasonic transducer. Higher resolution can be achieved by increasing the transducer's working frequency and bandwidth. IVPA imaging at 35 and 80-MHz by using newly designed integrated IVUS/IVPA probes were reported based on the development of 80-MHz IVUS Imaging, which is the first time IVPA has been achieved as high as 80-MHz.

The integrated IVUS/IVPA probe is composed of a parallel arranged, side-firing optical probe and side-viewing ultrasonic transducer, shown in Fig. 2.6. For the optical part, a 200-μm-core multimode fiber is used to deliver the 532 nm pulsed laser beams. At the distal end, a 45° polished microprism (0.25 × 0.25 × 2 mm^3; Bern Optics Inc., Westfield, MA) is connected to the fiber tip and sealed inside a glass capillary tube (0.4 mm inside diameter; 0.55 mm outside diameter). Air is trapped inside

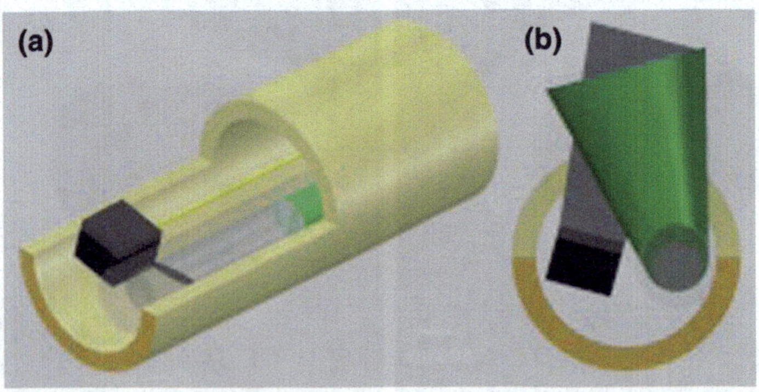

Fig. 2.6 Schematic of the integrated IVUS/IVPA probe: **a** top view and **b** front view; the light beam is in green and the acoustic beam in gray (Li et al. 2012)

the tube to form an air/glass interface at the prism polished surface to redirect laser beams by 90°, following the "total internal reflection" effects. The light divergence is measured to be 32°. For the acoustic part, 35- and 80-MHz ultrasonic transducers were fabricated in our laboratory and assembled into the IVUS/IVPA probes, respectively. In the 35-MHz transducer, a $[Pb(Mg_{1/3}Nb_{2/3})O_3]0.63[PbTiO_3]0.37$ (PMN-PT) crystal (HC Materials, Bolingbrook, IL) is used as the active piezoelectric element with a thickness of 55 μm and an aperture size of $0.4 \times 0.4\ mm^2$. In the 80-MHz transducer, the PMN-PT freestanding film, which is fabricated from a tape-casting method, is used as the piezoelectric layer with a thickness of 30 μm and the same aperture size as the 35-MHz transducer. All components are packaged in polyimide tubing (Small Parts, Inc., Miramar, FL) with 1.2 mm OD. The optical and acoustic beams are aligned to have an angle of approximately 20° to achieve optimal overlapping. The closer arrangement may improve the optical-acoustic overlapping, especially at the region close to the probe surface. By incorporating either 35 or 80-MHz ultrasonic transducers, we fabricated the integrated IVUS/IVPA probes working at each frequency range.

By using the integrated IVUS/IVPA probes, in vitro imaging of a normal rabbit aorta was conducted at both 35 and 80-MHz to demonstrate the probes' imaging feasibility. IVUS and IVPA images of a rabbit aorta at 35-MHz are shown in Fig. 2.7a, b. The IVUS image has a dynamic range of 50 dB, and the IVPA image has a dynamic range of 35 dB. Both images can be seen through the vessel wall. The 35-MHz IVPA imaging depth could be demonstrated up to 4 mm at the 12:00 to 2:00 o'clock position in Fig. 2.7b. The aorta vessel wall is composed of three-layer structures: intima, media, and adventitia. Intimal thickening is considered to be a manifestation of atherosclerosis (Fitzgerald et al. 1992). Since the aorta used in this study is from a healthy rabbit, the intima only displays as a very thin darker layer in the histology image, and the vessel wall has a relatively uniform composition, as shown in Fig. 2.7d. Therefore, it is not surprising that the IVUS and IVPA images have a relatively homogeneous appearance. In the IVPA image, due to the boundary buildup effect (Guo et al. 2009), the front and rear boundaries are more prominent than the middle. Figure 2.7c shows the fused IVUS/IVPA image, which demonstrates the co-registration of the two images.

The 80-MHz IVUS and IVPA images are shown in Fig. 2.8a, b. The IVUS image has a dynamic range of 50 dB, and the IVPA image has a dynamic range of 35 dB. Due to the improved axial resolution at 80-MHz, the profile of the vessel lumen is depicted more clearly in both IVUS and IVPA images than at 35-MHz. In the 80-MHz IVUS image, as shown in Fig. 2.8a, owing to the insufficient acoustic contrast between adventitia and soft tissue surrounding the aorta, the outer boundary of the vessel wall is not clearly displayed, especially from the 8:00 to 2:00 o'clock position, while in the IVPA image, as shown in Fig. 2.8b, the outer boundary is more evident. By fusing the IVUS and IVPA images together, as shown in Fig. 2.8c, the boundary between surrounding soft tissue and vessel wall can be easily seen. Compared to the 35-MHz IVPA image, the 80-MHz image displays a clearer boundary profile of the vessel wall, which demonstrates improved axial resolution at a higher frequency. Imaging results show that IVPA has superior contrast over IVUS in identifying the

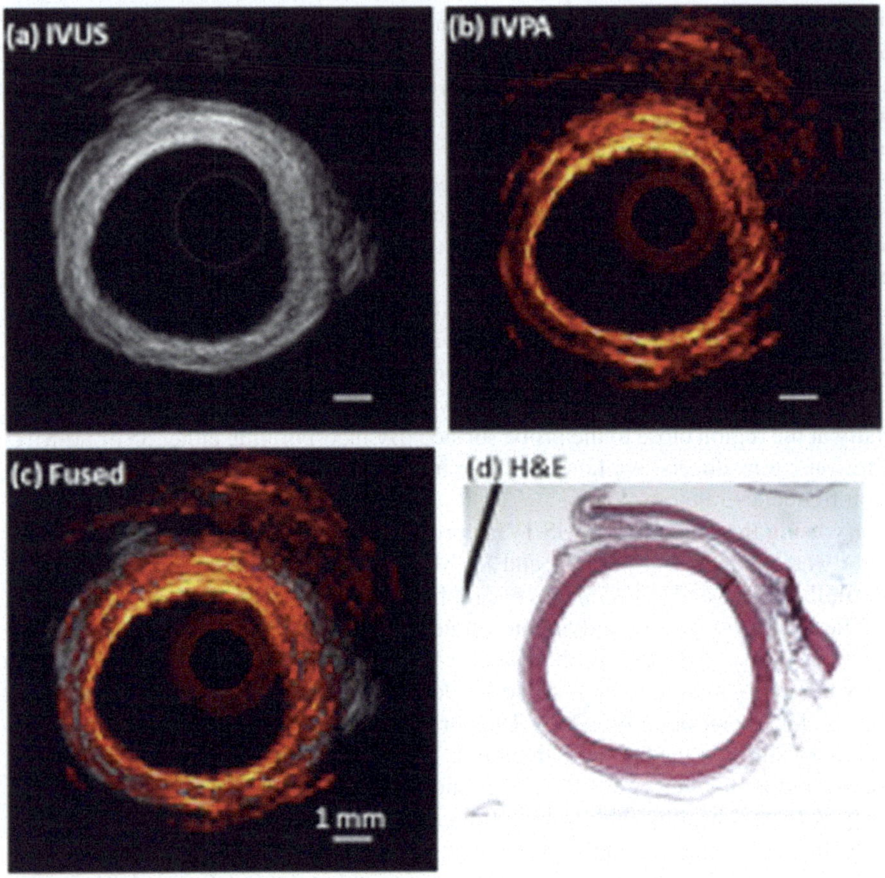

Fig. 2.7 Cross-sectional **a** IVUS, **b** IVPA, and **c** fused images of a healthy rabbit aorta at 35-MHz, and **d** hematoxylin-eosin (H&E)-stained histology image (Li et al. 2012)

arterial wall, and IVPA at 80-MHz demonstrates extraordinary resolution (35 μm) compared to 35-MHz.

Summary

In this section, we reviewed the recent development of high-frequency (80-MHz) IVUS application by using piezoelectric thin film technology. The reported results showed the 80-MHz PMN-PT thin-film IVUS transducer had superior resolution and sensitivity. An in vitro study was conducted with a healthy rabbit aorta. The 80-MHz IVUS image demonstrated improved resolution and contrast to allow a differentiation of the vascular wall and surrounding fatty tissue, which could not be achieved by a

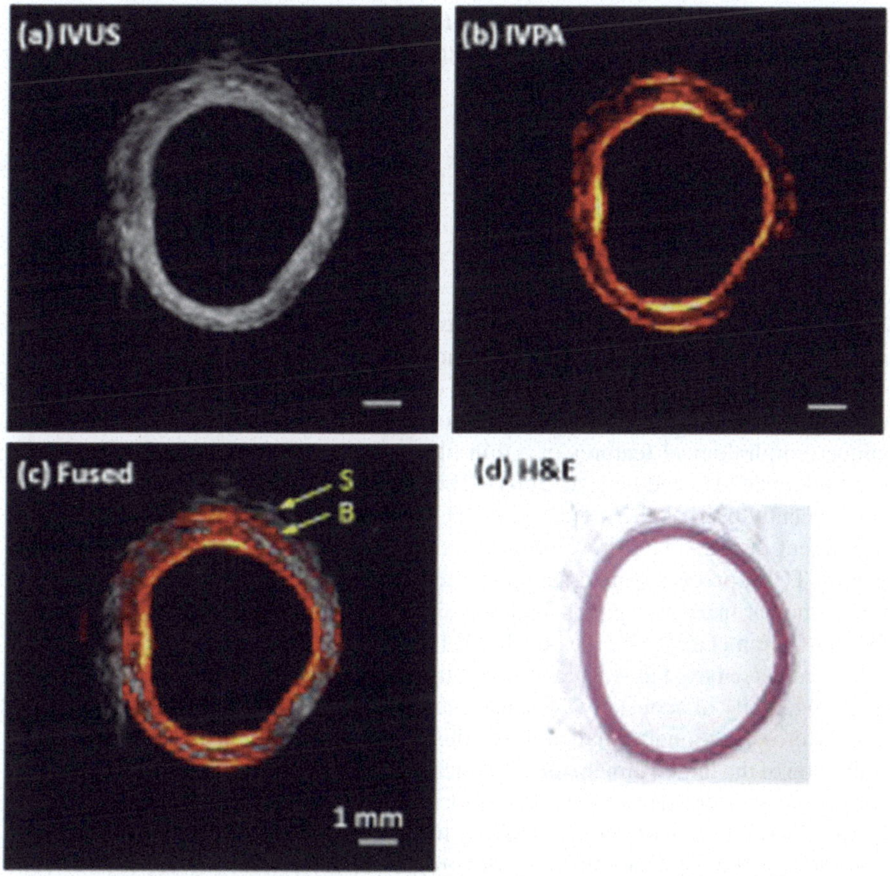

Fig. 2.8 Cross-sectional **a** IVUS, **b** IVPA, and **c** fused images of a healthy rabbit aorta at 80-MHz, and **d** H&E-stained histology image (Li et al. 2012)

35-MHz transducer. As expected, the imaging depth in the hypoechoic fatty tissue was inferior to that of a 35-MHz transducer. However, this capability is especially attractive in detecting a vulnerable plaque consisting of a lipid pool surrounded by a fibrous cap given the improved resolution and contrast at 80-MHz. Moreover, we also reviewed the miniature integrated IVUS/IVPA probes that could internally illuminate the vessel wall and provide IVUS and IVPA imaging simultaneously. The optical and acoustic components were arranged parallel to each other. 35 and 80-MHz ultrasonic transducers were incorporated into the integrated probes to perform IVPA imaging at each frequency. In vitro wire phantom and rabbit aorta experiments were successfully conducted to verify the feasibility of applying these probes for intravascular imaging. Compared to the 35-MHz IVPA imaging, the 80-MHz IVPA imaging demonstrated much finer resolution for delineating vessel boundaries. A further increase in frequency will allow the resolution and contrast to be further improved by sacrificing the depth of penetration.

Multi-frequency IVUS Imaging

Introduction

Vulnerable plaque has been hypothesized as an unstable atherosclerosis plaque building up in the walls of arteries, whose rupture is primarily responsible for acute coronary syndrome (ACS) and sudden cardiac death (Virmani et al. 2000). The thin-cap fibroatheroma (TCFA), which is the phenotype of vulnerable plaque, possesses the unique morphological features of a thin fibrous cap (<65 μm) and a large necrotic core underneath (Kolodgie et al. 2001). These morphological features in a tiny scale are frequently used as the indicators of TCFA and can be used to facilitate the risk assessment of ACS. High-resolution intravascular imaging techniques, which can identify TCFA precisely, are essentially needed to provide a detailed anatomic structure of the coronary artery wall with the aim of evaluating the plaque vulnerability (Garcia-Garcia et al. 2008; Puri et al. 2011).

The catheter-based grayscale intravascular ultrasound (IVUS) imaging, based on the echogenicity of acoustic waves, has been clinically available for 20 years to provide the cross-sectional visualization of the coronary artery wall and the quantitative evaluation of the lumen dimensions and the plaque area (Potkin et al. 1990; Aoki et al. 2005). The aperture size of the piezoelectric transducer incorporated at the catheter tip, usually smaller than 0.8 mm, is strictly limited by the confined size of the coronary artery lumen and the curvy structure of coronary artery system (Foster et al. 2000a). The typical center frequencies of IVUS transducers range from 20 to 40-MHz, providing 70–200 μm axial resolution, 200–400 μm lateral resolution and 5–10 mm imaging depth (Elliott and Thrush 1996; Brezinski et al. 1997). As a result, the spatial resolution of conventional IVUS is insufficient to measure the thin fibrous cap thickness, which is usually smaller than 65 μm. A newly developed radiofrequency backscatter spectrum analysis algorithm, virtual histology IVUS system (VH-IVUS), is capable of predicting the plaque vulnerability through characterizing plaque composition (Nair et al. 2002, 2007; Nasu et al. 2006). It is also a fact that data analyzed by VH is displayed on the order of 250 μm, which intrinsically downgrades the reliability to detect the thin fibrous cap of this technique (Garcia-Garcia et al. 2009). Super-harmonic IVUS imaging has been recently developed to image coronary vasa vasorum during the contrast agent injection; however, in vivo validation and long-term optimization of this technique are needed to prove its capability to perform accurate vulnerable plaque assessment (Ma et al. 2014). Optical coherence tomography (OCT), which utilizes back-scattered infrared light to generate high speed and high spatial resolution (10–30 μm) images of microstructures of blood vessels such as thin fibrous cap, is hypothesized to be a substitute of IVUS (Brezinski et al. 1996;

Regar et al. 2003). However, the penetration depth through blood and vascular tissues of OCT is shallow. A synergistic approach combining IVUS and OCT seems favorable by simultaneously providing the deep penetration depth of IVUS and the high spatial resolution of OCT (Li et al. 2014a). Several designs of integrated IVUS-OCT catheters have previously been evaluated both in vitro and in vivo to demonstrate the feasibility of this dual-modality intravascular imaging technique that allows more accurate plaque characterization than using IVUS or OCT alone (Li et al. 2010, 2014a; Yin et al. 2011; Bourantas et al. 2013). Nevertheless, technical requirements such as advancing IVUS-OCT catheter design, increasing the imaging speed and enhancing the accuracy of the co-registration between IVUS and OCT images need to be satisfied before this technique can be implemented in clinical practice (Bourantas et al. 2013). In addition, the manufacturing cost of such a hybrid system will be dramatically increased because two independent systems must be integrated into one single unit, which requires deliberation between translational strategies and long-term clinical utility (Maresca et al. 2014). Given the fact that OCT is the optical analog of ultrasound, a question is then raised: can ultrahigh-frequency ultrasonic imaging be considered as an alternative and cost-effective solution to replace OCT? Ultrahigh-frequency IVUS at 80 MHz has been previously investigated and proved to be able to improve artery characterization with much higher axial resolution (Li et al. 2011). Even though there is a foreseeable drawback that higher frequency ultrasound will have stronger attenuation in the blood and the vascular tissue, such transducers at 80-MHz center frequency can still provide an imaging depth of 2 mm that is comparable to the penetration depth of OCT. In addition, further increasing the frequency and bandwidth of the ultrasound transducer is a possible approach to reach the axial resolution level of OCT.

Although it is hard to identify the vulnerable plaques with the conventional IVUS because of its low resolution, it is still a reliable technique that cannot be easily substituted in the near future due to its accurate and deeply penetrated imaging capability (Puri et al. 2011). Therefore, compared to the integrated IVUS-OCT, incorporating another transducer at a higher frequency to the current IVUS system seems to be a simpler and more cost-effective approach to overcome the resolution limit of IVUS (Puri et al. 2011). In this paper, we present the development of a multi-frequency IVUS system by integrating a conventional transducer (35-MHz) and an ultrahigh-frequency transducer (80–150-MHz) into one single catheter. Three prototypes of the multi-frequency IVUS catheters with different frequency combinations (35/90, 35/120 and 35/150-MHz) were fabricated and evaluated. The two transducer elements were arranged in a back-to-back configuration to facilitate automatic image co-registration at the same longitudinal location of the artery. The flexible catheters were miniaturized to have an outer diameter of 0.95 mm and a front rigid length of 2 mm to enable the catheters' clinical compatibility. The multi-frequency IVUS imaging was assessed with a tissue-mimicking phantom with and without the presence of blood to evaluate the imaging depth, resolution and contrast-to-noise ratio (CNR) of the imaging system. In vitro intravascular imaging of human cadaver coronary arteries with atherosclerosis was conducted to demonstrate the merit of multi-frequency IVUS which possesses the complementary nature of both low-frequency and high-frequency transducers. The system provides the deep penetration to image the whole artery wall of the low-frequency transducer and the fine resolution in the lumen area of the high-frequency transducer, allowing more comprehensive visualization of the coronary artery. These results suggest that the multi-frequency IVUS system appears to be a feasible and more cost-effective approach to overcome the current limitations of the conventional IVUS system.

Design and Fabrication of Multi-frequency Catheter

The square element size of transducer was restricted to be less than 0.8 mm in the conventional IVUS catheter. In order to fit the two transducers in the small catheter, the transducer element size and thickness were fixed to be $0.5 \times 0.5\ \text{mm}^2$ and 0.3 mm for various center frequencies (35, 90, 120, and 150-MHz) in this study. Many different piezoelectric materials, including piezoceramics, piezoelectric polymers, and ferroelectric single-crystals can be used when designing a single-element transducer. Choices between the materials are made based on the piezoelectric properties and the specific requirements of the designed transducers such as center frequency, aperture size, and focal depth. Electromechanical coupling coefficient (k_t), piezoelectric coefficient (d_{33}) and dielectric permittivity ($\varepsilon_{33}^s/\varepsilon_0$), listed in Table 2.1, are the major parameters that determine the performance of a transducer (Li et al. 2014b). Since the transducer aperture size is already determined in this study, the effects caused by dielectric permittivity should be carefully taken into account during the material selection. The reason for that is the dielectric permittivity, together with the surface area of piezoelectric material, determines the electrical impedance of a transducer (Li et al. 2011). A properly matched electrical impedance of 50 Ω to the electronics will result in an optimized sensitivity in both transmitting and receiving. $Pb(Zr,Ti)O_3$ (PZT) ceramics have been widely used in the fabrication of IVUS transducers in the frequency range of 20–50-MHz (Lockwood et al. 1996). Compared to PZT, the recently developed single-crystal $Pb(Mg_{1/3}Nb_{2/3})$-$PbTiO_3$ (PMN-PT)

Table 2.1 Piezoelectric properties of materials for IVUS imaging application

Type	PZT-5H (Snook et al. 2002)	PMN-PT (Zhou et al. 2007a)	$LiNbO_3$ (Cannata et al. 2003)
	Ceramic	Single-crystal	Single-crystal
k_t	0.51	0.58	0.49
d_{33} (pm/V)	593	2000	6
$\varepsilon_{33}^s/\varepsilon_0$	1470	5229	39
v (m/s)	4580	4610	7340
Z_a (MRayl)	36.0	36.9	34
*Transducer fabrication in this study		35 and 90-MHz	120 and 150-MHz

(from H.C. Materials Inc) exhibits improved piezoelectric performance on account of its outstanding k_t and d_{33} (Table 2.1), making it a more favorable candidate for ultrasonic transducer fabrication (Zhou et al. 2007a). In addition, PMN-PT is more suitable for high sensitivity, small aperture ultrasound transducer application in a frequency range of 20–100-MHz, such as IVUS transducers, because of its superior dielectric permittivity value ($\varepsilon_{33}^{s}/\varepsilon_0$ ~ 5229) that makes the electrical impedance of the transducer to be better matched with the 50 Ω electronics within a tiny surface area. In consequence, PMN-PT was used in this study to fabricate the 35 and 90-MHz transducer elements for the multi-frequency IVUS catheter. Nevertheless, for the center frequency of over 100-MHz, the electrical impedance mismatch between the PMN-PT transducer and the acquisition platform will become significant for the same element size. And it is very challenging to further miniaturize the aperture size of element to be less than 0.3 mm. A need for materials with even lower dielectric permittivity rises for fabricating transducers with the center frequency of over 100-MHz. Lead-free single-crystal lithium niobate ($LiNbO_3$) provides a comparable k_t to PZT, a much lower dielectric permittivity, and high longitudinal sound speed (v ~ 7340 m/s). It is suitable for designing sensitive large aperture high-frequency single-element transducers (Cannata et al. 2003). Therefore, $LiNbO_3$ was selected to fabricate 120 and 150-MHz transducer elements.

The combination of two transducer elements in the multi-frequency IVUS catheter shares a similar principle with the integration of a transducer and optical probe in the IVUS-OCT catheter, as well as the hybrid catheter design for the other multi-modality intravascular imaging systems. The design scheme of the intravascular catheter should not only allow for the technical feasibility but also fulfill the clinical requirements. Several studies on the design of integrated IVUS-OCT catheter have been reported individually (Li et al. 2010, 2014a; Yin et al. 2011; Bourantas et al. 2013). Each of the designs carries its own features and limitations. Based on the curvy structure of the cardiovascular system, a catheter with small outer diameter (OD), short front rigid length, and high flexibility is demanded to enable the safe delivery of catheter to the coronary artery (Bourantas et al. 2013). Given the fact that the cardiovascular intervention procedure is time-limited, the multi-frequency IVUS system should be able to facilitate accurate fusion of two IVUS in a real-time manner. Based on the previous IVUS-OCT catheter designs, three design schemes, with different configurations of how the two transducers of the same shape are aligned, of the multi-frequency IVUS catheter are presented and summarized in Fig. 2.9. These design schemes include aligning the two transducers in the configurations of left-and-right, fore-and-aft, and back-to-back. For the left-and-right configuration (Fig. 2.9a), online image co-registration can be achieved by implementing a simple post-processing algorithm since the two transducers are imaging the same cross section within each image frame. However, this design scheme has an enlarged OD that is almost doubled as the single-element IVUS catheter. The fore-and-aft configuration (Fig. 2.9b) has a satisfactory OD that is the same size as the single-element IVUS catheter. But the increased front rigid length makes it less flexible when going through sharp turns in the cardiovascular system. Moreover, image co-registration has to be performed through pullback scanning because the two transducers have a

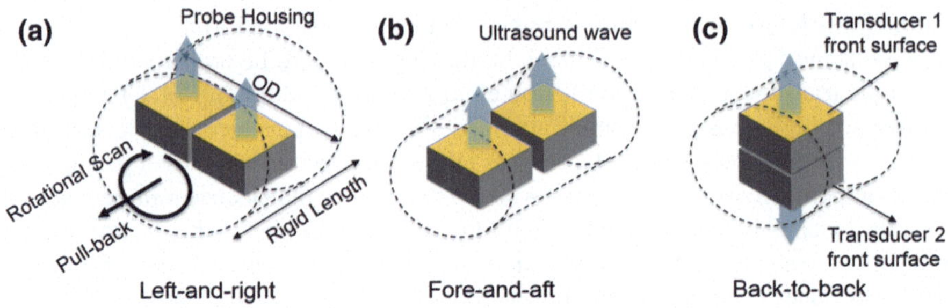

Fig. 2.9 Illustrations of design schemes of multi-frequency IVUS catheter: **a** left-and-right configuration; **b** fore-and-aft configuration; **c** back-to-back configuration (Ma et al. 2015b)

separation of more than 0.5 mm, which may result in inaccurate image co-registration for the artery walls that will be changing dynamically during each cardiac cycle. As shown in Fig. 2.9c, the back-to-back configuration provides a solution to the image co-registration problem in the fore-and-aft configuration, since the two transducers generate images of same cross section within one frame. This design scheme is preferred to the other two because it aligns the two transducers in the thickness direction, thus having the most effective use of space inside the cylindrical housing, and consequently offers a small OD and short front rigid length. Furthermore, based on the unique feature of back-to-back configuration, the real-time image co-registration at two frequencies can be accomplished by simply rotating one of the IVUS images of the same cross section by 180° apart.

In this study, three prototypes of back-to-back multi-frequency IVUS catheters with different frequency combinations (35/90, 35/120 and 35/150-MHz) were fabricated and evaluated. The general fabrication techniques described by Cannata et al. were used to prepare for the single-element IVUS transducer before assembling them into a catheter (Cannata et al. 2003). The PMN-PT single-crystal was lapped to 50 and 28 μm for the IVUS transducer operating at 35 and 90-MHz, respectively. $LiNbO_3$ wafer of 22 and 15 μm were also acquired to fabricate the transducer at 120 and 150-MHz. All the wafers were sputtered a 1500 Å chrome/gold on both sides to serve as the front and back electrodes. Due to the difficulties of lapping the piezoelectric material down to a thickness less than 5 μm, only the 35 and 90-MHz transducers had a matching layer made of the mixture of Insulcast 501 and 2–3 μm silver particle. The matching layer was cast on to the front electrode of piezoelectric material, and then lapped to the designed thickness listed in Table 2.2. Conductive epoxy (E-Solder 3022) was cast onto the back electrodes of the wafers, and all of the wafers were lapped to a thickness of 0.3 mm. The transducer element, together with the backing layer, was diced into a $0.5 \times 0.5\ mm^2$ square shape. The front electrode and back electrode were connected to the shielding wires and central wires of the coaxial cable insulated by the epoxy, respectively. A parylene layer was deposited on to the entire element of the transducers to serve as the second matching layer of 35 and 90-MHz transducer, and the first matching layer for the 120 and 150-MHz transducer. The thickness of the parylene layer for each transducer was

Table 2.2 Properties of the materials used in IVUS transducer fabrication (Cannata et al. 2003)

Material	Use	Density (g/cm^3)	Longitudinal sound speed (m/s)	Acoustic impedance (MRayl)
Insulcast 501 and 2–3 μm silver particles	First matching layer for 30-MHz (12 μm) 90-MHz (5 μm)	3.86	1900	7.3
Parylene	Second matching layer for 30-MHz (12 μm) 90-MHz (5 μm); First matching layer for 120-MHz (3 μm) 150-MHz (2 μm)	1.18	2200	2.6
E-Solder 3022	Backing layer	3.20	1890	5.9
EPO-TEK 301	Insolating epoxy	1.15	2650	3.1

listed in Table 2.2. The two elements of different center frequencies were carefully aligned into the back-to-back configuration before inserted into a two-window stainless steel cap (0.95 mm OD and 2 mm length). A polyimide layer was bonded by epoxy in between the two backings of the IVUS transducers so as to further insulate the electrical signals. A triple-wrapped torque coil (OD: 0.9 mm) was connected to the catheter cap, covering the two electrical wires of the transducers in order to provide the smooth torque translation. The two coaxial cables, a total of four electrical wires, were then connected to the separated channels of an eight-channel slip ring (SRH0317, Prosper M&E Tech Co, Hangzhou, China) to allow for the IVUS signal coupling for the two transducers during the rotational scan. The schematic and photograph of a back-to-back multi-frequency IVUS catheter were shown in Fig. 2.10a, b. The overall size of the multi-frequency IVUS catheter is 0.95 mm in diameter with a front rigid length of 2 mm, which is comparable to the size of a commercial IVUS catheter.

Characterization of Multi-frequency IVUS Imaging System

The experimental imaging setup is illustrated in Fig. 2.11. A low-frequency pulser (AVTECH AVB2-TB-C, Avtech Electrosystems Ltd, Ogdensburg, NY) and a high-frequency pulser (AVTECH AVB2-TC-C, Avtech Electrosystems Ltd, Ogdens-

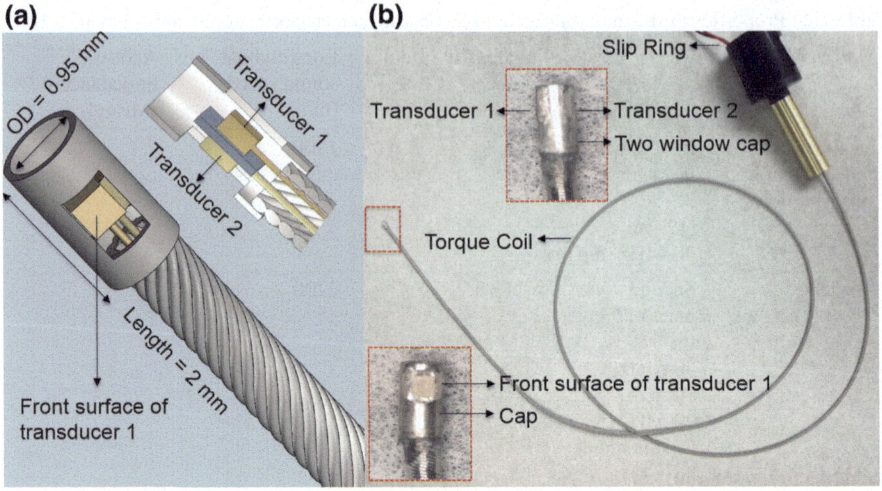

Fig. 2.10 **a** Diagram of back-to-back multi-frequency IVUS catheter. Middle: 3D drawing. Left-bottom: sectional drawing; **b** photograph of a multi-frequency IVUS catheter prototype. Enlarged photograph: side view (top) and front view (bottom) of catheter tip (Ma et al. 2015b)

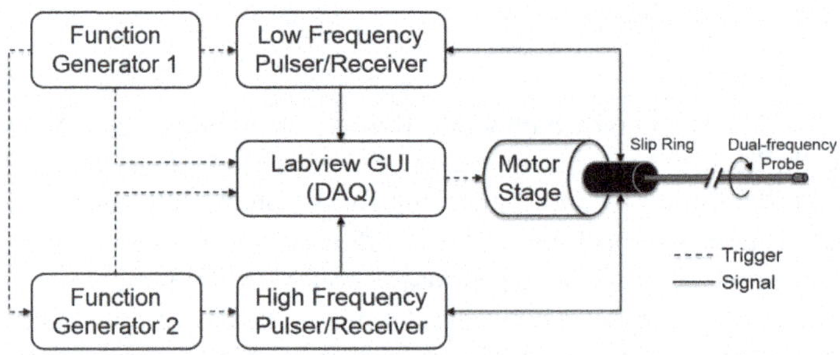

Fig. 2.11 A diagram of the multi-frequency IVUS imaging system (Ma et al. 2015b)

burg, NY) were used to excite the low-frequency transducer (35-MHz) and the high-frequency transducer (90, 120, and 150-MHz), respectively. The received radiofrequency (RF) signals were amplified by 33 dB (Miteq AU1114, MITEQ Inc, Hauppauge, NY) and digitized by a 12-bit data acquisition board (Alazar Technologies Inc, Pointe-Claire, Canada) at a sampling frequency of 1.8 GS/s. Two function generators were used to synchronize the pulser/receiver, digitizer, and the motor unit, and one of them has a 50-μs time delay in respect of the other. The saved RF data were processed and displayed using a custom-developed LabVIEW program (National Instruments, Austin, TX). During the imaging experiments, the catheter tip was positioned inside the central lumen of the phantom and the catheter was rotated by the motor unit while the phantom and the slip ring were kept stationary.

Pulse-echo testing (Cannata et al. 2003) was performed to evaluate the performance of the transducers. An X-cut quartz, placed at an approximate distance of 1.9 mm away from the transducers, was used as a reflecting target during the tests. The received echo signals in time-domain and the frequency responses of the 35-MHz transducer of the three catheters were recorded and only one of them, as a representative, is shown in Fig. 2.12a. The echo signals and the frequency responses of the 90, 120 and 150-MHz transducers from the pulse-echo testing are displayed in Fig. 2.12b–d, respectively. To evaluate the axial and lateral resolution of the transducers, 6-μm-diameter tungsten wire targets were imaged. The measured center frequency, −6 dB fractional bandwidth, axial resolution, and lateral resolution of these transducers are listed in Table 2.3.

The measured center frequencies of these transducers are in reasonably good agreement with the designed center frequencies. The comparisons in Fig. 2.12 also

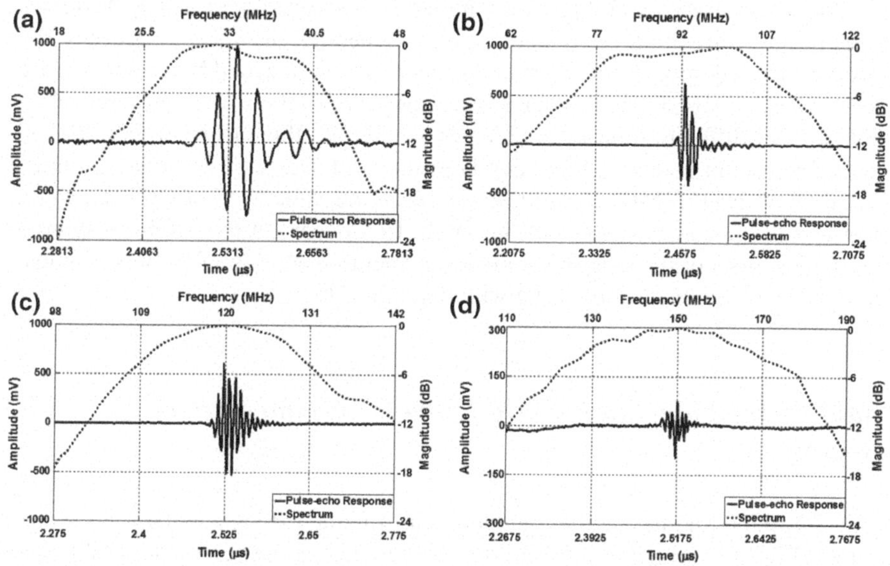

Fig. 2.12 Pulse-echo measurement results. Time-domain echo signals and frequency responses of a representative **a** 35-MHz transducer, **b** 90-MHz transducer, **c** 120-MHz transducer, and **d** 150-MHz transducer (Ma et al. 2015b)

Table 2.3 Measured center frequencies, bandwidths and resolutions of the representative transducers used in this study

Designed center frequency (MHz)	35	90	120	150
Measured center frequency (MHz)	33.6	91.2	120.0	149.7
−6 dB fractional bandwidth (%)	47.7	42.0	24.1	28.1
Axial resolution (μm)	46.0	21.5	25.7	17.2
Lateral resolution (μm)	231.5	123.5	105.3	87.3

show that as the center frequency of the transducer goes higher, the amplitude of the echo signals becomes smaller. These transducers exhibit various bandwidth characteristics. The −6 dB fractional bandwidths of 35 and 90-MHz transducer are 47.4 and 42.0%, respectively, which is attributed to the better piezoelectric properties of the PMN-PT single-crystal as well as the presence of the silver epoxy matching layer. However, due to the missing of the matching layer, the 120 and 150-MHz transducers made from $LiNbO_3$ exhibit narrower bandwidths of 24.1 and 28.1% respectively. Theoretically, a broader bandwidth will result in a shorter pulse length during the monocycle excitation, and then a better axial resolution of a transducer can be achieved owing to the broadband characteristics (Zhou et al. 2011b). It should be noted that the axial resolution of the 90-MHz transducer is even better than that of the 120-MHz transducer. This can be explained by the narrowband characteristics of the 120-MHz transducer. These results suggest that the high-frequency transducers ranging from 90 to 150-MHz in this study are potentially capable of resolving the thin fibrous cap in the axial direction. The lateral resolution is primarily dependent on the beamwidth within the imaging range. Generally, the center frequency of a transducer has a dominant influence on the lateral resolution, and hence a higher frequency transducer is able to provide a better lateral resolution. The measured lateral resolutions of different transducers are consistent with this theory, where the 150-MHz transducer provides the best lateral resolution of 87.3 μm. The axial resolution (17.2 μm) of the 150-MHz transducer can reach the same level of OCT. However, given that the lateral resolution of the IVUS transducers in this study is approximately four times of the axial resolution, the lateral resolution of IVUS is the least competitive aspect when comparing with that of OCT.

Phantom Imaging and Ex Vivo Human Coronary Artery Imaging

An agar-based tissue-mimicking phantom with silicon carbide powder as the sound scatters (Fig. 2.13), having a 3-mm-diameter lumen in the center, was fabricated to evaluate the CNR and the imaging depth of the multi-frequency IVUS catheter (Teirlinck et al. 1998; Ryan and Foster 1997). A polyimide tube with a 1.5-mm OD and a 50-μm wall thickness was placed right next to the central lumen. A graphite rod with a 0.5-mm OD was placed at a distance of 0.5 mm away from the border of central lumen. The polyimide tube wall was designed to be a resolution target, which was an analog to the thin fibrous cap. The graphite rod was a target with strong acoustic reflection.

To evaluate the imaging capability of the multi-frequency IVUS catheter, tissue-mimicking phantom images without the presence of blood were initially acquired. The central lumen and the polyimide tube were filled with DI water. A phantom image (Fig. 2.14a) captured by one of the three 35-MHz IVUS demonstrated an imaging depth of 5 mm at this frequency. The graphite rod was clearly distinguished from the

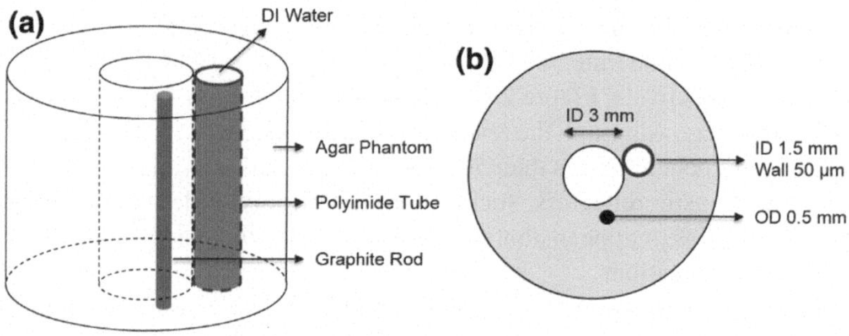

Fig. 2.13 Illustration of agar-based tissue-mimicking phantom: **a** 3D drawing; **b** sectional drawing (Ma et al. 2015b)

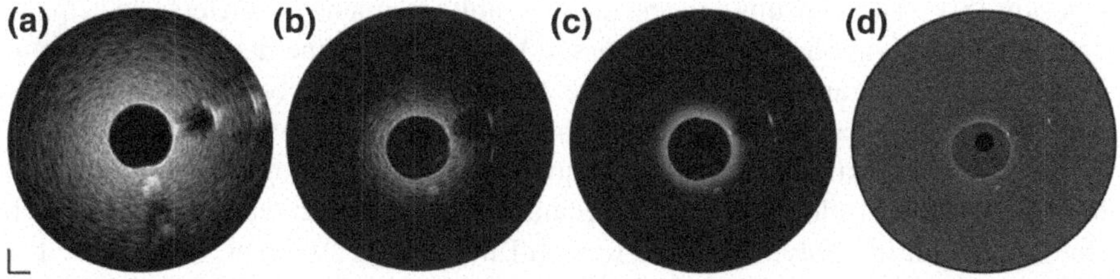

Fig. 2.14 Tissue-mimicking phantom images without the presence of blood at **a** 35-MHz, **b** 90-MHz, **c** 120-MHz, and **d** 150-MHz. Dynamic range: 45 dB; scale bar: 1 mm (Ma et al. 2015b)

surrounding agar with strong echo signal followed by an acoustic shadow. The front and back surfaces of the polyimide tube could be identified, but the actual thickness of the polyimide tube could not be precisely resolved due to insufficient resolution. Another important evaluation parameter of an IVUS image is the CNR, which is also critical for the detection of sub-resolution targets. The CNR was calculated by using the following equation (Li and O'Donnell 1994; Li et al. 2014b):

$$\mathrm{CNR} = \sqrt{\frac{(\mathrm{mean}_t - \mathrm{mean}_n)^2}{\mathrm{var}_t + \mathrm{var}_n}}$$

where mean_t and mean_n stand for the mean of the signal of the imaging target and the mean of the background noise; var_t and var_n represent the signal variance of the imaging target and the background noise. In this case, the imaging target is the agar and the background noise is the water filled lumen within the polyimide tube. The CNR of the phantom image of the 35-MHz transducer is 5.1. The phantom images of the 90 and 120-MHz transducers are shown in Fig. 2.14b, c, respectively. Because resolution has been improved, the front and back surfaces of the polyimide tube could be better differentiated. A more detailed structure around the boundary of the graphite rod could be visualized. Moreover, due to the stronger attenuation of higher frequency ultrasound, a significant decrease in imaging depths, about 2 and 1 mm compared to 5 mm with the 35-MHz transducer, was observed at 90 and 120-MHz, respectively. At the same time, the CNRs of the images at 90 and 120-MHz dropped to 2.1 and 1.6, respectively. Figure 2.14d shows the phantom image generated by the 150-MHz transducer. Although the resolution of the image was further improved, the image could only penetrate less than 500 μm in depth and provided weaker contrast for the strong acoustic reflectors, such as surface of polyimide tube and boundary of graphite rod. Thus, it appears that transducer at such frequency is not suitable for IVUS imaging applications.

Despite the decreased penetration depth and CNR, IVUS transducers of 90 and 120-MHz are still capable of providing valuable high-resolution information around the inner lumen area. The fused phantom images captured by 35/90-MHz catheter and 35/120-MHz catheter are shown in Fig. 2.15a, b, respectively. The image co-registration in each imaging pair is accomplished by simply rotating one of the IVUS images by 180°. These results suggest that the multi-frequency catheters can achieve a better characterization of the artery tissue by integrating the deep imaging depth of lower frequency transducer and high-resolution of higher frequency transducer.

To further investigate the feasibility of the multi-frequency catheters in a setup that is closer to the clinical settings, tissue-mimicking phantom imaging was performed with the presence of blood. The center lumen of the phantom was filled with fresh swine blood and the polyimide tube was still filled with DI water. As expected, a slightly decreased penetration depth of 4 mm and a reduced CNR value of 3.2 were obtained in the image at 35-MHz (Fig. 2.16a) due to the strong attenuation of the blood in the lumen. As shown in Fig. 2.16b–d, the much stronger attenuation of the higher frequency ultrasonic waves in the blood dramatically limits the imaging depth and degrades the image quality. These results indicate that to improve imaging depth with a high-frequency transducer may require the temporal removal of luminal blood during the in vivo imaging, similar to the flushing mechanism in OCT, which is an intrinsic drawback of multi-frequency IVUS imaging.

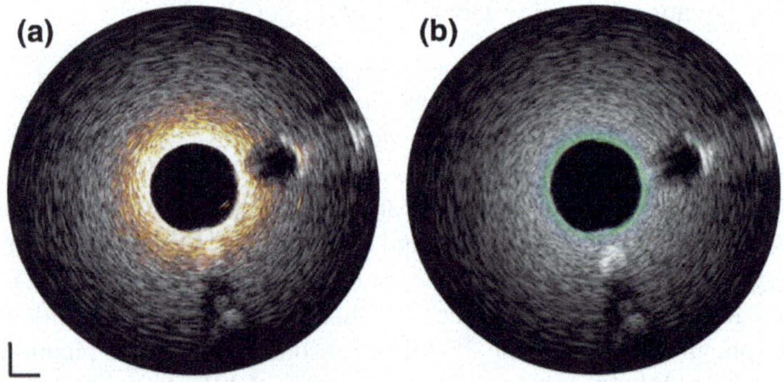

Fig. 2.15 Fused tissue-mimicking phantom images captured by **a** 35/90-MHz multi-frequency IVUS catheter and **b** 35/120-MHz multi-frequency IVUS catheter. White color: 35-MHz ultrasound image; orange color: 90-MHz ultrasound image; green color: 120-MHz ultrasound image; dynamic range: 45 dB; scale bar: 1 mm (Ma et al. 2015b)

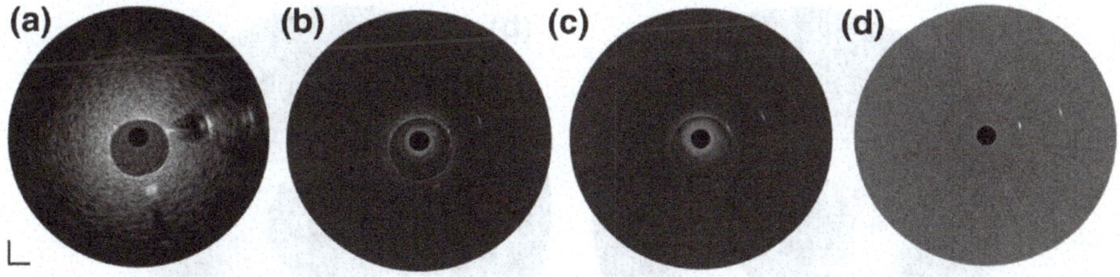

Fig. 2.16 Tissue-mimicking phantom images in the presence of blood at **a** 35-MHz, **b** 90-MHz, **c** 120-MHz, and **d** 150-MHz. Dynamic range: 45 dB; scale bar: 1 mm (Ma et al. 2015b)

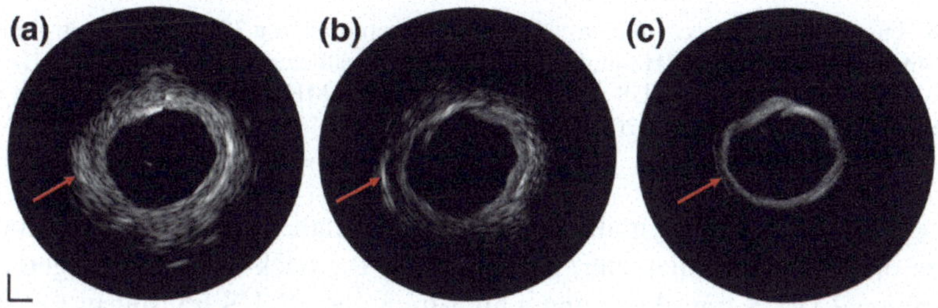

Fig. 2.17 IVUS images of the human coronary artery at **a** 35-MHz, **b** 90-MHz, and **c** 120-MHz. Dynamic range: 50 dB; scale bar: 1 mm (Ma et al. 2015b)

Human cadaver coronary artery images acquired at 35, 90, and 120-MHz are displayed in Fig. 2.17. The 35-MHz transducer provides more complete morphological information of the coronary artery owing to the deep penetration depth (Fig. 2.17a). With the 90-MHz transducer, the three-layer structure (intima, media and adventitia) pointed by the red arrow is identified in Fig. 2.17b because of the improved axial resolution. However, it is blurred and not apparent in Fig. 2.17b. In Fig. 2.17c, although the 120-MHz transducer can only image through the intima layer due to the limited penetration depth, an improvement in both axial resolution and lateral resolution is achieved, indicated by a reduced speckle size in the image of intima layer. The fused image pairs acquired by the 35/90 and 35/120-MHz integrated catheters were shown in Fig. 2.18a, b respectively. The integrated image pairs, combining the advantages of the high-resolution superficial image of lumen area offered by the high-frequency IVUS transducer (90 or 120-MHz) and the deep penetration image of the entire artery wall provided by the low-frequency transducer (35-MHz), yield a more comprehensive visualization of the artery wall and plaque volume.

Summary

We have successfully developed and prototyped multi-frequency IVUS imaging system catheters with three different frequency combinations prepared by PMN-PT and

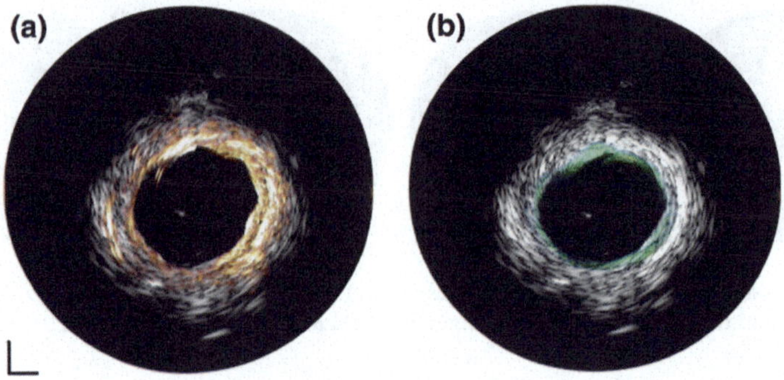

Fig. 2.18 Fused IVUS images of human coronary artery captured by **a** 35/90-MHz multi-frequency IVUS catheter and **b** 35/120-MHz multi-frequency IVUS catheter. White color: 35-MHz IVUS image; orange color: 90-MHz IVUS image; green color: 120-MHz IVUS image; dynamic range: 50 dB; Scale bar: 1 mm (Ma et al. 2015b)

$LiNbO_3$ single-crystals. The multi-frequency IVUS catheter, with a clinical compatible size of 0.95 mm in diameter, is featured by the back-to-back arrangement of a conventional IVUS transducer and a high-frequency IVUS transducer to achieve accurate co-registration of two IVUS images. The performance of the high-frequency IVUS transducer at different frequency ranges (90, 120, and 150-MHz) was evaluated and compared to find the optimal frequency range for the high-frequency transducer in the multi-frequency IVUS catheter, considering imaging depth, imaging resolution, and CNR. Tissue-mimicking phantom imaging with and without the presence of blood shows that the multi-frequency catheter carries the complementary strengths of the deep imaging depth of the conventional IVUS and the high-resolution of the high-frequency transducer. On the other hand, these results also point out a potential weakness of the multi-frequency catheter, which is flushing of the luminal blood is required to ensure the functionality of this technique. The in vitro human cadaver coronary artery imaging demonstrates the capability of the multi-frequency catheter to provide more comprehensive visualization of the vascular structure and to facilitate the assessment of the vulnerable plaque. Compared to other multi-modality intravascular imaging techniques, the multi-frequency IVUS imaging capitalizes the advantage of cost-effectiveness because only a moderate modification of the current commercial IVUS system is needed. Besides, the apparent clinical utility of this ultrasound-only technology can be promptly explored during the translational stage since most of the interventional physicians are familiar with the ultrasound technology. The future translation of the multi-frequency IVUS imaging into clinical use will not only consolidate the leading status of the IVUS technology in the interventional cardiology practice, but also prosperously lead to patient benefits.

Some Advances in High-Frequency IUVS

Introduction

The performance of the multi-frequency IVUS transducers used in Section "Multi--frequency IVUS Imaging" is summarized in Fig. 2.19, including center frequency, imaging depth, imaging contrast, and imaging resolution. The results are in agreement with the classic ultrasound theory that increasing the center frequency of the transducer results in better imaging resolution and decreased imaging depth. Even though different piezoelectric materials were selected to fabricate the IVUS transducers at different center frequencies, there was still some electrical impedance mismatch between the IVUS transducers and electrical system due to the fact that the transducer aperture size was set to be 0.5 mm. Future implementation of a matching circuit into the system will potentially solve this issue and improve the overall performance of multi-frequency IVUS catheter. There is a significant gap between the conventional IVUS and the competitive technology OCT, indicated by the red dot, regarding the image resolution and penetration depth. Combining a higher frequency IVUS transducer in a frequency range of 90–120-MHz with the conventional IVUS catheter is a more feasible and simpler solution to reduce the controversy caused by insufficient resolution and fill the gap between the conventional IVUS and OCT. However, the requirement of temporal removal of luminal blood with the aim of eliminating the stronger attenuation effect for higher frequency ultrasound is a weakness of this technology that needs to be addressed in the future. Although further increasing the center frequency of the transducer to 150-MHz or higher could theoretically

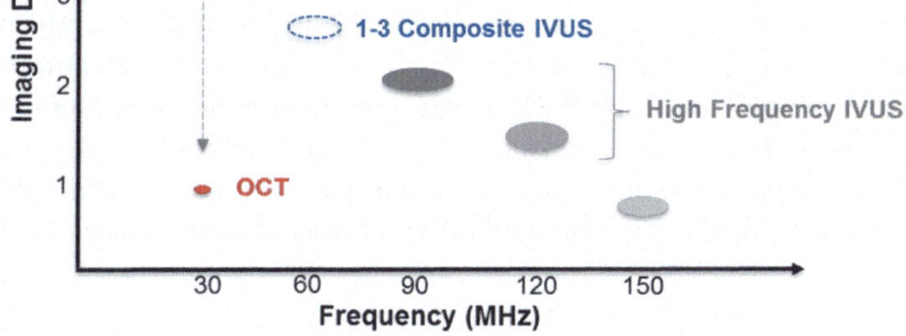

Fig. 2.19 Summaries of IVUS transducers' performances at different center frequencies in this study, including imaging depth, imaging contrast, and imaging resolution. Red dot: OCT; dashed blue dot: 1-3 composite IVUS transducer (Ma et al. 2015b)

reach the resolution level of OCT, the extremely shallow imaging depth and reduced imaging contrast will dramatically degrade the image quality. Thus, it may not be worthwhile to use the transducer with such high frequencies for IVUS applications. As discussed in Section "Multi-frequency IVUS Imaging", owing to its broadband characteristics, the 90-MHz transducer exhibits a better axial resolution than the 120-MHz transducer. This suggests that the integration of a broader bandwidth transducer is favorable to further renovate the multi-frequency IVUS catheters.

Therefore, one direction to advance IVUS technology is to implement the state-of-the-art composite materials. The newly developed micromachined PIN-PMN-PT single-crystal 1-3 composite transducer at 40-MHz for IVUS application, providing an improved axial resolution of 43 μm and a comparable penetration depth to conventional IVUS transducers, was investigated (Li et al. 2014a; Yuan et al. 2008). For the possible implementation of 1-3 composite material in multi-frequency IVUS imaging, the proposed 1-3 composite transducers in a frequency range of 60–90 MHz, indicated by the blue dashed dot in Fig. 2.19, serves as a more promising substitute for the high-frequency transducers in the multi-frequency IVUS catheters so that the enhanced resolution can be achieved without excessively sacrificing imaging depth. Moreover, it is the fact that lateral resolution of the multi-frequency IVUS catheter system is almost three to four times worse than the axial resolution, which is also the technical barrier and natural shortcoming of all other IVUS systems. The images quality of ultrahigh-frequency IVUS in the deeper region because of the further downgraded lateral and contrast-to-noise ration. Some beamforming techniques that combine virtual source synthetic aperture (VSSA) focusing and coherence factor weighting (CFW) are reported to have the possibility to improve the lateral resolution of IVUS imaging. Furthermore, as discussed previously, a further increase of operating frequency will allow for the improvement of axial resolution, while the penetration depth of IVUS will be sacrificed. Thus, an imaging method that improves the penetration depth of high-resolution IVUS would also be of major clinical importance. Modulated excitation imaging is known to allow ultrasound waves to penetrate further. Qiu et al. recently reported an ultrasound system specifically for modulated-excitation-based IVUS imaging (Qiu et al. 2017). The system incorporates a high-voltage waveform generator and an image processing board that is optimized for IVUS applications. The results show that the modulated excitation IVUS system was able to provide increases of 86.7% in penetration depth and 9.6 dB in the signal-to-noise ratio for 60-MHz IVUS. In this section, we will introduce three advances of ultrahigh-frequency IVUS technology including the implementation of broadband 1-3 composite IVUS transducer to enhance axial resolution, utilization of virtual source synthetic aperture focusing techniques to improve lateral resolution, as well as the modulated excitation method to increase the penetration depth.

Development of PIN-PMN-PT 1-3 Composite IVUS Transducer

Imaging resolution is inversely proportional to the frequency bandwidth of the transducers. Transducers with higher center frequency and broader bandwidth can provide better image resolution (Foster et al. 2000b). In our previous work, 80-MHz IVUS transducers were investigated and could provide axial resolution of 35 μm and penetration of 2 mm (Li et al. 2011). The fine resolution is critical for detecting certain vulnerable plaques such as thin-cap fibroatheroma (TCFA), which has a cap thickness less than 65 μm (Kolodgie et al. 2001). The drawback of working at 80-MHz is the shallow penetration depth resulting from the strong attenuation in blood and vessel wall. An alternative method to improve resolution without sacrificing penetration depth is to use transducers with broader bandwidth at relatively lower frequency (40-MHz). Moreover, a broadband transducer is advantageous for multi-frequency signal processing to differentiate blood and vessel (Li et al. 2008).

To achieve optimal imaging performance, the piezoelectric material of an IVUS transducer must be carefully chosen. Conventionally, $Pb(Zr,Ti)O_3$ (PZT)-based ceramics, such as PZT-5H, have been used for fabricating IVUS transducers. Recently, binary-relaxor-based ferroelectric single-crystal $Pb(Mg_{1/3}Nb_{2/3})$-$PbTiO_3$(PMN-PT) has also been investigated for fabricating high-frequency ultrasonic transducers (Rhee 2007; Yuan et al. 2006). Compared with PZT ceramics, PMN-PT crystal demonstrated improved sensitivity and bandwidth, resulting from its higher piezoelectric coefficient (d_{33} ~ 2000 pm/V), and electromechanical coupling coefficients (k_t ~ 0.58, k_{33} ~ 0.9) (Luo et al. 2000; Park and Shrout 1997b; Tian et al. 2007a; Zhang and Shrout 2010). Ceramic or crystal 1-3 composite material exhibits even broader bandwidth than monolithic materials because of composite's very high coupling coefficient and low acoustic impedance (Smith 1989). However, composites for transducers operating above 20-MHz cannot be made using traditional dice-and-fill technique because of the thickness limitation of the dicing saw blade, which is no less than 10 μm (Cheng et al. 2003; Zhang et al. 2011; Zhou et al. 2011a). Mechanical dicing can also produce highly stressed and damaged surface layers, thereby degrading the properties of 1-3 composites (Lee et al. 2010). For 1-3 composites, the frequency of the first lateral mode is empirically expressed as

$$f_1 = \frac{V_T}{2\sqrt{2}d_P}$$

where f_1 is the frequency of first lateral mode resonance, V_T is the shear wave velocity of filler, and d_P is the kerf width (Jiang et al. 2006; Tian et al. 2007a). However, previously reported micromachined 1-3 composites were fabricated from PMN-PT crystal. The binary PMN-PT crystal has relatively low coercive field (E_c ~ 2.5 kV/cm), depoling temperature ($T_{R/T}$ ~ 60 to 95 °C), and Curie temperature (T_c ~ 130 to 178 °C) (Park and Shrout 1997a; Tian et al. 2007b; Ye and Dong 2000; Zhang et al. 2008). Higher E_c would allow transducers to be driven in a higher electrical field and higher $T_{R/T}$ for a broader temperature usage range (Tian et al. 2007b; Zhang et al. 2008). During the fabrication process, transducers are frequently incubated at elevated temperatures (40–90 °C) for epoxy curing, wax bonding, electrode sputtering, etc., which can cause PMN-PT depoling. Depoling can make testing transducer during fabrication inconvenient. In addition, an IVUS catheter must go through a sterilization process before entering the market. Some sterilization processes, such as steam sterilization, work at relatively high temperature (90 °C). After sterilization, it is difficult to repole the IVUS transducer without taking the catheter out of the sterilized pouch. Therefore, thermal instability of PMN-PT may degrade transducers' performance during fabrication and sterilization processes. By adding lead indium niobate (PIN) into the PMN-PT system, the ternary crystal PIN-PMN-PT was found to have superior electrical and thermal stability (E_c ~ 5.5 to 6 kV/cm, $T_{R/T}$ ~ 120 to 130 °C, T_c ~ 160 to 200 °C) to overcome the aforementioned drawbacks, while maintaining the excellent piezoelectric properties (d_{33} ~ 900 to 1900 pm/V, k_{33} ~ 0.83 to 0.92) (Tian et al. 2007b; Zhang et al. 2008). PIN-PMN-PT has been investigated for high-frequency ultrasonic applications and demonstrated reduced temperature dependence of performance and minimal depolarization during fabrication (Chen et al. 2012; Sun et al. 2009). Composites based on PIN-PMN-PT crystal inherit the improvements of the thermal and electrical properties (Tian et al. 2010). Considering the merits of higher usage temperature range and coercive field, PIN-PMN-PT could be more advantageous than PMN-PT to fabricate micromachined 1-3 composite for high-frequency IVUS applications. In this paper, we report the use of micromachined PIN-PMN-PT 1-3 composite material for IVUS application. Miniature side-looking needle transducers and flexible IVUS catheters were built from the composite to operate at 40-MHz. Ex vivo experiments were conducted to demonstrate the superiority of the composite for IVUS application.

PIN-PMN-PT crystal 1-3 composite was fabricated using inductively coupled plasma (ICP)-enhanced DRIE technique by H. C. Materials Corp. (Tian et al. 2010). Polished PIN-PMN-PT crystal plate was first coated with Cr/Au as the seed layer for subsequent nickel electroplating. Photolithography was used to lay out the etching pattern over the Cr/Au layer. A 4- to 5-μm-thick nickel mask was then electroplated with the inverse pattern of the photoresist. Crystal parts with the patterned nickel mask were loaded in ICP-enhanced plasma (Oxford Instruments, PlasmaLab 100, Abingdon, Oxford, UK) for DRIE. Following etching, kerfs in the etched crystals were filled with epoxy (EPO-TEK 301, Epoxy Technology Inc. Billerica, MA). After epoxy cures, the composite was lapped from both sides to the final thickness. The composite plate was sputtered with Cr/Au electrodes on both sides and then deposited with a thick conductive backing material. A cross-sectional micrograph of the composite on top of a thick conductive backing material is shown in Fig. 2.20. The thickness of the composite is around 27 μm and the kerf width is less than 5 μm. DRIE produces a near vertical etched profile with a side wall angle greater than 85°.

Side-viewing needle-type IVUS transducers and flexible catheter-type IVUS transducers were fabricated based on the PIN-PMN-PT 1-3 composite. The measured f_c was 41-MHz and BW was 86%. BW is comparable to the reported PMN-PT

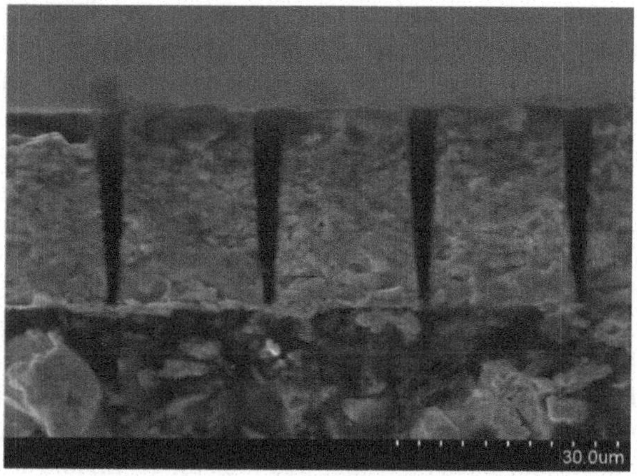

Fig. 2.20 PIN-PMN-PT 1-3 composite with conductive backing (Li et al. 2014b)

composite transducer (77–90%) (Yuan et al. 2006, 2008; Sun et al. 2010; Jiang et al. 2008), but almost doubled that of a PMN-PT crystal transducer (45%) (Zhou et al. 2007b). Tungsten wire targets with 6-μm OD were imaged to determine axial and lateral resolutions of the transducer, as shown in Fig. 2.21a. For comparison, the wire targets were also imaged by a 40-MHz PMN-PT single-crystal transducer (two matching layers; one backing layer; 0.5×0.4 mm^2 aperture; unfocused; 47% BW (Zhou et al. 2007b), as shown in Fig. 2.21b. The axial and lateral resolutions were determined from the −6 dB envelope width from the wire located at 1.3 mm, which is close to the transducer's natural focus (1.1 mm). For the PIN-PMN-PT composite

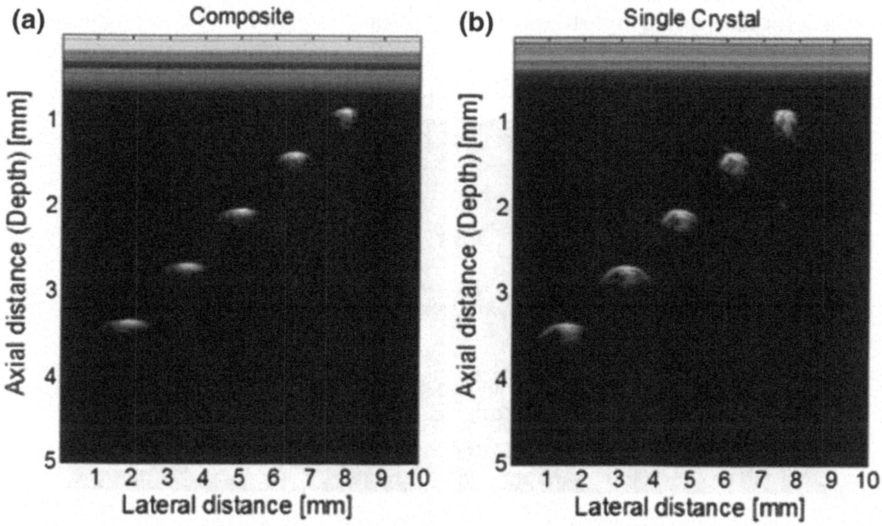

Fig. 2.21 Wire phantom images from **a** PIN-PMN-PT composite transducer; and **b** PMN-PT single-crystal transducer (Li et al. 2014b)

transducer, axial and lateral resolutions were 43 and 226 μm, respectively. For the PMN-PT single-crystal transducer, axial and lateral resolutions were 62 and 278 μm. As shown in Fig. 2.21, the wire phantom images of the composite transducer were much thinner than those of the single-crystal transducer because of the improved axial resolution. Images are displayed with 50 dB dynamic range.

An ex vivo IVUS image from the composite transducer is shown in Fig. 2.22a. For comparison, an image of the artery from the PMN-PT crystal transducer of the same cross section is shown in Fig. 2.22b. Both images are capable of differentiating the thickened intima (*I*), media (*M*), and adventitia (*A*) layers, as well as visualizing the calcified plaques (Ca, high-intensity areas followed by acoustic shadow). However, because of the improved axial resolution of composite transducer, Fig. 2.22a displays higher clarity (finer speckles) of all the three layers. The boundaries of IM and MA (denoted by green arrows) were identifiable by the composite transducer in Fig. 2.22a but are blurred in Fig. 2.22b and even unidentifiable at 9 to 11 o'clock. The improved resolution is also helpful to accurately determine the percentage of calcified plaque over the entire lumen circumference.

The micromachined PIN-PMN-PT single-crystal 1-3 composite material demonstrated improved effective electromechanical coupling coefficient k_t (eff) compared with monolithic single-crystal. The usage temperature of the composite is at least 30 °C higher than PMN-PT. The acoustic impedance is as low as 20 MRayl. Using the composite, needle-type and flexible-type IVUS transducers were fabricated and tested. 41-MHz transducers with 86% bandwidth were achieved, which resulted in 43-μm axial resolution. The characterization results of PIN-PMN-PT crystal composite showed superior piezoelectric properties which were comparable to that of PMN-PT-based crystal composite, as well as improved thermal property compared with that of PMN-PT based composite. PIN-PMN-PT crystal can be an alternative approach for fabricating high-frequency composite. Ex vivo IVUS imaging was con-

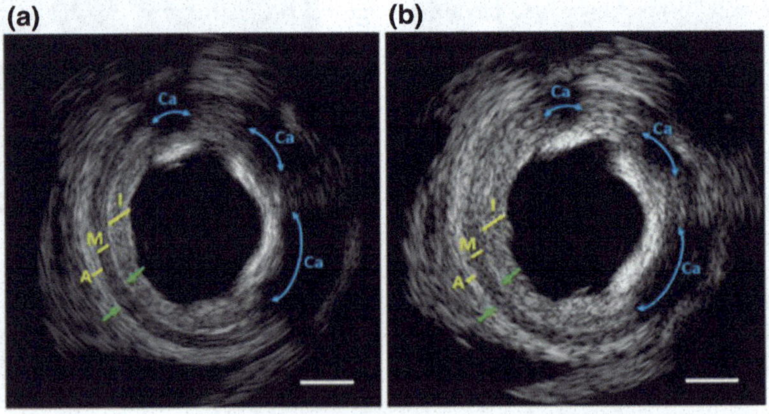

Fig. 2.22 IVUS image of human coronary artery at 40-MHz from **a** PIN-PMN-PT composite transducer; and **b** PMN-PT single-crystal transducer. *I* = intima; *M* = media; *A* = adventitia; Ca = calcified plaque. Green arrows denote the boundaries of IM and MA (Li et al. 2014b)

ducted to demonstrate the superiority of improved axial resolution. The composite transducer was able to identify the three layers of a human coronary artery specimen with high definition, whereas the image from a single-crystal transducer showed lower clarity of the layered structures.

IVUS with Virtual Source Synthetic Aperture Focusing and Coherence Factor Weighting

As a widely recognized clinical tool for intravascular imaging and the irreplaceable technical basis of most multi-modality intravascular imaging techniques, however, IVUS itself still requires further improvement in the overall performance including spatial resolution and contrast-to-noise ratio (CNR) over the whole range of interest. Several approaches that have been reported to improve the IVUS image quality are illustrated in Fig. 2.23 by simulating the two-way beam pattern of single-element IVUS transducers with a 0.5-mm active aperture size in PZFlex (Weidlinger Associates, Cupertino, CA). Figure 2.23a shows the two-way beam pattern of a conventional flat 40-MHz IVUS transducer that is naturally focused at 1.7 mm. Increasing

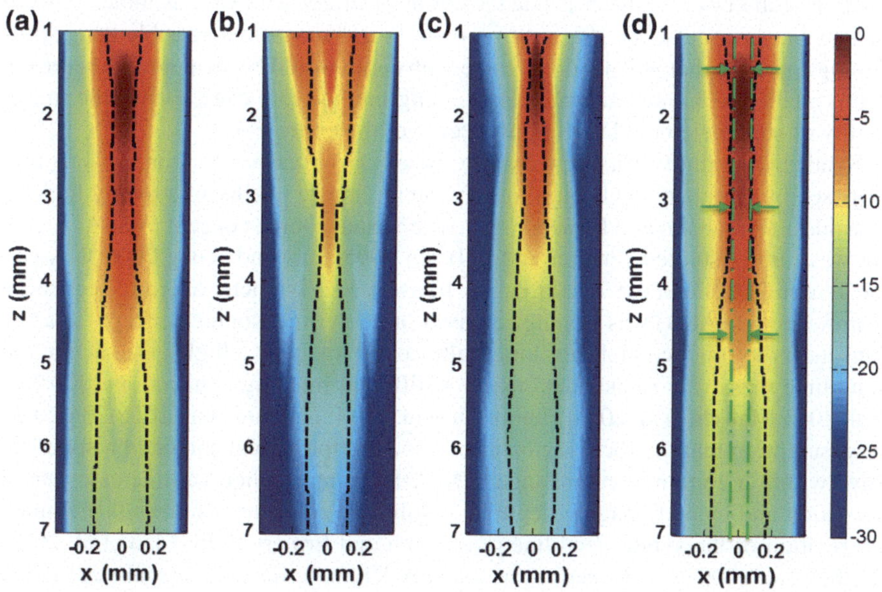

Fig. 2.23 Simulated two-way beam patterns and −6 dB beamwidth contour (black dashed line) of IVUS transducers with 0.5 mm active aperture size: **a** 40-MHz IVUS transducer with flat aperture, **b** 80-MHz IVUS transducer with flat aperture, **c** 40-MHz IVUS transducer with mechanical focusing at 3 mm. **d** Simulated beam pattern from (**a**) with expected −6 dB beamwidth contour (green dash–dot line) after VSSA&CFW processing. The *x*-axis is lateral direction, and the *z*-axis is axial direction (Yu et al. 2017)

the operational frequency of IVUS provides better spatial resolution within limited depth while sacrificing the CNR, and penetration depth of IVUS images beyond that depth (Ma et al. 2015b). The radiation pattern of a flat 80-MHz IVUS transducer is shown in Fig. 2.23b. It can provide good resolution at 3–4 mm, but the intensity is significantly attenuated beyond the natural focus, causing degradation in CNR.

Mechanical focusing of the IVUS transducer is another approach to achieve higher resolution within the focus region (Lee et al. 2016). The beam pattern of a 40-MHz IVUS transducer with a mechanical focus at 3 mm is presented in Fig. 2.23c. Since the mechanical focal point of 3 mm is beyond the natural focal point of 1.7 mm, further mechanical focusing will only pull the focal zone toward the transducer surface, which will in turn decrease the sensitivity and lateral resolution in the rest of field of view (FOV) (Lee et al. 2016). As shown in Fig. 2.23c, narrow beamwidth can be achieved at 1–2 mm, but the beam diverges beyond that region, where more artery structure information lies. Moreover, it is practically very challenging to geometrically focus the transducer of very small aperture size (e.g., $0.5 \times 0.5\ mm^2$).

Given the fact that increasing the center frequency and mechanical focusing method could only improve the lateral resolution and enhance the CNR within a certain region of FOV while downgrading the image quality outside of the focal region, it requires an accurate alignment between the focal zone of IVUS transducer and the region of interest. However, it is unrealizable to precisely control the exact position of the IVUS catheter inside the coronary artery with various lumen sizes in an actual intravascular imaging environment. In this scenario, a feasible image processing approach that will not downgrade with depth would be invaluable. Therefore, in this study, we applied an image processing method that can enhance the image quality of single-element IVUS transducer over the entire FOV.

Synthetic aperture focusing technique based on virtual source concept was first proposed by Passman and Ermert to improve the image quality of strongly focused transducer (Passmann and Ermert 1996). The virtual source concept (VSSA) focusing was further studied (Frazier and O'Brien 1998; Bae and Jeong 2000; Nikolov and Jensen 2002) and extended to array transducers. Synthetic aperture sequential beamforming (SASB) was developed based on the virtual source concept for array transducers to produce high-resolution ultrasound image at a high speed and at the same time massively reduce data load for different imaging applications (Kortbek et al. 2013; Hansen et al. 2014; Hemmsen et al. 2014; Di Ianni et al. 2016). Based on the assumption that the focal region can be seen as the virtual source of a spherical wave within a certain angular range, VSSA-focusing method can increase lateral resolution and extend imaging depth by combining adjacent scanlines that contain information of the synthesized line (Passmann and Ermert 1996; Liao et al. 2004; Kortbek et al. 2007). This inspires us to apply VSSA to the mechanical-state IVUS imaging for improving lateral resolution and SNR.

Even though the VSSA-focusing method improves the resolution by synthesizing a large aperture, it brings in higher sidelobes, which can cause unwanted artifact (Frazier and O'Brien 1998; Karaman et al. 1995). Amplitude apodization applied on the received data can suppress the sidelobe level and increase contrast resolution but broadening the main lobe, or sacrificing lateral resolution (Frazier and O'Brien 1998).

Coherence factor based on the coherence of delayed RF signal has been used as a focusing criterion to quantitatively evaluate the image quality (Mallart and Fink 1994; Hollman et al. 1999). Higher coherence factor value indicates delayed RF signals are more coherent among each other. It has been demonstrated that weighting the delayed beams adaptively using the coherent factor can reduce the sidelobe level of the VSSA-focused radiation pattern (Li et al. 2004) and to further improve the resolution (Liao et al. 2004). This method is called coherence factor weighting (CFW). The expected -6 dB beamwidth contour after VSSA-CFW processing is illustrated by the green dash–dot line in Fig. 2.23d. The application of this combined method to conventional IVUS transducer has not yet been investigated.

In this study, the VSSA-CFW beamforming method was extended to IVUS application for improving lateral resolution and CNR through the whole FOV. The applicability of this VSSA-CFW method was first demonstrated through simulations in Field II (Jensen and Svendsen 1992). Then, the efficacy was explored experimentally on tungsten wire phantom imaging, homogeneous agar-based phantom imaging, anechoic cyst phantom imaging and ex vivo human artery imaging. Both simulated and experimental results of wire target imaging demonstrate improvements in lateral resolution. CNR improvement is also clearly shown in the anechoic cyst phantom images. Image depth extension was quantitatively measured through in vitro imaging of the homogeneous phantom. In the human artery image, extended image depth, and better layer definition are achieved after image processing. To the best of our knowledge, this is the first time VSSA-CFW being used for IVUS imaging. This image processing method can be integrated into a current imaging system to improve the image quality of current IVUS imaging catheters. Moreover, it can further refine the multi-modality intravascular imaging techniques to provide more accurate and comprehensive information for vulnerable plaque identification.

The experiment results in Fig. 2.24 of rotational scanning also show an increase in the lateral resolution over the whole imaged range. The resolution results are summarized in Fig. 2.25. The VSSA-focusing alone does not improve the resolution of rotational scan mode since the moderate expansion of the synthesized aperture cannot compensate the beam divergence. However, after applying CFW, the lateral

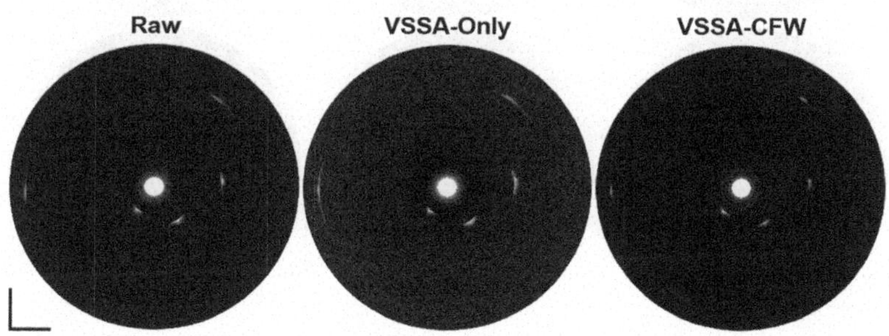

Fig. 2.24 Experimental results of wire phantom imaging by rotational scanning. Dynamic range: 50 dB; scale bar: 2 mm (Yu et al. 2017)

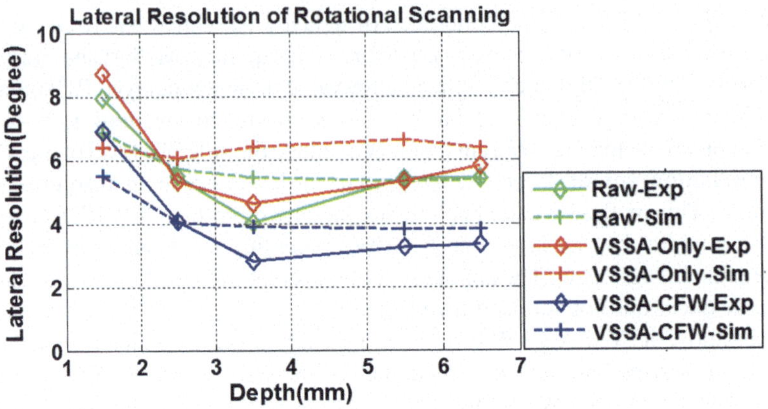

Fig. 2.25 −6 dB lateral resolution results of wire phantom images by rotational scanning (Yu et al. 2017)

resolutions increase by up to 42% for both simulation and experimental results. The beam profile of the wire phantom clearly shows the improvement after VSSA-CFW processing. For instance, 11 dB sidelobe suppression and 2.5° resolution increase were obtained for the simulated point target at 3.5 mm after CFW.

The experimental result of agar-based anechoic cyst phantom imaging is shown in Fig. 2.26. The calculated CNR values are summarized in Fig. 2.27. The CNR

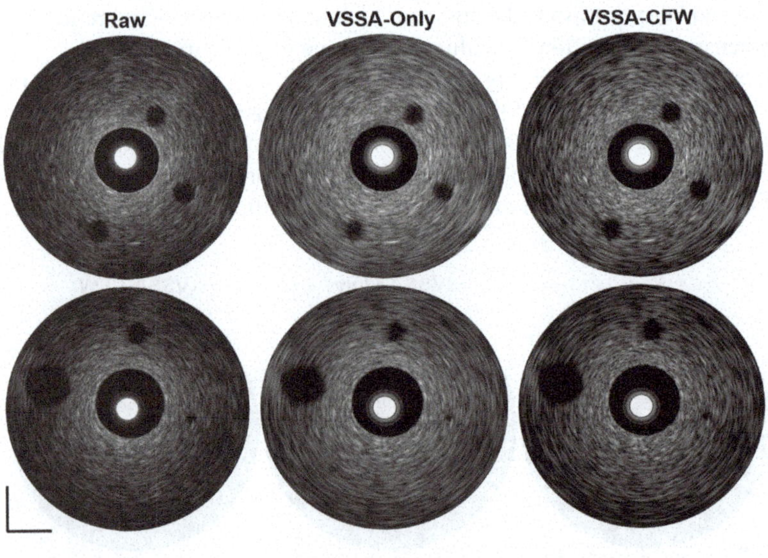

Fig. 2.26 Experimental imaging result of the two agar-based anechoic cyst phantom imaging: top row shows phantom with 1-mm cyst at different depth; bottom row shows phantom with different cysts at 3 mm. Dynamic range: 50 dB; scale bar: 2 mm (Yu et al. 2017)

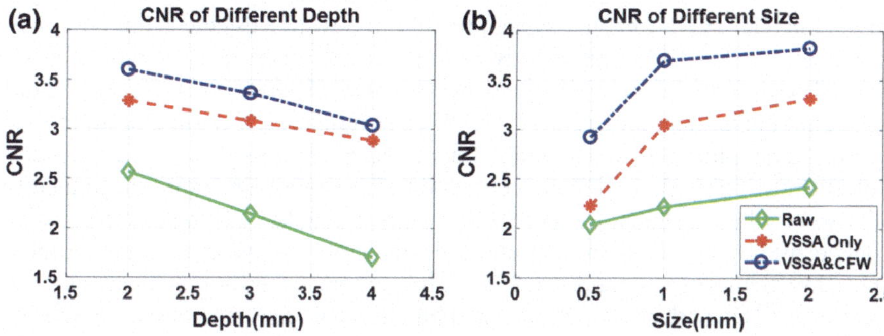

Fig. 2.27 Contrast-to-noise ratio (CNR) results of the two anechoic cyst phantom images. **a** CNR of 1-mm cyst at different depth. **b** CNR of different cysts at 3 mm (Yu et al. 2017)

improves after applying VSSA. However, the higher sidelobe still creates artifacts within the cyst region. After applying CFW onto the VSSA focused image, the CNR further improves, which demonstrates that CFW can effectively suppress the sidelobes and result in a higher CNR for cyst of different sizes and at different depths. In the VSSA-CFW image, superior CNR can be observed through the more distinct boundary between anechoic area and surrounding agar phantom, as well as at the boundary of the central lumen. IVUS images of the human artery are shown in Fig. 2.28. The suppressed noise floor and enhanced resolution help to better delineate media layer structure of the artery wall, as indicated by the yellow arrowheads.

This study demonstrates that the beamforming method of combining virtual source synthetic aperture and coherence factor weighting can effectively improve the overall quality of IVUS images, including lateral resolution, imaging depth and CNR. This is for the first time, to our best knowledge, that the VSSA-CFW is applied to single-element-based IVUS imaging. The VSSA-CFW beamforming method holds a promising potential for clinical translation since it can be integrated into the cur-

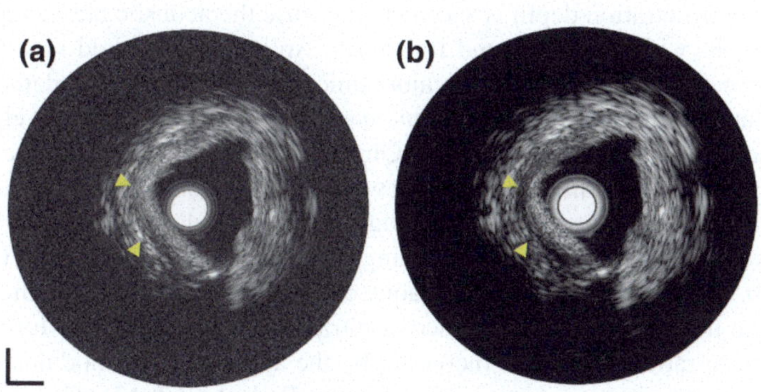

Fig. 2.28 Experimental result of ex vivo human artery imaging: **a** raw image, **b** beamformed image using VSSA-CFW. Dynamic range: 50 dB; scale bar: 1 mm (Yu et al. 2017)

rent commercial IVUS imaging system readily to post-process the acquired IVUS. In summary, VSSA-CFW can effectively improve the image quality in both cross-sectional and longitudinal plane beyond the natural focal point, which ensures the full coverage of the region of interest for vessels of different sizes. It can also be applied to intravascular photoacoustic imaging, for which lower noise level and better lateral resolution are highly desired (Liao et al. 2004). The improvement in image quality will also be a benefit for those increasingly miniaturized IVUS transducers, which is indispensable for advancing current multi-modality intravascular imaging techniques. IVUS images with improved lateral resolution, extended imaging depth, and enhanced CNR in both cross-sectional and longitudinal view can assist cardiologists to acquire more accurate plaque volume measurement, layer structure delineation, and artery remodeling evaluation.

IVUS Imaging with a Modulated Excitation

Traditional IVUS technology uses ultrasound with a center frequency of 20–45 MHz, but the resulting imaging spatial resolution makes it impossible to delineate the thin cap or microvessels. A multi-modality intravascular imaging method combining ultrasound with optical coherence tomography (OCT) was employed to improve both imaging resolution and penetration (Li et al. 2014a), but this had the disadvantage of the high costs of the catheter and the overall system. Recent progress of novel IVUS imaging techniques has been proposed using micromachined 1-3 composite (Li et al. 2014b; Yuan et al. 2008), and dual frequency IVUS acoustic angiography (Ma et al. 2014; Ma et al. 2015a). In addition, high-frequency ultrasound was proposed for improving the imaging resolution of the IVUS method (Ma et al. 2015b). Increasing the ultrasound frequency to more than 100-MHz enables the resolution of IVUS to approach that of optical imaging methods. This means that IVUS could be an alternative and cost-effective solution to replace OCT-based intravascular imaging. However, the penetration depth is sacrificed because the acoustic attenuation in the tissue increases with the ultrasound frequency. An imaging method for improving the depth of penetration would be of major clinical interest. It has been demonstrated that modulated excitation imaging can increase the penetration depth by elongating the ultrasound wave (Mamou et al. 2008; Qiu et al. 2013), which extends the view for acquiring structural information of vessel walls (Maresca et al. 2012; Shekhar et al. 2016). In addition, contrast-enhanced imaging with modulated excitation was proposed for visualizing the vasa vasorum using a conventional IVUS catheter (Maresca et al. 2013). The vulnerability of the plaque could be quantified by the microvasculature within the plaque in the presence of ultrasound microbubbles. Moreover, the signal-to-noise ratio (SNR) was increased by the application of modulated excitation to pulse-inversion tissue harmonic imaging (Park et al. 2013). The modulated-excitation-based IVUS (ME-IVUS) imaging technique could increase the efficacy of ultrasound in assessments of cardiovascular diseases. However, no previous study has specifically applied high-frequency (>50-MHz) ultrasound to improve the penetration depth in IVUS. Moreover, no study has proposed a design for a bespoke ME-IVUS system. Published studies of modulated excitation imaging have mostly implemented systems that combine different equipment such as commercial functional generators (from suppliers such as Agilent Technologies, Santa Clara, CA, and B&K Precision Instruments, Yorba Linda, CA) and power amplifiers (from suppliers such as Electronics and Innovation, Rochester, NY and Amplifier Research Corporation, Souderton, PA) (Maresca et al. 2012, 2013; Park et al. 2013; Shekhar et al. 2016). These evaluation setups were noisy, bulky, and expensive, and so an ultrasound system specifically optimized for ME-IVUS still needs to be developed.

Qiu et al. proposed an ultrasound system specifically designed for ME-IVUS (Qiu et al. 2017). An arbitrary-waveform generator was developed using a power amplifier with a custom-built circuit for switching it off. A digital-to-analog converter (DAC) and data acquisition circuitry were incorporated for pulse generation and data processing. A normal swine thoracic aorta specimen was used to evaluate the system in vitro. The transducer was inserted into the specimen to allow cross-sectional imaging. Ultrasound images of an aorta fixed in a water tank are shown in Fig. 2.29, which indicate that the penetration depth was greatly improved by using the chirp-based modulated excitation technique at both ultrasound frequencies (Increases of

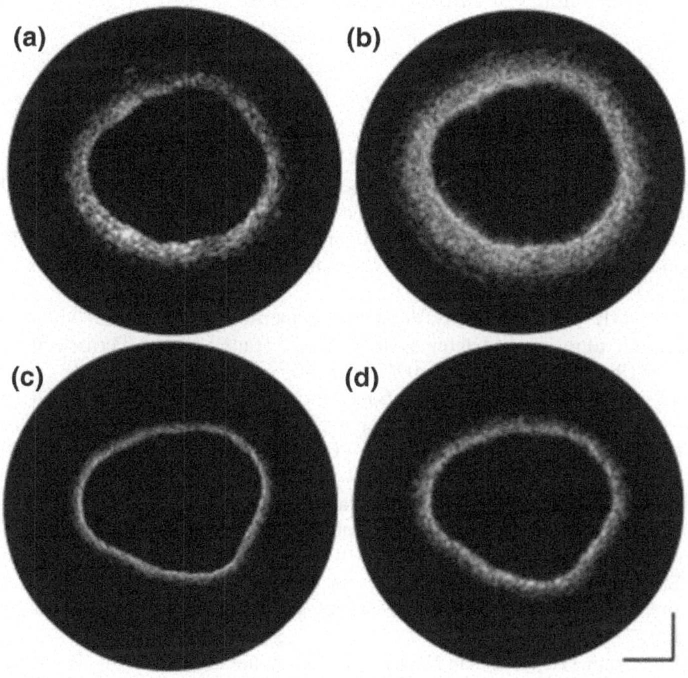

Fig. 2.29 Images of a swine aorta specimen in vitro acquired using the proposed system: **a** 30-MHz short pulse, **b** 30-MHz chirp, **c** 60-MHz short pulse, and **d** 60-MHz chirp. Scale bars: 2 mm; dynamic range: 45 dB (Qiu et al. 2017)

more than 41 and 63% in penetration depth for 30 and 60-MHz IVUS, respectively), while the quality of the images was maintained.

This work presents a new IVUS system based on a modulated excitation imaging method. Quantitative measurements demonstrated that the SNR and penetration depth were increased significantly: there was an improvement in the SNR of about 12 dB, while the penetration depth increased by 47.1% for the 30-MHz chirp ultrasound and 86.7% for the 60-MHz chirp ultrasound, clearly demonstrating good system performance. It would be an alternative method to see deep of the tissue with the state-of-the-art catheter. Hydrophone scan was performed with a 3D ultrasound intensity measurement system (UMS3, Precision acoustics, Dorchester, UK). The peak acoustic pressure for 30-MHz IVUS transducer is about 9 kPa. There is no data for 60-MHz probe as the hydrophone in our laboratory can only cover up to 40-MHz ultrasound frequency. It can be predicted that the acoustic pressure is lower than 9 kPa for 60-MHz transducer. A novel ultrasound system has been proposed and evaluated specifically for ME-IVUS, with the underlying methods, system design, and imaging results presented in detail. The SNR and penetration depth are increased significantly when using the proposed system. Test results show that the proposed system is flexible, and suitable for different applications with a modulated excitation intravascular imaging method, which could potentially increase the usefulness of ultrasound in assessments of cardiovascular diseases.

References

Aoki J, Abizaid AC, Serruys PW, Ong AT, Boersma E, Sousa JE, Bruining N (2005) Evaluation of four-year coronary artery response after sirolimus-eluting stent implantation using serial quantitative intravascular ultrasound and computer-assisted grayscale value analysis for plaque composition in event-free patients. J Am Coll Cardiol 46(9):1670–1676. https://doi.org/10.1016/j.jacc.2005.06.076

Bae MH, Jeong MK (2000) A study of synthetic-aperture imaging with virtual source elements in B-mode ultrasound imaging systems. IEEE Trans Ultrason Ferroelectr Freq Control 47(6):1510–1519. https://doi.org/10.1109/58.883540

Bourantas CV, Garcia-Garcia HM, Naka KK, Sakellarios A, Athanasiou L, Fotiadis DI, Michalis LK, Serruys PW (2013) Hybrid intravascular imaging: current applications and prospective potential in the study of coronary atherosclerosis. J Am Coll Cardiol 61(13):1369–1378. https://doi.org/10.1016/j.jacc.2012.10.057

Bourantas CV, Serruys PW (2014) Hybrid intravascular imaging: the key for a holistic evaluation of plaque pathology. EuroIntervJ EuroPCR Collab Work Group Interv Cardiol Eur Soc Cardiol 10(3):296–298. https://doi.org/10.4244/EIJV10I3A51

Brezinski ME, Tearney GJ, Bouma BE, Boppart SA, Hee MR, Swanson EA, Southern JF, Fujimoto JG (1996) Imaging of coronary artery microstructure (in vitro) with optical coherence tomography. Am J Cardiol 77(1):92–93

Brezinski ME, Tearney GJ, Weissman NJ, Boppart SA, Bouma BE, Hee MR, Weyman AE, Swanson EA, Southern JF, Fujimoto JG (1997) Assessing atherosclerotic plaque morphology: comparison of optical coherence tomography and high frequency intravascular ultrasound. Heart 77(5):397–403

Brugaletta S, Garcia-Garcia HM, Serruys PW, Maehara A, Farooq V, Mintz GS, de Bruyne B, Marso SP, Verheye S, Dudek D, Hamm CW, Farhat N, Schiele F, McPherson J, Lerman A, Moreno PR,

Wennerblom B, Fahy M, Templin B, Morel MA, van Es GA, Stone GW (2012) Relationship between palpography and virtual histology in patients with acute coronary syndromes. JACC Cardiovasc Imaging 5(3 Suppl):S19–S27. https://doi.org/10.1016/j.jcmg.2011.02.026

Buschman HP, Marple ET, Wach ML, Bennett B, Schut TC, Bruining HA, Bruschke AV, van der Laarse A, Puppels GJ (2000) In vivo determination of the molecular composition of artery wall by intravascular Raman spectroscopy. Anal Chem 72(16):3771–3775

Calzada ML, Alguero M, Ricote J, Santos A, Pardo L (2007) Preliminary results on sol-gel processing of <100> oriented $Pb(Mg_{1/3}Nb_{2/3})O_3$-$PbTiO_3$ thin films using diol-based solutions. J Sol-Gel Sci Techn 42(3):331–336. https://doi.org/10.1007/s10971-006-0203-9

Cannata JM, Ritter TA, Chen WH, Silverman RH, Shung KK (2003) Design of efficient, broadband single-element (20–80 MHz) ultrasonic transducers for medical imaging applications. IEEE Trans Ultrason Ferroelectr Freq Control 50(11):1548–1557. https://doi.org/10.1109/Tuffc.2003.1251138

Chen RM, Wu JC, Lam KH, Yao LH, Zhou QF, Tian J, Han PD, Shung KK (2012) Thermal-Independent properties of PIN-PMN-PT single-crystal linear-array ultrasonic transducers. IEEE Trans Ultrason Ferroelectr Freq Control 59(12):2777–2784. https://doi.org/10.1109/Tuffc.2012.2519

Cheng KC, Chan HLW, Choy CL, Yin QR, Luo HS, Yin ZW (2003) Single crystal PMN-0.33PT/epoxy 1-3 composites for ultrasonic transducer applications. IEEE Trans Ultrason Ferroelectr Frequency Control 50(9):1177–1183. https://doi.org/10.1109/tuffc.2003.1235328

Dawson P (1989) Cardiovascular effects of contrast agents. Am J Cardiol 64(9):2E–9E

Deleaval F, Bouvier A, Finet G, Cloutier G, Yazdani SK, Le Floc'h S, Clarysse P, Pettigrew RI, Ohayon J (2013) The intravascular ultrasound elasticity-palpography technique revisited: a reliable tool for the in vivo detection of vulnerable coronary atherosclerotic plaques. Ultrasound Med Biol 39(8):1469–1481. https://doi.org/10.1016/j.ultrasmedbio.2013.03.001

Di Ianni T, Hemmsen MC, Jensen JA (2016) Vector velocity estimation for portable ultrasound using directional transverse oscillation and synthetic aperture sequential beamforming. 2016 IEEE Int Ultrason Symp (IUS)

Elliott MR, Thrush AJ (1996) Measurement of resolution in intravascular ultrasound images. Physiol Meas 17(4):259–265

Finn AV, Nakano M, Narula J, Kolodgie FD, Virmani R (2010) Concept of vulnerable/unstable plaque. Arterioscler Thromb Vasc Biol 30(7):1282–1292. https://doi.org/10.1161/ATVBAHA.108.179739

Fitzgerald PJ, St Goar FG, Connolly AJ, Pinto FJ, Billingham ME, Popp RL, Yock PG (1992) Intravascular ultrasound imaging of coronary arteries. Is three layers the norm? Circulation 86(1):154–158. https://doi.org/10.1161/01.cir.86.1.154

Foster FS, Pavlin CJ, Harasiewicz KA, Christopher DA, Turnbull DH (2000) Advances in ultrasound biomicroscopy. Ultrasoun Med Biol 26(1):1–27. https://doi.org/10.1016/S0301-5629(99)00096-4

Frazier CH, O'Brien WR (1998) Synthetic aperture techniques with a virtual source element. IEEE Trans Ultrason Ferroelectr Freq Control 45(1):196–207. https://doi.org/10.1109/58.646925

Garcia-Garcia HM, Gonzalo N, Granada JF, Regar E, Serruys PW (2008) Diagnosis and treatment of coronary vulnerable plaques. Expert Rev Cardiovasc Therapy 6(2):209–222. https://doi.org/10.1586/14779072.6.2.209

Garcia-Garcia HM, Mintz GS, Lerman A, Vince DG, Margolis MP, van Es GA, Morel MA, Nair A, Virmani R, Burke AP, Stone GW, Serruys PW (2009) Tissue characterisation using intravascular radiofrequency data analysis: recommendations for acquisition, analysis, interpretation and reporting. EuroInterv J EuroPCR Collab Work Group Interv Cardiol Eur Soc Cardiol 5(2):177–189

Go AS, Mozaffarian D, Roger VL, Benjamin EJ, Berry JD, Blaha MJ, Dai S, Ford ES, Fox CS, Franco S, Fullerton HJ, Gillespie C, Hailpern SM, Heit JA, Howard VJ, Huffman MD, Judd SE, Kissela BM, Kittner SJ, Lackland DT, Lichtman JH, Lisabeth LD, Mackey RH, Magid DJ, Marcus GM, Marelli A, Matchar DB, McGuire DK, Mohler ER, Moy CS, Mussolino ME,

Neumar RW, Nichol G, Pandey DK, Paynter NP, Reeves MJ, Sorlie PD, Stein J, Towfighi A, Turan TN, Virani SS, Wong ND, Woo D, Turner MB, American Heart Association Statistics C, Stroke Statistics S (2014) Executive summary: heart disease and stroke statistics–2014 update: a report from the American Heart Association. Circulation 129 (3):399–410. https://doi.org/10.1161/01.cir.0000442015.53336.12

Guo ZJ, Li L, Wang LHV (2009) On the speckle-free nature of photoacoustic tomography. Med Phys 36(9):4084–4088. https://doi.org/10.1118/1.3187231

Hansen PM, Hemmsen M, Brandt A, Rasmussen J, Lange T, Krohn PS, Lonn L, Jensen JA, Nielsen MB (2014) Clinical evaluation of synthetic aperture sequential beamforming ultrasound in patients with liver tumors. Ultrasoun Med Biol 40(12):2805–2810. https://doi.org/10.1016/j.ultrasmedbio.2014.07.008

Hemmsen MC, Rasmussen JH, Jensen JA (2014) Tissue harmonic synthetic aperture ultrasound imaging. J Acoust Soc Am 136(4):2050–2056. https://doi.org/10.1121/1.4893902

Hollman KW, Rigby KW, O'Donnell M (1999) Coherence factor of speckle from a multi-row probe. In: 1999 IEEE ultrasonics symposium proceedings, vols 1 and 2, pp 1257–1260

Honda Y, Fitzgerald PJ (2008) Frontiers in intravascular imaging technologies. Circulation 117(15):2024–2037. https://doi.org/10.1161/CIRCULATIONAHA.105.551804

Jansen K, van Soest G, van der Steen AF (2014) Intravascular photoacoustic imaging: a new tool for vulnerable plaque identification. Ultrasoun Med Biol 40(6):1037–1048. https://doi.org/10.1016/j.ultrasmedbio.2014.01.008

Jensen JA, Svendsen NB (1992) Calculation of pressure fields from arbitrarily shaped, apodized, and excited ultrasound transducers. IEEE Trans Ultrason Ferroelectr Freq Control 39(2):262–267. https://doi.org/10.1109/58.139123

Jiang X, Yuan JR, Cheng A, Snook K, Cao PJ, Rehrig PW, Hackenberger WS, Lavalelle G, Geng X, Shrout TR (2006) Microfabrication of piezoelectric composite ultrasound transducers (PC-MUT). Ultrason 922–+

Jiang X, Snook K, Cheng A, Hackenberger WS, Geng X (2008) Micromachined PMN-PT single crystal composite transducers-15–75 MHz PC-MUT. In: 2008 IEEE Ultrasonics Symposium, vols 1–4 and Appendix: 164–+ . https://doi.org/10.1109/ultsym.2008.0040

Karaman M, Li PC, Odonnell M (1995) Synthetic-aperture imaging for small-scale systems. IEEE Trans Ultrason Ferroelectr Freq Control 42(3):429–442. https://doi.org/10.1109/58.384453

Kolodgie FD, Burke AP, Farb A, Gold HK, Yuan J, Narula J, Finn AV, Virmani R (2001) The thin-cap fibroatheroma: a type of vulnerable plaque: the major precursor lesion to acute coronary syndromes. Curr Opin Cardiol 16(5):285–292

Kortbek J, Jensen JA, Gammelmark KL (2007) Synthetic aperture focusing applied to imaging using a rotating single element transducer. In: 2007 IEEE ultrasonics symposium proceedings, vols 1–6, pp 1504–1507. https://doi.org/10.1109/ultsym.2007.378

Kortbek J, Jensen JA, Gammelmark KL (2013) Sequential beamforming for synthetic aperture imaging. Ultrasonics 53(1):1–16. https://doi.org/10.1016/j.ultras.2012.06.006

Kuscer D, Skalar M, Holc J, Kosec M (2009) Processing and properties of 0.65Pb(Mg1/3Nb2/3)O-3-0.35PbTiO(3) thick films. J Eur Ceram Soc 29(1):105–113. https://doi.org/10.1016/j.jeurceramsoc.2008.06.010

Lee DJ, Sigel B, Swami VK, Justin JR, Gahtan V, O'Brien SP, Dwyer-Joyce L, Feleppa EJ, Roberts AB, Berkowitz HD (1998) Determination of carotid plaque risk by ultrasonic tissue characterization. Ultrasoun Med Biol 24(9):1291–1299

Lee HJ, Zhang SJ, Shrout TR (2010) Scaling effects of relaxor-PbTiO3 crystals and composites for high frequency ultrasound. J Appl Phy 107(12). Artn 124107. https://doi.org/10.1063/1.3437068

Lee J, Jang J, Chang JH (2016) Oblong-shaped focused transducers for intravascular ultrasound imaging. IEEE Trans Biomed Eng. https://doi.org/10.1109/tbme.2016.2572182

Li ML, Guan WJ, Li PC (2004) Improved synthetic aperture focusing technique with applications in high-frequency ultrasound imaging. IEEE Trans Ultrason Ferroelectr Freq Control 51(1):63–70

Li PC, O'Donnell M (1994) Elevational spatial compounding. Ultrason Imaging 16(3):176–189

Li WG, Carrillo R, Yuan J, Teo TJ, Thomas L (2008) Multi-frequency processing for lumen enhancement with wideband intravascular ultrasound. In: 2008 IEEE ultrasonics symposium, vols 1–4 and Appendix: 371–374

Li X, Yin J, Hu C, Zhou Q, Shung KK, Chen Z (2010) High-resolution coregistered intravascular imaging with integrated ultrasound and optical coherence tomography probe. Appl Phys Lett 97(13):133702. https://doi.org/10.1063/1.3493659

Li X, Wu W, Chung Y, Shih WY, Shih WH, Zhou Q, Shung KK (2011) 80-MHz intravascular ultrasound transducer using PMN-PT free-standing film. IEEE Trans Ultrason Ferroelectr Freq Control 58(11):2281–2288. https://doi.org/10.1109/TUFFC.2011.2085

Li X, Wei W, Zhou QF, Shung KK, Chen ZP (2012) Intravascular photoacoustic imaging at 35 and 80 MHz. J Biomed Opt 17(10). Artn 106005. https://doi.org/10.1117/1.jbo.17.10.106005

Li J, Li X, Mohar D, Raney A, Jing J, Zhang J, Johnston A, Liang S, Ma T, Shung KK, Mahon S, Brenner M, Narula J, Zhou Q, Patel PM, Chen Z (2014a) Integrated IVUS-OCT for real-time imaging of coronary atherosclerosis. JACC Cardiovasc Imaging 7(1):101–103. https://doi.org/10.1016/j.jcmg.2013.07.012

Li X, Ma T, Tian J, Han P, Zhou Q, Shung K (2014b) Micromachined PIN-PMN-PT crystal composite transducer for high-frequency intravascular ultrasound (IVUS) imaging. IEEE Trans Ultrason Ferroelectr Freq Control 61(7):1171–1178. https://doi.org/10.1109/TUFFC.2014.3016

Li J, Minami H, Steward E, Ma T, Mohar D, Robertson C, Shung K, Zhou Q, Patel P, Chen Z (2015) Optimal flushing agents for integrated optical and acoustic imaging systems. J Biomed Opt 20(5):56005. https://doi.org/10.1117/1.JBO.20.5.056005

Liao CK, Li ML, Li PC (2004) Optoacoustic imaging with synthetic aperture focusing and coherence weighting. Opt Lett 29(21):2506–2508

Libby P (1995) Molecular bases of the acute coronary syndromes. Circulation 91(11):2844–2850

Lockwood GR, Turnbull DH, Christopher DA, Foster FS (1996) Beyond 30 MHz—applications of high-frequency ultrasound imaging. IEEE Eng Med Biol Mag 15(6):60–71. https://doi.org/10.1109/51.544513

Luo HS, Xu GS, Xu HQ, Wang PC, Yin ZW (2000) Compositional homogeneity and electrical properties of lead magnesium niobate titanate single crystals grown by a modified Bridgman technique. Jpn J Appl Phys Part 1 Regul Papers Short Notes Rev Papers 39 (9b):5581–5585. https://doi.org/10.1143/jjap.39.5581

Luo H, Shih WY, Shih WH (2007) Double precursor solution coating approach for low-temperature sintering of $[Pb(Mg_{1/3}Nb_{2/3})O_3]_{0.63}[PbTiO_3]_{0.37}$ solids. J Am Ceram Soc 90(12):3825–3829. https://doi.org/10.1111/j.1551-2916.2007.02072.x

Ma J, Martin K, Dayton PA, Jiang X (2014) A preliminary engineering design of intravascular dual-frequency transducers for contrast-enhanced acoustic angiography and molecular imaging. IEEE Trans Ultrason Ferroelectr Freq Control 61(5):870–880. https://doi.org/10.1109/TUFFC.2014.6805699

Ma JG, Martin KH, Li Y, Dayton PA, Shung KK, Zhou QF, Jiang XN (2015a) Design factors of intravascular dual frequency transducers for super-harmonic contrast imaging and acoustic angiography. Phys Med Biol 60(9):3441–3457. https://doi.org/10.1088/0031-9155/60/9/3441

Ma T, Yu M, Li J, Munding CE, Chen Z, Fei C, Shung KK, Zhou Q (2015b) Multi-frequency intravascular ultrasound (IVUS) imaging. IEEE Trans Ultrason Ferroelectr Freq Control 62(1):97–107. https://doi.org/10.1109/TUFFC.2014.006679

Maehara A, Mintz GS, Weissman NJ (2009) Advances in intravascular imaging. Circ Cardiovasc Interv 2(5):482–490. https://doi.org/10.1161/CIRCINTERVENTIONS.109.868398

Mallart R, Fink M (1994) Adaptive focusing in scattering media through sound-speed inhomogeneities—the Vancittert-Zernike approach and focusing criterion. J Acoust Soc Am 96(6):3721–3732. https://doi.org/10.1121/1.410562

Mamou J, Ketterling JA, Silverman RH (2008) Chirp-coded excitation imaging with a high-frequency ultrasound annular array. IEEE Trans Ultrason Ferroelectr Freq Control 55(2):508–513. https://doi.org/10.1109/Tuffc.2008.670

Maresca D, Jansen K, Renaud G, van Soest G, Li X, Zhou Q, de Jong N, Shung KK, van der Steen AFW (2012) Intravascular ultrasound chirp imaging. Appl Phys Lett 100(4). Artn 043703. https://doi.org/10.1063/1.3679375

Maresca D, Renaud G, van Soest G, Li X, Zhou QF, Shung KK, de Jong N, van der Steen AFW (2013) Contrast-enhanced intravascular ultrasound pulse sequences for bandwidth-limited transducers. Ultrasoun Med Biol 39(4):706–713. https://doi.org/10.1016/j.ultrasmedbio.2012.10.020

Maresca D, Adams S, Maresca B, van der Steen AF (2014) Mapping intravascular ultrasound controversies in interventional cardiology practice. PLoS ONE 9(5):e97215. https://doi.org/10.1371/journal.pone.0097215

Moreno PR (2010) Vulnerable plaque: definition, diagnosis, and treatment. Cardiol Clin 28(1):1–30. https://doi.org/10.1016/j.ccl.2009.09.008

Moreno PR, Lodder RA, Purushothaman KR, Charash WE, O'Connor WN, Muller JE (2002) Detection of lipid pool, thin fibrous cap, and inflammatory cells in human aortic atherosclerotic plaques by near-infrared spectroscopy. Circulation 105(8):923–927

Nair A, Kuban BD, Tuzcu EM, Schoenhagen P, Nissen SE, Vince DG (2002) Coronary plaque classification with intravascular ultrasound radiofrequency data analysis. Circulation 106(17):2200–2206

Nair A, Margolis MP, Kuban BD, Vince DG (2007) Automated coronary plaque characterisation with intravascular ultrasound backscatter: ex vivo validation. EuroIntery J EuroPCR Collab Work Group Interv Cardiol Eur Soc Cardiol 3(1):113–120

Nasu K, Tsuchikane E, Katoh O, Vince DG, Virmani R, Surmely JF, Murata A, Takeda Y, Ito T, Ehara M, Matsubara T, Terashima M, Suzuki T (2006) Accuracy of in vivo coronary plaque morphology assessment: a validation study of in vivo virtual histology compared with in vitro histopathology. J Am Coll Cardiol 47(12):2405–2412. https://doi.org/10.1016/j.jacc.2006.02.044

Nikolov SI, Jensen JA (2002) Virtual ultrasound sources in high-resolution ultrasound imaging. P Soc Photo-Opt Ins 4687:395–405. https://doi.org/10.1117/12.462178

Noritomi T, Sigel B, Gahtan V, Swami V, Justin J, Feleppa E, Shirouzu K (1997a) In vivo detection of carotid plaque thrombus by ultrasonic tissue characterization. J Ultrasoun Med 16(2):107–111

Noritomi T, Sigel B, Swami V, Justin J, Gahtan V, Chen X, Feleppa EJ, Roberts AB, Shirouzu K (1997b) Carotid plaque typing by multiple-parameter ultrasonic tissue characterization. Ultrasoun Med Biol 23(5):643–650

Park J, Li X, Zhou QF, Shung KK (2013) Combined chirp coded tissue harmonic and fundamental ultrasound imaging for intravascular ultrasound: 20–60 MHz phantom and ex vivo results. Ultrasonics 53(2):369–376. https://doi.org/10.1016/j.ultras.2012.07.003

Park SE, Shrout TR (1997a) Characteristics of relaxor-based piezoelectric single crystals for ultrasonic transducers. IEEE Trans Ultrason Ferroelectr Freq Control 44(5):1140–1147. https://doi.org/10.1109/58.655639

Park SE, Shrout TR (1997b) Ultrahigh strain and piezoelectric behavior in relaxor based ferroelectric single crystals. J Appl Phys 82(4):1804–1811. https://doi.org/10.1063/1.365983

Passmann C, Ermert H (1996) A 100-MHz ultrasound imaging system for dermatologic and ophthalmologic diagnostics. IEEE Trans Ultrason Ferroelectr Freq Control 43(4):545–552. https://doi.org/10.1109/58.503714

Potkin BN, Bartorelli AL, Gessert JM, Neville RF, Almagor Y, Roberts WC, Leon MB (1990) Coronary artery imaging with intravascular high-frequency ultrasound. Circulation 81(5):1575–1585

Puri R, Worthley MI, Nicholls SJ (2011) Intravascular imaging of vulnerable coronary plaque: current and future concepts. Nat Rev Cardiol 8(3):131–139. https://doi.org/10.1038/nrcardio.2010.210

Qiu WB, Yu YY, Tsang FK, Zheng HR, Sun L (2013) A novel modulated excitation imaging system for microultrasound. IEEE Trans Biomed Eng 60(7):1884–1890. https://doi.org/10.1109/Tbme.2013.2244887

Qiu WB, Wang XY, Chen Y, Fu Q, Su M, Zhang LN, Xia JJ, Dai JY, Zhang YN, Zheng HR (2017) Modulated excitation imaging system for intravascular ultrasound. IEEE Trans Biomed Eng 64(8):1935–1942. https://doi.org/10.1109/Tbme.2016.2631224

Regar E, Schaar JA, Mont E, Virmani R, Serruys PW (2003) Optical coherence tomography. Cardiovascular radiation medicine 4(4):198–204. https://doi.org/10.1016/j.carrad.2003.12.003

Rhee S (2007) High frequency (IVUS) ultrasound transducer technology—applications and challenges. IEEE Int Ferro 850–851

Ross R (1993) The pathogenesis of atherosclerosis: a perspective for the 1990s. Nature 362(6423):801–809. https://doi.org/10.1038/362801a0

Ross R (1999) Atherosclerosis—an inflammatory disease. N Engl J Med 340(2):115–126. https://doi.org/10.1056/NEJM199901143400207

Ryan LK, Foster FS (1997) Tissue equivalent vessel phantoms for intravascular ultrasound. Ultrasound Med Biol 23(2):261–273

Schaar JA, de Korte CL, Mastik F, Baldewsing R, Regar E, de Feyter P, Slager CJ, van der Steen AF, Serruys PW (2003) Intravascular palpography for high-risk vulnerable plaque assessment. Herz 28(6):488–495. https://doi.org/10.1007/s00059-003-2488-6

Schaar JA, van der Steen AF, Mastik F, Baldewsing RA, Serruys PW (2006) Intravascular palpography for vulnerable plaque assessment. J Am Coll Cardiol 47(8 Suppl):C86–C91. https://doi.org/10.1016/j.jacc.2006.01.035

Shekhar H, Huntzicker S, Awuor I, Doyley MM (2016) Chirp-coded ultraharmonic imaging with a modified clinical intravascular ultrasound system. Ultrason Imaging 38(6):403–419. https://doi.org/10.1177/0161734615618639

Shih WY, Luo HY, Li HD, Martorano C, Shih WH (2006) Sheet geometry enhanced giant piezoelectric coefficients. Appl Phys Lett 89(24). Artn 242913. https://doi.org/10.1063/1.2408633

Shin ES, Garcia-Garcia HM, Ligthart JM, Witberg K, Schultz C, van der Steen AF, Serruys PW (2011) In vivo findings of tissue characteristics using iMap IVUS and virtual histology IVUS. EuroInterv Journal EuroPCR Collab Work Group Interv Cardiol Eur Soc Cardiol 6(8):1017–1019. https://doi.org/10.4244/EIJV6I8A175

Smith WA (1989) The role of piezocomposites in ultrasonic transducers. In: IEEE 1989 ultrasonics symposium: proceedings, vols 1 and 2, pp 755–766

Snook KA, Zhao JZ, Alves CH, Cannata JM, Chen WH, Meyer RJ Jr, Ritter TA, Shung KK (2002) Design, fabrication, and evaluation of high frequency, single-element transducers incorporating different materials. IEEE Trans Ultrason Ferroelectr Freq Control 49(2):169–176

Stephens DN, Park J, Sun Y, Papaioannou T, Marcu L (2009) Intraluminal fluorescence spectroscopy catheter with ultrasound guidance. J Biomed Opt 14(3):030505. https://doi.org/10.1117/1.3146813

Sun P, Zhou QF, Zhu BP, Wu DW, Hu CL, Cannata JM, Tian J, Han PD, Wang GF, Shung KK (2009) Design and fabrication of PIN-PMN-PT single-crystal high-frequency ultrasound transducers. IEEE Trans Ultrason Ferroelectr Freq Control 56(12):2760–2763. https://doi.org/10.1109/Tuffc.2009.1367

Sun P, Wang GF, Wu DW, Zhu BP, Hu CH, Liu CG, Djuth FT, Zhou QF, Shung KK (2010) High frequency PMN-PT 1-3 composite transducer for ultrasonic imaging application. Ferroelectrics 408:120–128. https://doi.org/10.1080/00150193.2010.485546

Sun Y, Chaudhari AJ, Lam M, Xie H, Yankelevich DR, Phipps J, Liu J, Fishbein MC, Cannata JM, Shung KK, Marcu L (2011) Multimodal characterization of compositional, structural and functional features of human atherosclerotic plaques. Biomed Opt Express 2(8):2288–2298. https://doi.org/10.1364/BOE.2.002288

Teirlinck CJ, Bezemer RA, Kollmann C, Lubbers J, Hoskins PR, Ramnarine KV, Fish P, Fredeldt KE, Schaarschmidt UG (1998) Development of an example flow test object and comparison of five of these test objects, constructed in various laboratories. Ultrasonics 36(1–5):653–660

Tian J, Han P, Payne DA (2007a) Measurements along the growth direction of PMN-PT crystals: dielectric, piezoelectric, and elastic properties. IEEE Trans Ultrason Ferroelectr Freq Control 54(9):1895–1902. https://doi.org/10.1109/Tuffc.2007.474

Tian J, Han PD, Huang XL, Pan HX, Carroll JF, Payne DA (2007b) Improved stability for piezoelectric crystals grown in the lead indium niobate-lead magnesium niobate-lead titanate system. Appl Phys Lett 91(22). Artn 222903. https://doi.org/10.1063/1.2817743

Tian J, Meneou K, Stone B, Han P (2010) Piezoelectric crystal composite for high frequency ultrasound application. In: 2010 IEEE international ultrasonics symposium, 11–14 Oct 2010, pp 65–67. https://doi.org/10.1109/ultsym.2010.5935822

Ursic H, Skarabot M, Hrovat M, Holc J, Skalar M, Bobnar V, Kosec M, Musevic I (2008) The electrostrictive effect in ferroelectric 0.65Pb(Mg1/3Nb2/3)O-3-0.35PbTiO(3) thick films. J Appl Phys 103(12). Artn 124101. https://doi.org/10.1063/1.2938848

van de Poll SW, Romer TJ, Puppels GJ, van der Laarse A (2002) Imaging of atherosclerosis. Raman spectroscopy of atherosclerosis. J Cardiovasc Risk 9(5):255–261

Virmani R, Kolodgie FD, Burke AP, Farb A, Schwartz SM (2000) Lessons from sudden coronary death: a comprehensive morphological classification scheme for atherosclerotic lesions. Arterioscler Thromb Vasc Biol 20(5):1262–1275

Virmani R, Burke AP, Kolodgie FD, Farb A (2003) Pathology of the thin-cap fibroatheroma: a type of vulnerable plaque. J Interv Cardiol 16(3):267–272

Wang B, Su JL, Karpiouk AB, Sokolov KV, Smalling RW, Emelianov SY (2010) Intravascular photoacoustic imaging. IEEE J Quantum Electron 16(3):588–599. https://doi.org/10.1109/JSTQE.2009.2037023

Waxman S, Dixon SR, L'Allier P, Moses JW, Petersen JL, Cutlip D, Tardif JC, Nesto RW, Muller JE, Hendricks MJ, Sum ST, Gardner CM, Goldstein JA, Stone GW, Krucoff MW (2009) In vivo validation of a catheter-based near-infrared spectroscopy system for detection of lipid core coronary plaques: initial results of the SPECTACL study. JACC Cardiovasc Imaging 2(7):858–868. https://doi.org/10.1016/j.jcmg.2009.05.001

Ye ZG, Dong M (2000) Morphotropic domain structures and phase transitions in relaxor-based piezo-/ferroelectric $(1-x)Pb(Mg_{1/3}Nb_{2/3})O_3-xPbTiO_3$ single crystals. J Appl Phys 87(5):2312–2319. https://doi.org/10.1063/1.372180

Yin J, Li X, Jing J, Li J, Mukai D, Mahon S, Edris A, Hoang K, Shung KK, Brenner M, Narula J, Zhou Q, Chen Z (2011) Novel combined miniature optical coherence tomography ultrasound probe for in vivo intravascular imaging. J Biomed Opt 16(6):060505. https://doi.org/10.1117/1.3589097

Yoo H, Kim JW, Shishkov M, Namati E, Morse T, Shubochkin R, McCarthy JR, Ntziachristos V, Bouma BE, Jaffer FA, Tearney GJ (2011) Intra-arterial catheter for simultaneous microstructural and molecular imaging in vivo. Nat Med 17(12):1680–1684. https://doi.org/10.1038/nm.2555

Yu MY, Li Y, Ma T, Shung KK, Zhou QF (2017) Intravascular ultrasound imaging with virtual source synthetic aperture focusing and coherence factor weighting. IEEE Trans Med Imaging 36(10):2171–2178. https://doi.org/10.1109/Tmi.2017.2723479

Yuan JR, Jiang X, Cao IPJ, Sadaka A, Bautista R, Snook K, Rehrig PW (2006) High frequency piezo composites microfabricated ultrasound transducers for intravascular imaging. Ultrason:264–+. https://doi.org/10.1109/ultsym.2006.83

Yuan J, Rhee S, Jiang XN, IEEE (2008) 60 MHz PMN-PT based 1-3 composite transducer for IVUS imaging. In: 2008 IEEE ultrasonics symposium, vols 1–4 and Appendix, pp 682–685. https://doi.org/10.1109/ultsym.2008.0164

Zhang SJ, Luo J, Hackenberger W, Shrout TR (2008) Characterization of $Pb(In_{1/2}Nb_{1/2})O_3–Pb(Mg_{1/3}Nb_{2/3})O_3–PbTiO_3$ ferroelectric crystal with enhanced phase transition temperatures. J Appl Phys 104(6). Artn 064106. https://doi.org/10.1063/1.2978333

Zhang SJ, Shrout TR (2010) Relaxor-PT single crystals: observations and developments. IEEE Trans Ultrason Ferroelectr Freq Control 57(10):2138–2146. https://doi.org/10.1109/Tuffc.2010.1670

Zhang YY, Zhao XY, Wang W, Ren B, Liu DA, Luo HS (2011) Fabrication of PIMNT/epoxy 1-3 composites and ultrasonic transducer for nondestructive evaluation. IEEE Trans Ultrason Ferroelectr Freq Control 58(9):1774–1781. https://doi.org/10.1109/Tuffc.2011.2014

Zhou Q, Xu X, Gottlieb EJ, Sun L, Cannata JM, Ameri H, Humayun MS, Han P, Shung KK (2007) PMN-PT single crystal, high-frequency ultrasonic needle transducers for pulsed-wave Doppler application. IEEE Trans Ultrason Ferroelectr Freq Control 54(3):668–675. https://doi.org/10.1109/Tuffc.2007.290

Zhou D, Cheung KF, Chen Y, Lau ST, Zhou QF, Shung KK, Luo HS, Dai JY, Chan HLW (2011a) Fabrication and performance of endoscopic ultrasound radial arrays based on PMN-PT single crystal/epoxy 1-3 composite. IEEE Trans Ultrason Ferroelectr Freq Control 58(2):477–484. https://doi.org/10.1109/Tuffc.2011.1825

Zhou Q, Lau S, Wu D, Shung KK (2011b) Piezoelectric films for high frequency ultrasonic transducers in biomedical applications. Prog Mater Sci 56(2):139–174. https://doi.org/10.1016/j.pmatsci.2010.09.001

Chapter 3
The Integration of IVUS and OCT

Jiawen Li, Teng Ma, Qifa Zhou and Zhongping Chen

Introduction

Ultrasound imaging provides a tomographic view inside the body by detecting echoes. Similarly, optical coherence tomography (OCT) is also capable of noninvasive cross-sectional imaging of an internal structure. By using near-infrared light instead of ultrasound and analyzing the signal with an interferometric technique, OCT enables higher spatial resolution than ultrasound (Boppart et al. 1998; Leitgeb et al. 2004; Werkmeister et al. 2013). Ultrasound and OCT share many similarities (non-ionic, label-free, and cross-sectional imaging capability), and they both have been applied in similar fields (Radhakrishnan et al. 2005; Zagaynova et al. 2008; Bezerra et al. 2009). In cardiovascular applications, ultrasound and OCT provide

J. Li
Adelaide Medical School, Australian Research Council Centre of Excellence for Nanoscale Biophotonics, Institute for Photonics and Advanced Sensing, The University of Adelaide, Adelaide, SA 5005, Australia
e-mail: jiawen.li01@adelaide.edu.au

T. Ma
Paul C. Lauterbur Research Center for Biomedical Imaging, Institute of Biomedical and Health Engineering, Shenzhen Institutes of Advanced Technology, Chinese Academy of Sciences, Shenzhen 518055, China
e-mail: teng.ma@siat.ac.cn

Q. Zhou
Roski Eye Institute, University of Southern California, Los Angeles, CA 90033, USA
e-mail: qifazhou@usc.edu

Department of Biomedical Engineering, University of Southern California, Los Angeles, CA 90089, USA

Z. Chen (✉)
Beckman Laser Institute and Department of Biomedical Engineering, University of California, Irvine, Irvine, CA 92697, USA
e-mail: z2chen@uci.edu

Q. Zhou and Z. Chen (eds.), *Multimodality Imaging*,
https://doi.org/10.1007/978-981-10-6307-7_3

complementary capabilities: ultrasound has high penetration depth but low resolution; OCT has high resolution but low penetration depth. Ultrasound is superior in visualizing the deep structure (up to 7 mm) while OCT is better at imaging and accurately quantifying small features that are close (within 1–3 mm) to the surface. A hybrid ultrasound-OCT system would not only enable clear visualization of subsurface microstructure (e.g., thin fibrous cap and endothelial erosion) but also reveal features deep in the tissue (e.g., necrotic core). Such a "two-in-one" capability may be clinically significant in the diagnosis of cardiovascular diseases (Bourantas et al. 2016).

The research of integrating these two imaging modalities has been driven by the strong clinical need to image atherosclerotic plaques, specifically to identify vulnerable plaques inside coronary arteries. When the first integrated intravascular ultrasound (IVUS)-OCT system was built, images of a rabbit aorta were obtained ex vivo (Yin et al. 2010). After four years of endeavor on improving the clinical adaptability of the system, in vivo intracoronary US-OCT imaging was successfully achieved (Li et al. 2014a). To further improve the safety of the system by reducing the time needed for imaging, an ultrafast system (72 frames per second, fps) was developed and safely imaged live animals (Li et al. 2015b). Meanwhile, the feasibility of using IVUS-OCT to detect vulnerable plaques was also demonstrated (Li et al. 2015b).

This chapter begins with elucidating fundamentals of IVUS-OCT technology and goes on to review advances that make the in vivo utilization of IVUS-OCT technically possible. It is followed by a review of in vitro and in vivo validations of IVUS-OCT and introducing other studies related to IVUS-OCT.

IVUS-OCT Fundamentals

In this section, we introduce the fundamentals of integrated ultrasound-OCT technology, which is vital for both building an ultrasound-OCT system and understanding the development of IVUS-OCT.

As shown in Fig. 3.1, a representative IVUS-OCT system consists of an ultrasound sub-system, an OCT sub-system, a motion control unit, a data acquisition (DAQ) system, and an IVUS-OCT imaging catheter. This section describes major considerations of building these constituents, presents important measures of an IVUS-OCT system, and briefly discusses how the system can be improved.

Ultrasound Sub-system

The main component for the ultrasound system is a pulser/receiver. It generates a high-voltage pulse to excite the transducer. After the echoed acoustic wave is

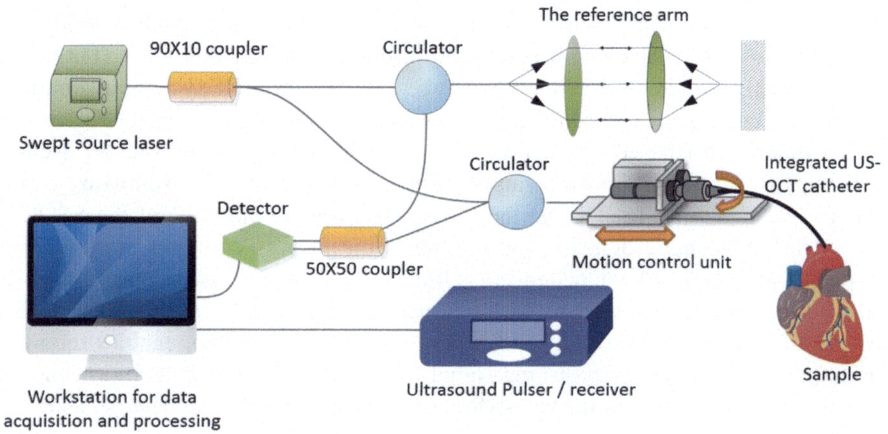

Fig. 3.1 Schematic of an integrated IVUS-OCT system

received and converted to an electrical signal by a transducer, the pulser/receiver amplifies and filters the signal.

Many factors will influence the signal-to-noise ratio (SNR) of the ultrasound system, including (1) piezoelectric material used in the transducer; (2) beam diameter generated by the transducer; (3) impedance match between each component of the imaging system; and (4) selection of filters.

OCT Sub-system

Since OCT was first reported (Huang et al. 1991), two generations of OCT systems have been developed. These two generations applied different techniques to obtain signals at different depths. The first-generation OCT (Time domain OCT, TDOCT) directly scans different depths by moving the reference arm while the second-generation OCT (Fourier Domain OCT, FDOCT) accomplishes depth-resolved imaging by using Fourier transform of the acquired spectra (Leitgeb et al. 2003). FDOCT surpasses TDOCT by its faster imaging speed and higher SNR (Choma et al. 2003). Thus, previously reported IVUS-OCT systems all employ the FDOCT technique. To perform fast Fourier transform (FFT) and reconstruct an axial scan as a function of depth (i.e., an A-line) in FDOCT, measuring the intensity distribution as a function of wavelength is needed. There are two main techniques to accomplish this measurement: spectral-domain OCT (SDOCT) applies a spectrometer to separate different wavelength components in space; swept source OCT (SSOCT) uses a frequency scanning laser to separate different wavelength components in time. Although SDOCT technique, particularly the micro-OCT technique, could realize a higher spatial resolution and reveal many significant microstructures (Liu et al. 2011), SSOCT utilizes a simpler optical setup and can achieve higher imaging speed than SDOCT at around 1310 nm, the center wavelength commonly used for intravascular OCT (IVOCT). Thus, clinically used IVOCT and previously reported IVUS-OCT systems all employ the SSOCT technique. Accordingly, we now limit our discussion to SSOCT.

A standard fiber-based SSOCT system includes a swept laser source, a coupler/couplers, a reference arm, a sample arm, and a detector. The following components, as shown in Fig. 3.1, can be utilized to improve the SNR of the SSOCT system: (a) a balanced detector can compress the useless common mode signal (Bouma et al. 2008); (b) when shot-noise limit is reached, a coupler that splits more lights to the sample arm than to the reference arm can usually improve the SNR; and (c) a circulator, with low insertion loss (each port ~ 0.6 dB), enables the system to make full use of the returning signal from each arm. A more comprehensive discussion on factors that influence SNR can be found in the section 'Key Measures'.

Motion Control Unit

To form three-dimensional (3D) images from the obtained A-lines, scanning the light and sound beam by a motion control unit is necessary. In an IVUS-OCT system, a 3D scan is conducted by pulling back and rotating the imaging probe. The pullback can be accomplished by a linear translational stage, as indicated by the arrow in Fig. 3.1. To perform a rotational scan, two mechanisms have been proposed: the proximal and distal rotations.

Proximal rotation is achieved by using a rotational motor outside the imaging probe to actuate the proximal end of the probe and a torque coil around the imaging probe to transfer the rotation all the way to the distal end of the probe. To achieve this scanning mechanism, an optical rotary joint and an electrical slip ring are also needed. They are connected in between the imaging probe and ultrasound/OCT sub-systems for facilitating transmissions of optical and electrical signals between the rotary and stationary parts (Li et al. 2014c). As the motor is outside of the probe, the diameter and the rigid part of the probe can be kept small. However, this design is not suitable for a system that requires a rotary speed higher than a couple hundred revolutions per second (rps). At such a high speed, the transmission of rotation from the proximal end to the distal end becomes inaccurate and severe non-uniform rotational distortion can be observed (Wang et al. 2013).

To rotate over 200 rps, the distal scanning mechanism that uses a micromotor (Wang et al. 2013; Yin et al. 2009) at the distal end of the imaging probe is more suitable. Yet the speed is increased at the cost of the imaging catheter profile. To the best of our knowledge, the outer diameter (OD) of the smallest micromotor-based IVOCT catheter to date is 1.1 mm (Wang et al. 2013), much larger than the OD that can be achieved by the proximal scanning design (Moon et al. 2013). Thus, the distal scanning mechanism is suggested to be utilized in applications where the miniaturization of the probe is less vital than the imaging speed.

DAQ and Signal Processing

To ensure the OCT and ultrasound signals are acquired simultaneously, a dual-channel DAQ card (Yin et al. 2011) is often used. One channel obtains the OCT signal, and the other obtains the ultrasound signal. Data acquisitions of both channels are synchronized by the same trigger signal.

The captured ultrasound data undergo a bandpass filter to isolate the echo signal, a Hilbert transform, and a logarithmic scaling to construct an ultrasound A-scan. In parallel, the following steps are undertaken to create an OCT A-scan: resampling/calibration (Bouma et al. 2008), FFT, and logarithmic scaling. Apart from these steps, extra steps can be added to improve the image quality. For example, applying digital dispersion compensation (Hoang et al. 2009) can usually improve the actual axial resolution, since dispersions generated in the reference and sample arms are not the same. As each A-line requires the same processing procedures, parallel computing using a graphic processing unit (GPU) can improve the data processing speed (Yin et al. 2011). The speed for DAQ and signal processing affects the speed a system can achieve. With the advancement of the dual-channel digitizer, gigabits speed solid-state drive (SSD), powerful GPUs and data processing algorithms, the speed of IVUS-OCT imaging has increased 72-fold from 1 fps to 72 fps. (Yin et al. 2010; Li et al. 2015b).

IVUS-OCT Imaging Catheter Design

An IVUS-OCT catheter is made up of an imaging probe (IVUS and OCT sub-probes) and a catheter outer sheath. Accordingly, we describe these components one by one in the following paragraphs.

IVUS Sub-probe Design

All IVUS-OCT probes reported to date (Li et al. 2010, 2013a, 2014a, c, 2015b; Yin et al. 2010, 2011) used single-element transducers, although a phase array system has been used to make a commercial IVUS catheter.

A single-element transducer usually consists of three layers. The vital layer in the center is the piezoelectric material which converts the US signal to rapid changing electrical potential and vice versa. Many polymers, ceramics, and crystals display superior piezoelectric properties, such as lead zirconate titanate (PZT), lithium niobate ($LiNbO_3$), and lead magnesium niobate–lead titanate (PMN–PT). The thickness of the piezoelectric material determines the central frequency of a transducer. The key parameters representing the material's piezoelectric properties include electromechanical coupling coefficient (k_t), piezoelectric strain constant (d_{33}), and relative clamped dielectric constant ($\varepsilon_s/\varepsilon_0$); k_t and d_{33} are related to the energy conversion efficiency and sensitivity of a transducer while the electrical impedance of a

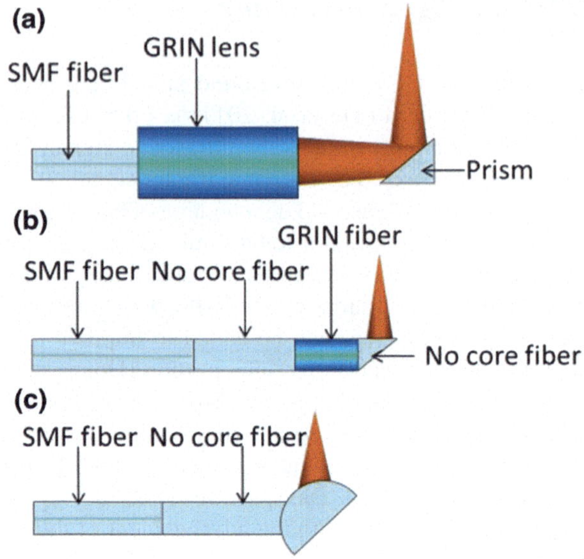

Fig. 3.2 Designs of OCT sub-probes. **a** GRIN lens probe. **b** GRIN fiber probe. **c** Ball lens probe

transducer is proportional to $\frac{\text{thickness}}{\varepsilon_s \times \text{Area}}$. For a miniaturized transducer that has a small surface area, such as a transducer used for intravascular imaging, selecting a material with a high $\varepsilon^s/\varepsilon_0$ is critical for maintaining electrical impedance matching of all electronics and high transmitting sensitivity. PMN–PT has higher $\varepsilon_s/\varepsilon_0$ (Kelly et al. 1997) than PZT [1350 for PZT-5H (Cannata et al. 2003)] and $LiNbO_3$ [43 (Cannata et al. 2003)] and thus is suitable to make miniaturized ultrasound transducers. Good acoustic impedance matching also requires selecting optimal thicknesses of the layers below (backing layer) and above (matching layer) the piezoelectric layer. The optimal values of thicknesses of these two layers can be calculated based on the Krimholtz, Leedom, and Matthaei (KLM) equivalent circuit model.

OCT Sub-probe Design

Different designs of OCT sub-probes (Fig. 3.2) can be used in making an IVUS-OCT catheter, such as gradient refractive index (GRIN) lens design (Li et al. 2010; Yin et al. 2010, 2011), GRIN fiber design (Mao et al. 2010), and ball-lens design (Li et al. 2013b; Tan et al. 2012).

The diameters of a GRIN fiber (without the buffer layer) and a ball lens are 125 and 200–250 μm, respectively. However, off-the-shelf GRIN lenses are usually with an OD larger than 250 μm. Thus, the GRIN fiber or the ball-lens probe design is usually chosen to make miniature probes due to its small OD.

On the other hand, a GRIN lens design is commonly used where the image performance is more important because the lateral resolution of a GRIN lens probe is often better than that of a GRIN fiber or ball-lens probe. The reasons include firstly, as the aperture of the GRIN lens is larger than GRIN fiber probe, the theoretical lateral resolution of a GRIN lens probe is usually better; secondly, a GRIN lens usually performs with a higher-quality gradient refractive index profile than that of a GRIN fiber and less spherical aberration than a ball lens, which makes its lateral resolution even better.

After separately fabricating the OCT sub-probe and the ultrasound transducer, these two components can be fixed inside a stainless steel (Yin et al. 2010) or polyimide (Yin et al. 2011) tube. On the tube, a window needs to be cut to allow the light beam and the sound wave to exit. The different arrangements of the transducer and the OCT sub-probe have been proposed, to minimize the size of the integrated probe while preserving its imaging performance. As they are all designed for intracoronary applications, we review them later in this chapter.

Outer Sheath Design

An outer sheath can help avoid cross-contamination between the imaging probe and the biological tissue. Furthermore, the sheath is designed to remain stationary when the probe rotates for scanning. Such a design can prevent the rotating probe from contacting and damaging the tissue.

The design of an IVUS-OCT sheath is similar to that of a clinically used IVOCT catheter, except that special attention needs to be paid to the selection of its material. The sheath material has to allow both light beams and sound waves to pass through without high attenuation or reflection (Li and Chen 2016).

Since the sheath is a cylindrical tube, its curved surface makes the light beam diverge faster in one axis than in the other axis. Consequently, astigmatism is generated to the light beam leaving the sheath. Various methods of astigmatism correction can be used, such as purging (Katwal and Lopez 2015) and using the ellipse ball-lens design (Tan et al. 2012). Purging means injecting a high refractive index medium (such as a saline solution or a contrast agent) into the lumen between the imaging probe and the sheath. Since the media in both sides of the curved interface have similar high refractive indices, the amount of undesired focusing is reduced. In a later design (Tan et al. 2012), a ball lens with different radii in two axes was employed so as to focus light to different extents and compensate astigmatism caused by the curved surface of the sheath.

Key Measures

To characterize the imaging performance of the system we built, two key measures, resolution and SNR, are commonly used. Here, we present equations that relate the imaging performance, the system speed and the probe size, and briefly discuss their implications. Technical approaches to maintain high performance without significantly increasing the system complexity or reducing the imaging speed are summarized later in this section.

Resolution

Resolution is defined as the minimal resolvable distance between two-point objects.
Transverse resolution For both ultrasound and OCT technologies, the transverse resolution is determined by the beam spot exiting the imaging sensor.

In ultrasound, the theoretical lateral resolution is typically expressed as (Biomedical 2006; Ng 2011):

$$R_{\mathrm{L}} = \frac{V}{f_0} \cdot F\# \tag{1}$$

where V is the speed of ultrasound, f_0 is central frequency, $F\#$ is the quotient of the focal length $l_{\mathrm{f},}$ and the outer diameter D of the transducer aperture.

In OCT, the theoretical lateral resolution is related to the numerical aperture (NA) of the imaging optics:

$$R_{\mathrm{L}} = 0.37\frac{\lambda_0}{\mathrm{NA}} \tag{2}$$

where λ_0 is the center wavelength of the OCT light source.

In reality, focusing aberration usually occurs. It deteriorates the beam quality and, thus, makes the actual transverse resolutions worse than the theoretical ones calculated by Eqs. (1) and (2).
Axial resolution Unlike transverse resolution, axial resolution is independent of the focusing ability of an imaging sensor.

In ultrasound, the axial resolution can be estimated by the following equation (Biomedical 2006; Ng 2011):

$$R_{\mathrm{A}} = \frac{V}{2 \cdot f_{\mathrm{o}} \cdot \mathrm{BW}} \tag{3}$$

where BW is the fractional frequency bandwidth.

For an OCT sub-system with a Gaussian spectrum light source, the axial resolution is determined by the center frequency and the bandwidth of the light source $\Delta\lambda$:

$$R_{\mathrm{A}} = \frac{2\mathrm{In}2}{n_{\mathrm{s}}\pi}\left(\frac{\lambda_0^2}{\Delta\lambda}\right) \tag{4}$$

where n_{s} is the refractive index of the sample.

For an OCT sub-system with a non-Gaussian spectrum light source and/or with chromatic dispersion between the sample and reference arms, estimation of its axial resolution by the above equation can be inaccurate. A measurement of the full-width-at-half-maximum (FWHM) of the axial point spread function is usually used to obtain the actual axial resolution.
SNR Classically, the SNR of an imaging system is the ratio of the peak signal to the root-mean-square of the noise level.

In an ultrasound sub-system, the SNR is related to the pulse duration, which is inversely proportional to the imaging speed, and the lateral beam width (Werkmeister et al. 2013):

$$\mathrm{SNR} = \sqrt{\frac{16}{\rho v w_x w_y \Delta t} \frac{A(f_0)}{n(f_0)}} \tag{5}$$

where w_x and w_y are the beam width in the x-and y-axis, respectively, v is sound speed in the medium, Δt is the pulse duration, $A(f_0)$ is the amplitude of the signal at the center frequency, and $n(f_0)$ is the noise at the center frequency.

In shot-noise limit FDOCT (i.e., the system used in the integrated IVUS-OCT system) the SNR can be expressed by the following equation (Bouma et al. 2008):

$$\mathrm{SNR} = \frac{\eta P_\mathrm{s}}{h\upsilon f_\mathrm{A}} \tag{6}$$

where f_A is the A-line speed, P_s is the signal power of the sample arm, $h\upsilon$ is the single-photon energy, and η is the detector sensitivity. As illustrated by this equation, the power of the sample arm is usually positively related to the SNR. Accordingly, an OCT sub-system usually uses a 90:10 or 80:20 coupler that transmits more lights to the sample arm than to the reference arm, so as to improve the SNR.

Similarly, as can be seen from Eqs. (5) and (6), SNRs of both ultrasound and OCT sub-systems will drop as the imaging speed goes up. Thus, many challenges must be overcome so as to achieve an ultrafast high-SNR IVUS-OCT system, which is introduced in the section 'Technical Advances: high-speed imaging system'. In addition, the aperture size of an imaging sensor influences its lateral resolution and the SNR according to Eqs. (1), (2), and (5). A summary on approaches to overcome the trade-off between the performance and the size of an integrated probe is provided in the section 'Technical Advances: miniaturized IVUS-OCT probe'.

Case Study: IVUS-OCT to Detect Vulnerable Plaques

Background

It was briefly mentioned in the introduction that the development of integrated IVUS-OCT technology was driven by the need for identifying vulnerable plaques. A thin-cap fibroatheroma (TCFA), the most common type of vulnerable plaques, is prone to rupture and trigger thrombus formation. Thrombus can block blood flow within coronary arteries and subsequently results in acute coronary syndrome (ACS). However, the majority of plaques in arteries are stable and do not cause ACS (Farooq et al. 2009; Sanz and Fayad 2008; Nakano et al. 2012; Ferrante et al. 2010). Treating a plaque without assessing its vulnerability may lead to overtreatment of stable plaques, which is not recommended by clinical guidelines, and could result in an increase of complications (Waxman et al. 2006; Falk et al. 2013; Montalescot et al. 2013). On the contrary, identifying vulnerable plaques and optimizing therapies accordingly (Fassa et al. 2005; Kereiakes et al. 2003; Wykrzykowska et al. 2012) are expected to improve clinical outcomes. A tool that has both a high positive predictive value and a high negative predictive value in detecting vulnerable plaques, i.e., tell cardiologists to locally treat a plaque or not (Mintz 2016), is needed clinically. It holds great promise in preventing plaque rupture and life-threatening sequelae (Fleg et al. 2012; Waxman et al. 2006). Many imaging techniques have been used to image plaques (Lusis 2000; Narula and Strauss 2007), including conventional fluoroscopic angiography, magnetic resonance imaging (MRI), X-ray or computed tomography (CT), OCT, IVUS with virtual histology (VH), angioscopy and near-infrared fluorescence (NIRF) imaging (Puri et al. 2011; Sanz and Fayad 2008; Narula and Strauss 2007). However, none of these currently available technologies is sufficient to accurately evaluate the vulnerability of a plaque (Sanz and Fayad 2008; Puri et al. 2011).

Particularly, a stand-alone intravascular imaging modality can visualize a certain feature of a TCFA, but not all key features. IVUS, with its deep penetration, can image the large necrotic pool and the total plaque volume of a TCFA. The limitation of IVUS is its resolution (~100 μm). On the contrary, OCT, with its high resolution, permits a definite measurement of the cap thickness of a plaque. This critical parameter for characterizing TCFA is otherwise not accessible by other in vivo imaging techniques (Farooq et al. 2009). However, the primary limitation of OCT for this application is its penetration depth (~1.5 mm in tissue) which is not enough to reveal the necrotic core size. Clinical evidence has also proven these limitations. A large cohort IVUS trial confirmed that IVUS alone is not sufficient in identifying vulnerable plaques (Stone et al. 2011, 2012). Similarly, clinical data (Mintz 2016) also revealed that the diagnostic accuracy of OCT to identify vulnerable plaque is limited mainly due to its shallow penetration depth and low molecular specificity (Virmani 2011; Prati et al. 2010; Bezerra et al. 2009).

As illustrated above, the characteristics of TCFA highlight the necessity of an imaging technique with a high resolution to identify the thin cap and with a deep enough penetration depth to visualize the necrotic core simultaneously. Moreover, several clinical studies were reported and validated the usefulness of a combined use of IVUS and OCT in characterizing plaque compositions (Kawasaki et al. 2006; Rieber et al. 2006) and detecting vulnerable plaques (Sawada et al. 2008; Zhang et al. 2014; Fujii et al. 2015; Mintz 2016). All these findings motivate the development of a fully integrated intracoronary ultrasound-OCT system.

Technical Advances

In this section, we review the technical development of the IVUS-OCT technology for imaging vulnerable plaques in coronary arteries. Apart from common considerations of an integrated IVUS-OCT system previously described, there are several particular technical challenges that need to be addressed to translate this technology for clinical intravascular imaging. These challenges include miniaturizing the IVUS-OCT probe, increasing the imaging speed and optimizing the flushing process.

Miniaturized IVUS-OCT Probe

The inner diameter of a human coronary artery is 2–5 mm. Patients with atherosclerosis may have coronary arteries that are even narrower. Accordingly, an intracoronary imaging catheter (including the imaging probe and the outer sheath) is usually made with an OD smaller than 1 mm and with a soft tip to avoid scraping and damaging the coronary artery wall during imaging procedures. However, over-reducing the size (either the thickness or the aperture area) of an OCT or ultrasound probe will sacrifice the performance of the integrated imaging system. The aperture size of a sensor (either a transducer or an OCT lens) limits the lateral resolution and SNR that a system can achieve, based on Eqs. (1), (2), and (5). In addition, as described previously, the thicknesses of the piezoelectric element of a transducer affect its central frequency, while thicknesses of the matching and backing layers influence the sensitivity and the impedance of a transducer. Thus, the challenge of making an IVUS-OCT catheter is to maintain good imaging quality of IVUS and OCT without making the crossing profile of the IVUS-OCT catheter larger than 1 mm. To address this technical challenge, an innovative probe arrangement that can make good use of the space has to be applied.

In 2010, our groups published the first demonstration of an integrated IVUS-OCT probe (Yin et al. 2010). This probe applied a side-by-side arrangement and achieved an OD of 2.4 mm, see Fig. 3.3a. The size of the GRIN lens used in the OCT probe had an OD of 0.5 mm, and the size of the PZT transducer was around $0.4 \times 0.4\ mm^2$. To better use space and provide coaxial IVUS-OCT imaging ability, an inside-outside arrangement was proposed and achieved (Li et al. 2010) by placing a focused ring transducer (made by $LiNbO_3$) around a 0.7 mm-OD GRIN lens, see Fig. 3.3b. Although the OD of this prototype was the same as the former design (2.4 mm), apertures for the OCT and IVUS sensors were both larger, enabling higher lateral resolutions. Particularly with a large-aperture focused transducer, the measured lateral resolution of the IVUS sub-system was as high as 22 μm.

In 2011, we reported the first IVUS-OCT probe suitable for in vivo imaging in a coronary-sized artery (Yin et al. 2011). A sequential arrangement was applied to reduce the probe profile. The OD of the integrated probe was 0.68 mm. It utilized a relatively large GRIN lens (0.35 mm OD) to maintain high lateral resolution of the OCT sub-system. PMN-PT material, which has high k_t, $\varepsilon_s/\varepsilon_0$, and d_{33}, was used to fabricate a miniaturized transducer while maintaining high sensitivity of the ultrasound sub-system. However, there is a 2-mm longitudinal offset between the transducer and the OCT sensor due to the sequential arrangement, see Fig. 3.3c. Although the obtained ultrasound and OCT signal could be co-registered by post-processing, a real-time automatically co-registered arrangement would still be ideal for providing real-time guidance for interventions in the cardiac catheter laboratory

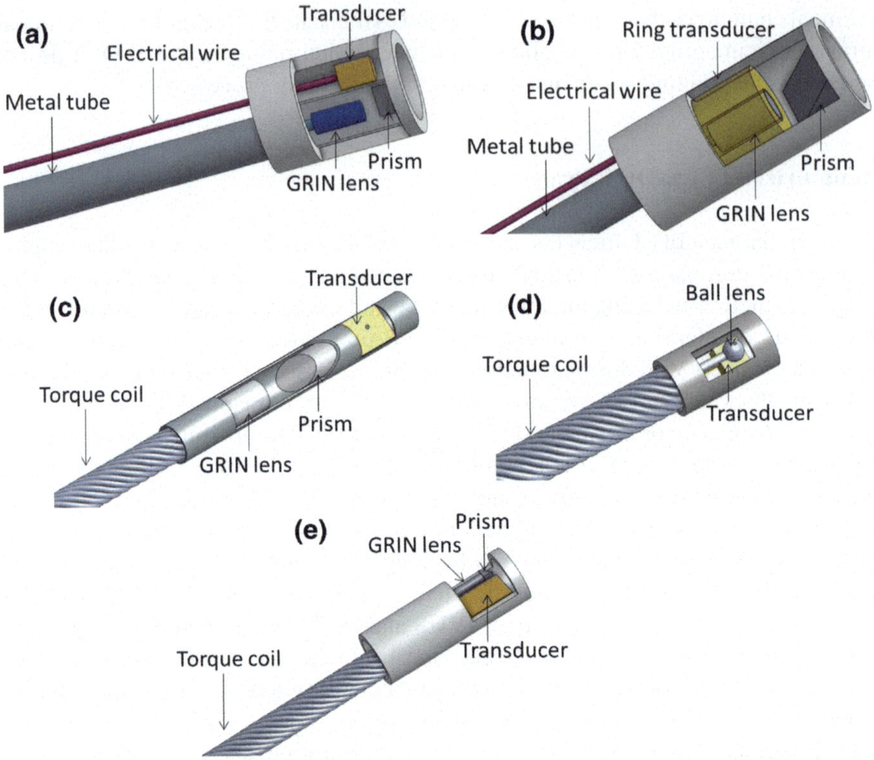

Fig. 3.3 Schematics of different IVUS-OCT probe designs. **a** Side-by-side arrangement (Yin et al. 2010); **b** inside–outside arrangement (Li et al. 2010); **c** sequential arrangement (Yin et al. 2011); **d** back-to-back arrangement (Li et al. 2013b); **e** side-by-side 90° apart design (Li et al. 2013a)

in a hospital. Moreover, the offset made this probe have a long rigid part [5-mm long, similar to previously reported designs (Li et al. 2010; Yin et al. 2010)]. It would be difficult to navigate this probe through a coronary artery that has a narrow lumen and sharp bends. To overcome these drawbacks, a modified version was reported in 2013 (Li et al. 2013b). This probe applied a "back-to-back" arrangement, see Fig. 3.3d, to ensure a coaxial positioning of the transducer and the OCT sensor, while keep a small OD. In addition, a ball lens (OD: 0.28 mm) instead of a miniaturized GRIN lens (OD: 0.35 mm) was used to further reduce the size of the OCT probe. The OD of this probe was 0.9 mm, and the rigid part of this probe was only 2.5 mm long, which was the shortest ever reported. Apart from these designs proposed by our group, the ultrasound group led by F. Stuart Foster and the OCT group led by Victor X. D. Yang, in Canada have also reported an IVUS-OCT probe design, see Fig. 3.3e. The probe used a side-by-side 90° apart design and a ultrasmall GRIN lens (OD 0.14 mm) to reduce the size of the IVUS-OCT probe (Li et al. 2013a). The OD of this probe was 1 mm. F. Foster and V. Yang also briefly reported a new probe design in a review paper (Bourantas et al. 2016). Although limited information was provided, it was stated that this new IVUS-OCT probe allowed optics to reside within the ultrasound transducer to facilitate miniaturization and the IVUS and OCT beams travel in the same direction. These four designs (Yin et al. 2011; Li et al. 2013a, b; Bourantas et al. 2016) described above ensure that the OD of the probe is smaller than 1 mm and are likely to provide safe access to coronary arteries.

High-Speed Imaging System

During intracoronary imaging, the imaging catheter stays inside a coronary artery and may partially block the normal blood flow. Coronary arteries supply blood to cardiac muscle that pumps oxygen-rich blood throughout the body. Due to this critical role of the coronary artery, the time needed for imaging the artery, i.e., the time to keep a catheter inside it, should be as short as possible. Furthermore, a short imaging time will also reduce the risk of flushing-induced complications. Because blood causes high optical attenuation, flushing an artery is often necessary to generate clear IVOCT images (Li et al. 2015a). However, the longer it takes to image, the more flushing agents will be injected, which can be harmful for the patient (Ozaki et al. 2012). To ensure the safety of using an IVUS-OCT catheter in patients, a vast amount of effort was put in reducing the time needed to image a coronary artery and, accordingly, increasing the imaging speed.

By innovatively applying state-of-the-art devices that were originally developed for other fields (including optical communication and computer engineering) in the development of an IVUS-OCT system, the imaging speed of IVUS-OCT has been greatly increased. The first reported IVUS-OCT system (Yin et al. 2010) was able to image at 1 fps by using separated DAQ cards for US and OCT sub-systems. Utilizing a more powerful dual-channel DAQ card and a GPU for parallel computing, 4 fps was accomplished (Yin et al. 2011). Subsequently, a 20-fps system was achieved (Li et al. 2014a, c) by using a modified probe design that ensures a more uniform rotation and smoother transmission of the rotation from the proximal to the distal end. With a recently developed ultrasound pulser and a more advanced design of a customized slip ring, the achievement of 72 fps was reported (Li et al. 2015b). By using such a system, IVUS-OCT imaging of a coronary artery can be conducted within 4 s. As a clinically used IVOCT system takes 3–6 s to image a coronary artery, it is reasonable to believe that the imaging speed of this IVUS-OCT system is sufficient for clinical applications.

Flushing

Intracoronary imaging is conducted in an environment naturally filled with blood. As blood is a strong attenuation source for OCT and high-frequency ultrasound signals, finding an optimal flushing agent that works well for both IVUS and OCT can improve in vivo imaging quality.

Flushing media with high viscosity can result in efficient blood displacement and clear OCT images (Suter et al. 2015). However, flushing media with too high viscosity will reduce the SNR of an ultrasound sub-system (Li et al. 2015a). This is because friction and thermal consumption of energy are induced when sound propagates through such media (Stokes 1849). To evaluate the effectiveness of different flushing agents and speeds and optimize flushing for both ultrasound and OCT sub-systems, we performed a well-controlled quantitative test in an in vitro model that mimics the human arterial system and in vivo in a rabbit (Li et al. 2015a).

Apart from considering the effectiveness of a flushing medium, toxicity is also a crucial concern. As the flushing agent is injected into blood and can be circulated through the whole body of the patient, we need to avoid using a medium with high toxicity (McCullough 2008; Ozaki et al. 2012). In consideration of both flushing effectiveness and toxicity, it was found that Dextran is the best among three flushing media we used (mannitol, Iohexol, and dextran). Other OCT flushing media, such as glucose solution and propylene glycol (Tuchin et al. 2002) and iodixanol (Suter et al. 2015), may work too but need to be validated with an IVUS-OCT system in vivo.

These above developments have made the intracoronary application of IVUS-OCT technically possible. Meanwhile, experiments in animal and human specimens have been conducted to evaluate the clinical adaptability and the safety of these developments which are reviewed below.

In Vivo *and* Ex Vivo *Validations*

In Vivo Validations

Rabbit and porcine models are commonly used (Ohtsuki et al. 2001) prior to human experiments to validate the design of an intravascular imaging system. The aortas of rabbits and the coronary arteries of pigs are comparable to the caliber of human coronary arteries. In addition, rabbits and pigs can grow lesions similar to human atherosclerotic plaques (Kolodgie et al. 2003; Schoenhagen et al. 2001).

The first in vivo demonstration of an integrated IVUS-OCT probe was achieved in a rabbit (Yin et al. 2011). IVUS-OCT image pairs of a rabbit aorta were acquired with a 4-fps system and a 0.68 mm-OD probe. The first in vivo intracoronary IVUS-OCT imaging was conducted in a pig (Li et al. 2014a). Imaging was performed in a swine left anterior descending artery by using a 20-fps system and a 3.7 F IVUS-OCT catheter (Fig. 3.4). Successively, an in vivo rabbit experiment validated the effectiveness of using dextran as the flushing agent for simultaneous IVUS-OCT imaging (Li et al. 2015a). In 2015, the safe use of an ultrafast IVUS-OCT system (72 fps) was demonstrated in live rabbits (Li et al. 2015b).

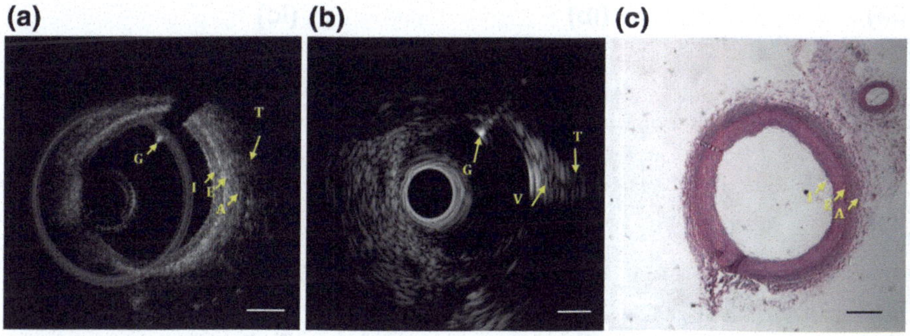

Fig. 3.4 An OCT-IVUS image pair obtained in a normal swine coronary artery in vivo. **a** OCT image, **b** IVUS image, **c** corresponding H&E histology. G: guide wire. T: tissue. V: vessel. I: intima. E: EEL. A: adventitia. Scale bar: 1 mm. Reprint from (Li et al. 2014a) with permission

Ex Vivo Validation

Although animal experiments can evaluate the clinical adaptability and the safety of these developed systems, few TCFA-like plaques can be found in animal models (Brezinski 2006; Rekhter 2002; Kolodgie et al. 2003; Rekhter et al. 1998). Thus, human cadaver samples were used to evaluate whether a device can acurately detect TCFA. In 2015, we reported our work on demonstrating the usefulness of IVUS-OCT in distinguishing a TCFA from false TCFAs (Li et al. 2015b), i.e., plaques that are easily misdiagnosed by stand-alone IVUS or OCT. Fibrous caps were shown as signal high regions in the OCT images (Fig. 3.5Ia, IIa, and IIIa). In IVUS and OCT images, the signal-low regions behind the caps revealed the presence of necrotic or lipid pools. As anticipated, the OCT imaging capability of the IVUS-OCT catheter differentiated thin cap (Fig. 3.5Ia) and thick cap (Fig. 3.5IIa) whereas the ultrasound imaging capability in the IVUS-OCT catheter differentiated large (Fig. 3.5Ib) and small (Fig. 3.5IIIb) necrotic cores. IVUS-OCT with simultaneous dual-imaging capabilities clearly discerned differences between TCFA and false TCFA (plaques with a thick cap or a small necrotic core).

Other Studies Related to IVUS-OCT

IVUS-OCT for Angioplasty Planning and Follow-Up

Apart from being used for the detection of TCFA, IVUS-OCT may also play an important role in angioplasty planning and follow-up.

Angioplasty is a commonly used non-surgical procedure to treat arteries, such as coronary or peripheral arteries (Arthurs et al. 2010), that were narrowed by atherosclerotic plaques. During this procedure, stents that open up arteries are usually

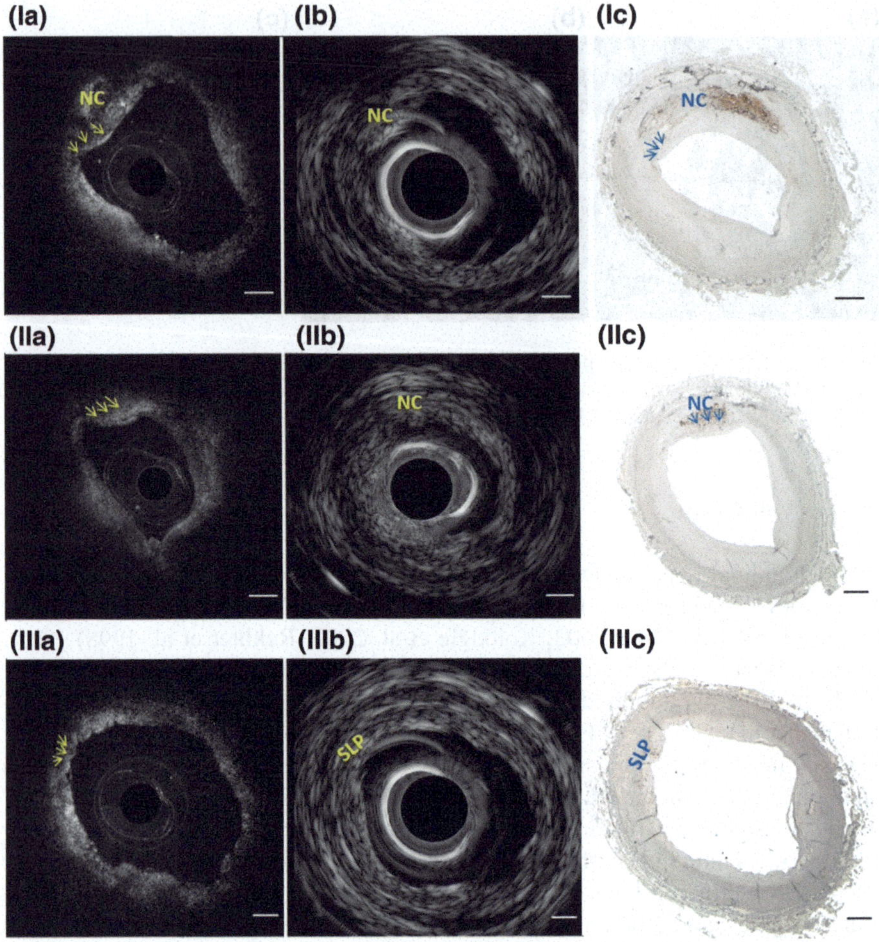

Fig. 3.5 Characterizing human atherosclerotic plaques by the ultrafast IVUS-OCT system. First row: Example of a TCFA. (**Ia**) OCT image in which arrows point at the fibrous cap; (**Ib**) corresponding IVUS image indicates the location of necrotic core; (**Ic**) photograph of the corresponding histology slide with CD 68 stain, highlighting macrophages and necrotic core. Middle row: A false-positive case of TCFA diagnosis based on IVUS only (**IIb**) was produced due to the insufficient resolution and sensitivity. Size of the thick cap can be determined by the corresponding OCT (**IIa**) and CD 68 histology (**IIc**). Bottom row: A false-positive case of TCFA diagnosis based on OCT only (**IIIa**) was produced due to OCT's limited penetration depth. A small lipid pool can be determined by IVUS (**IIIb**) and CD 68 histology (**IIIc**). Arrows denote the fibrous cap. NC: necrotic core; SLP: small lipid pool. Scale bar: 0.5 mm. Reprint from the reference Li et al. (2015b) with permission

installed under X-ray guidance. However, X-ray doesn't provide sufficient information for optimizing angioplasty procedures (Gonzalo et al. 2011; Chamié et al. 2013; Bezerra 2016). On the contrary, an imaging modality that can accurately measure and characterize plaques is able to address this need (Habara et al. 2012). In light of the encouraging results obtained by stand-alone IVUS (Roy et al. 2008; Sonoda et al. 2004) or OCT (Gatto et al. 2013; Prati et al. 2012) in guiding angioplasty, it is reasonable to believe that IVUS-OCT will be useful in facilitating angioplasty planning.

Studies showed that the outcomes of angioplasty can be optimized by sensible selections of the lesion preparation technique and stent landing zones, and these selections can be guided by accurate characterization of coronary plaques (Gonzalo et al. 2011; Bezerra 2016). Accordingly, research evaluating the accuracy of using IVUS-OCT to characterize plaques has been conducted. In 2013, F. Stuart Foster and Victor X. D. Yang et al. published their work on imaging 31 arterial segments from 11 cadaver human coronaries using a hybrid IVUS-OCT catheter and system (Li et al. 2013a). IVUS-OCT image pairs of calcified plaques, a plaque with a lipid pool and a fibrous cap, a plaque with a large necrotic pool and a fibrous cap were reported and these classifications were validated by histological analysis. To quantitatively evaluate the advantages of using the integrated IVUS-OCT over stand-alone IVUS or OCT in plaque classification, a larger sample size study (Li et al. 2014b) was completed. Over 240 regions of interests from 25 cadavers were imaged. The sensitivity and specificity of differentiating three main plaque types, calcified, lipid, and fibrotic, were calculated.

In addition to the potential value of using IVUS-OCT for angioplasty planning, IVUS-OCT may also provide clinically significant information on angioplasty follow-up. According to clinical guidelines, the use of IVUS is reasonable for evaluating stent restenosis (Levine et al. 2011). The effectiveness of OCT in angioplasty follow-up was also evaluated through clinical trials (Zivelonghi et al. 2014). Moreover, stand-alone IVUS and OCT have demonstrated complementary diagnostic values in imaging stent-tissue interactions (Alfonso et al. 2012; Tahara et al. 2010; Alfonso et al. 2012). Thus, it can be anticipated that the utility of an integrated IVUS-OCT catheter may benefit angioplasty follow-up. Three-dimensional IVUS-OCT imaging of a stent in a human coronary artery has been conducted (Li and Chen 2016). Small degrees of incomplete stent coverage were revealed in the OCT images whereas these features were not clearly shown in the IVUS images. On the other hand, deep invisible tissues in OCT images were visualized in the corresponding IVUS images.

Tri-modality Imaging System

Despite the synergetic advantages of IVUS-OCT illustrated above, IVUS-OCT has its limitation: the lack of a biochemical contrast and thus the inability to make a definite diagnosis of tissue types (Thim et al. 2010; Li et al. 2014b; Kawasaki et al. 2006;

Rieber et al. 2006). For example, homogenous plaque components that cause reduced light and sound backscattering may be misidentified as plaque components that cause an increase of absorption (Manfrini et al. 2006; Phipps et al. 2016). Although IVUS-VH was developed to provide an illustration of plaque components, its original validation was achieved by using necropsy specimens from only 51 patients (Nair et al. 2002). Moreover, an in vivo animal experiment demonstrated there is no relationship between a VH-IVUS identified necrotic core and one classified by histology (Nair et al. 2002).

To enhance the molecular specificity of IVUS-OCT for identifying tissue composition, fluorescence imaging capability was added and a tri-modality imaging system has been developed (Liang et al. 2014) and is detailed in Chapter 8. The system is able to acquire IVUS, OCT, and fluorescence images simultaneously. It may improve the diagnostic accuracy of IVUS-OCT.

References

Alfonso F, Dutary J, Paulo M, Gonzalo N, Pérez-Vizcayno MJ, Jiménez-Quevedo P, Escaned J, Bañuelos C, Hernández RMC (2012) Combined use of optical coherence tomography and intravascular ultrasound imaging in patients undergoing coronary interventions for stent thrombosis. Heart 98(16):1213–1220

Arthurs ZM, Bishop PD, Feiten LE, Eagleton MJ, Clair DG, Kashyap VS (2010) Evaluation of peripheral atherosclerosis: a comparative analysis of angiography and intravascular ultrasound imaging. J Vasc Surg 51(4):933–939

Bezerra H (2016) Intravascular OCT in PCI. American College of Cardiology. http://www.acc.org/latest-in-cardiology/articles/2016/06/13/10/01/intravascular-oct-in-pci

Bezerra HG, Costa MA, Guagliumi G, Rollins AM, Simon DI (2009) Intracoronary optical coherence tomography: a comprehensive review: clinical and research applications. JACC: Cardiovasc Interv 2(11):1035–1046

Boppart SA, Bouma BE, Pitris C, Southern JF, Brezinski ME, Fujimoto JG (1998) In vivo cellular optical coherence tomography imaging. Nat Med 4(7):861–865

Bouma BE, Tearney GJ, Vakoc BJ, Yun SH (2008) Optical frequency domain imaging. In: Drexler W, Fujimoto JG (eds) Optical coherence tomography: technology and applications. Springer, Berlin, Heidelberg, pp 209–237. https://doi.org/10.1007/978-3-540-77550-8_7

Bourantas CV, Jaffer FA, Gijsen FJ, Van Soest G, Madden SP, Courtney BK, Fard AM, Tenekecioglu E, Zeng Y, Van Der Steen AFW (2016) Hybrid intravascular imaging: recent advances, technical considerations, and current applications in the study of plaque pathophysiology. Eur Heart J 38(6):400–412

Brezinski ME (2006) OCT in cardiovascular medicine: animal models. In: Optical coherence tomography: principles and applications. p 414

Cannata JM, Ritter TA, Wo-Hsing C, Silverman RH, Shung KK (2003) Design of efficient, broadband single-element (20–80 MHz) ultrasonic transducers for medical imaging applications. IEEE Trans Ultrason Ferroelectr Freq Control 50(11):1548–1557. https://doi.org/10.1109/TUFFC.2003.1251138

Chamié D, Bezerra HG, Attizzani GF, Yamamoto H, Kanaya T, Stefano GT, Fujino Y, Mehanna E, Wang W, Abdul-Aziz A (2013) Incidence, predictors, morphological characteristics, and clinical outcomes of stent edge dissections detected by optical coherence tomography. JACC: Cardiovasc Interv 6(8):800–813

Choma MA, Sarunic MV, Yang C, Izatt JA (2003) Sensitivity advantage of swept source and Fourier domain optical coherence tomography. Opt Express 11(18):2183–2189

Cobbold RSCC (2007) Foundations of biomedical ultrasound. Oxford University Press, USA

Falk E, Nakano M, Bentzon JF, Finn AV, Virmani R (2013) Update on acute coronary syndromes: the pathologists' view. Eur Heart J 34(10):719–728

Farooq MU, Khasnis A, Majid A, Kassab MY (2009) The role of optical coherence tomography in vascular medicine. Vasc Med 14:63–71

Fassa AA, Wagatsuma K, Higano ST, Mathew V, Barsness GW, Lennon RJ, Holmes DR Jr, Lerman A (2005) Intravascular ultrasound-guided treatment for angiographically indeterminate left main coronary artery disease: a long-term follow-up study. J Am Coll Cardiol 45:204–211

Ferrante G, Nakano M, Prati F, Niccoli G, Mallus MT, Ramazzotti V, Montone RA, Kolodgie FD, Virmani R, Crea F (2010) High levels of systemic myeloperoxidase are associated with coronary plaque erosion in patients with acute coronary syndromes. Circulation 122:2505–2513

Fleg JL, Stone GW, Fayad ZA, Granada JF, Hatsukami TS, Kolodgie FD, Ohayon J, Pettigrew R, Sabatine MS, Tearney GJ, Waxman S, Domanski MJ, Srinivas PR, Narula J (2012) Detection of high-risk atherosclerotic plaque: report of the NHLBI Working Group on current status and future directions. JACC Cardiovasc Imaging 5(9):941–955

Fujii K, Hao H, Shibuya M, Imanaka T, Fukunaga M, Miki K, Tamaru H, Sawada H, Naito Y, Ohyanagi M (2015) Accuracy of OCT, grayscale IVUS, and their combination for the diagnosis of coronary TCFA: an ex vivo validation study. JACC: Cardiovasc Imaging 8(4):451–460

Gatto L, Chisari A, La Manna A, Burzotta F, Di Vito L, Mallus MT, Cremonesi A, Pappalardo A, Albertucci M, Prati F (2013) TCT-555 comparative incidence of optical coherence tomography features indicative of uncorrected stent deployment in patients with and without major adverse cardiac events in the OCT guided arm of the CLI-OPCI study. J Am Coll Cardiol 62(18_S1):B167

Gonzalo N, Serruys PW, Okamura T, Shen ZJ, Garcia-Garcia HM, Onuma Y, van Geuns RJ, Ligthart J, Regar E (2011) Relation between plaque type and dissections at the edges after stent implantation: an optical coherence tomography study. Int J Cardiol 150(2):151–155

Habara M, Nasu K, Terashima M, Kaneda H, Yokota D, Ko E, Ito T, Kurita T, Tanaka N, Kimura M, Ito T, Kinoshita Y, Tsuchikane E, Asakura K, Asakura Y, Katoh O, Suzuki T (2012) Impact of frequency-domain optical coherence tomography guidance for optimal coronary stent implantation in comparison with intravascular ultrasound guidance. Circ Cardiovasc Interv 5(2):193–201

Hoang KC, Edris A, Su J, Mukai DS, Mahon S, Petrov AD, Kern M, Ashan C, Chen Z, Tromberg BJ (2009) Use of an oxygen-carrying blood substitute to improve intravascular optical coherence tomography imaging. J Biomed Opt 14(3):034028

Huang D, Swanson EA, Lin CP, Schuman JS, Stinson WG, Chang W, Hee MR, Flotte T, Gregory K, Puliafito CA et al (1991) Optical coherence tomography. Science 254(5035):1178–1181

Katwal AB, Lopez JJ (2015) Technical considerations and practical guidance for intracoronary optical coherence tomography. In: Price MJ (ed) Intravascular imaging: OCT and IVUS, an issue of interventional cardiology clinics, vol 4. Elsevier Health Sciences, pp 239–250

Kawasaki M, Bouma BE, Bressner J, Houser SL, Nadkarni SK, MacNeill BD, Jang I-K, Fujiwara H, Tearney GJ (2006) Diagnostic accuracy of optical coherence tomography and integrated backscatter intravascular ultrasound images for tissue characterization of human coronary plaques. J Am Coll Cardiol 48(1):81–88

Kelly J, Leonard M, Tantigate C, Safari A (1997) Effect of composition on the electromechanical properties of (1-x)Pb(Mg1/3Nb2/3)O3− XPbTiO3 ceramics. J Am Ceram Soc 80(4):957–964. https://doi.org/10.1111/j.1151-2916.1997.tb02927.x

Kereiakes DJ, Szyniszewski AM, Wahr D, Herrmann HC, Simon DI, Rogers C, Kramer P, Shear W, Yeung AC, Shunk KA, Chou TM, Popma J, Fitzgerald P, Carroll TE, Forer D, Adelman DC (2003) Phase I drug and light dose-escalation trial of motexafin lutetium and far red light activation (phototherapy) in subjects with coronary artery disease undergoing percutaneous coronary intervention and stent deployment: procedural and long-term results. Circulation 108(11):1310–1315

Kolodgie FD, Petrov A, Virmani R, Narula N, Verjans JW, Weber DK, Hartung D, Steinmetz N, Vanderheyden JL, Vannan MA, Gold HK, Reutelingsperger CP, Hofstra L, Narula J (2003) Targeting of apoptotic macrophages and experimental atheroma with radiolabeled annexin V: a technique with potential for noninvasive imaging of vulnerable plaque. Circulation 108:3134–3139

Leitgeb R, Hitzenberger CK, Fercher AF (2003) Performance of fourier domain vs. time domain optical coherence tomography. Opt Express 11(8):889–894

Leitgeb R, Drexler W, Unterhuber A, Hermann B, Bajraszewski T, Le T, Stingl A, Fercher A (2004) Ultrahigh resolution Fourier domain optical coherence tomography. Opt Express 12(10):2156–2165

Levine GN, Bates ER, Blankenship JC, Bailey SR, Bittl JA, Cercek B, Chambers CE, Ellis SG, Hollenberg SM, Khot UN (2011) 2011 ACCF/AHA/SCAI guideline for percutaneous coronary intervention. Circulation 124(23):2574–2609

Li J, Chen Z (2016) Integrated intravascular ultrasound and optical coherence tomography technology: a promising tool to identify vulnerable plaques [INVITED PAPER]. J Biomed Photonics Eng 1(4):209–224

Li X, Yin J, Hu C, Zhou Q, Shung KK, Chen Z (2010) High-resolution coregistered intravascular imaging with integrated ultrasound and optical coherence tomography probe. Appl Phys Lett 97(13):133702

Li BH, Leung ASO, Soong A, Munding CE, Lee H, Thind AS, Munce NR, Wright GA, Rowsell CH, Yang VXD (2013a) Hybrid intravascular ultrasound and optical coherence tomography catheter for imaging of coronary atherosclerosis. Catheter Cardiovasc Interv 81(3):494–507

Li J, Ma T, Jing J, Zhang J, Patel PM, Shung KK, Zhou Q, Chen Z (2013b) Miniature optical coherence tomography-ultrasound probe for automatically coregistered three-dimensional intracoronary imaging with real-time display. J Biomed Opt 18(10):100502–100502

Li J, Li X, Mohar D, Raney A, Jing J, Zhang J, Johnston A, Liang S, Ma T, Shung KK (2014a) Integrated IVUS-OCT for real-time imaging of coronary atherosclerosis. JACC: Cardiovasc Imaging 7(1):101–103

Li J, Ma T, Mohar D, Correa A, Minami H, Jing J, Zhou Q, Patel PM, Chen Z (2014b) Diagnostic accuracy of integrated intravascular ultrasound and optical coherence tomography (IVUS-OCT) system for coronary plaque characterization. In: SPIE BiOS, International Society for Optics and Photonics, pp 892635–892636

Li X, Li J, Jing J, Ma T, Liang S, Zhang J, Mohar D, Raney A, Mahon S, Brenner M (2014c) Integrated IVUS-OCT imaging for atherosclerotic plaque characterization. IEEE J Sel Top Quantum Electron 20(2):196–203

Li J, Minami H, Steward E, Ma T, Mohar D, Robertson C, Shung K, Zhou Q, Patel P, Chen Z (2015a) Optimal flushing agents for integrated optical and acoustic imaging systems. J Biomed Opt 20(5):056005

Li J, Ma T, Mohar D, Steward E, Yu M, Piao Z, He Y, Shung KK, Zhou Q, Patel PM, Chen Z (2015b) Ultrafast optical-ultrasonic system and miniaturized catheter for imaging and characterizing atherosclerotic plaques in vivo. Sci Rep 5:18406

Liang S, Ma T, Jing J, Li X, Li J, Shung KK, Zhou Q, Zhang J, Chen Z (2014) Trimodality imaging system and intravascular endoscopic probe: combined optical coherence tomography, fluorescence imaging and ultrasound imaging. Opt Lett 39(23):6652–6655

Liu L, Gardecki JA, Nadkarni SK, Toussaint JD, Yagi Y, Bouma BE, Tearney GJ (2011) Imaging the subcellular structure of human coronary atherosclerosis using micro–optical coherence tomography. Nat Med 17:1010–1014

Lusis AJ (2000) Atherosclerosis. Nature 497:233–241

Manfrini O, Mont E, Leone O, Arbustini E, Eusebi V, Virmani R, Bugiardini R (2006) Sources of error and interpretation of plaque morphology by optical coherence tomography. Am J Cardiol 98(2):156–159

Mao Y, Chang S, Flueraru C (2010) Fiber lenses for ultra-small probes used in optical coherent tomography. J Biomed Sci Eng 3:27–34

McCullough PA (2008) Contrast-induced acute kidney injury. J Am Coll Cardiol 51(15):1419–1428

Mintz GS (2016) Understanding why and when optical coherence tomography does not detect vulnerable plaques is it important? Circ Cardiovasc Interv 9(7):e004144

Montalescot G, Sechtem U, Achenbach S, Andreotti F, Arden C, Budaj A, Bugiardini R, Crea F, Cuisset T, Di Mario C (2013) 2013 ESC guidelines on the management of stable coronary artery disease. Eur Heart J 34(38):2949–3003

Moon S, Piao Z, Kim C-S, Chen Z (2013) Lens-free endoscopy probe for optical coherence tomography. Opt Lett 38(12):2014–2016

Nair A, Kuban BD, Tuzcu EM, Schoenhagen P, Nissen SE, Vince DG (2002) Coronary plaque classification with intravascular ultrasound radiofrequency data analysis. Circulation 106(17):2200–2206

Nakano M, Kolodgie FD, Otsuka F, Yazdani SK, Ladich ER, Virmani R (2012) Vulnerable plaque. In: Coronary heart disease. pp 192–195

Narula J, Strauss HW (2007) The popcorn plaques. Nat Med 13:532–534

Ng A (2011) Resolution in ultrasound imaging. Contin Educ Anaesth Crit Care Pain 11(5):186–192

Ohtsuki K, Hayase M, Akashi K, Kopiwoda S, Strauss HW (2001) Detection of monocyte chemoattractant protein-1 receptor expression in experimental atherosclerotic lesions an autoradiographic study. Circulation 104:203–208

Ozaki Y, Kitabata H, Tsujioka H, Hosokawa S, Kashiwagi M, Ishibashi K, Komukai K, Tanimoto T, Ino Y, Takarada S, Kubo T, Kimura K, Tanaka A, Hirata K, Mizukoshi M, Imanishi T, Akasaka T (2012) Comparison of contrast media and low-molecular-weight dextran for frequency-domain optical coherence tomography. Circ J 76(4):922–927

Phipps JE, Hoyt T, Vela D, Wang T, Michalek JE, Buja LM, Jang I-K, Milner TE, Feldman MD (2016) Diagnosis of thin-capped fibroatheromas in intravascular optical coherence tomography images effects of light scattering. Circ Cardiovasc Interv 9 (7):e003163

Prati F, Regar E, Mintz GS, Arbustini E, Di Mario C, Jang I-K, Akasaka T, Costa M, Guagliumi G, Grube E (2010) Expert review document on methodology, terminology, and clinical applications of optical coherence tomography: physical principles, methodology of image acquisition, and clinical application for assessment of coronary arteries and atherosclerosis. Eur Heart J 31(4):401–415

Prati F, Di Vito L, Biondi-Zoccai G, Occhipinti M, La Manna A, Tamburino C, Burzotta F, Trani C, Porto I, Ramazzotti V (2012) Angiography alone versus angiography plus optical coherence tomography to guide decision-making during percutaneous coronary intervention: the Centro per la Lotta contro l'Infarto-Optimisation of Percutaneous Coronary Intervention (CLI-OPCI) study. EuroIntervention 8(7):823–829

Puri R, Worthley MI, Nicholls SJ (2011) intravascular imaging of vulnerable coronary plaque: current and future concepts. Nat Rev 8:131–139

Radhakrishnan S, Goldsmith J, Huang D, Westphal V, Dueker DK, Rollins AM, Izatt JA, Smith SD (2005) Comparison of optical coherence tomography and ultrasound biomicroscopy for detection of narrow anterior chamber angles. Arch Ophthalmol 123(8):1053–1059

Rekhter MD (2002) How to evaluate plaque vulnerability in animal models of atherosclerosis? Cardiovasc Res 54:36–41

Rekhter MD, Hicks GW, Brammer DW, Work CW, Kim JS, Gordon D, Keiser JA, Ryan MJ (1998) Animal model that mimics atherosclerotic plaque rupture. Circ Res 83(7):705–713

Rieber J, Meissner O, Babaryka G, Reim S, Oswald M et al (2006) Diagnostic accuracy of optical coherence tomography and intravascular ultrasound for the detection and characterization of atherosclerotic plaque composition in ex-vivo coronary specimens: a comparison with histology. Diagn Methods 17(5):425–430

Roy P, Steinberg DH, Sushinsky SJ, Okabe T, Slottow TLP, Kaneshige K, Xue Z, Satler LF, Kent KM, Suddath WO (2008) The potential clinical utility of intravascular ultrasound guidance in patients undergoing percutaneous coronary intervention with drug-eluting stents. Eur Heart J 29(15):1851–1857

Sanz J, Fayad ZA (2008) Imaging of atherosclerotic cardiovascular disease. Nature 45:953–957

Sawada T, Shite J, Garcia-Garcia HM, Shinke T, Watanabe S, Otake H, Matsumoto D, Tanino Y, Ogasawara D, Kawamori H (2008) Feasibility of combined use of intravascular ultrasound radiofrequency data analysis and optical coherence tomography for detecting thin-cap fibroatheroma. Eur Heart J 29(9):1136–1146

Schoenhagen P, Ziada KM, Vince DG, Nissen SE, Tuzcu EM (2001) Arterial remodeling and coronary artery disease: the concept of "dilated" versus "obstructive" coronary atherosclerosis. J Am Coll Cardiol 38(2):297–306

Sonoda S, Morino Y, Ako J, Terashima M, Hassan AHM, Bonneau HN, Leon MB, Moses JW, Yock PG, Honda Y (2004) Impact of final stent dimensions on long-term results following sirolimus-eluting stent implantation: serial intravascular ultrasound analysis from the sirius trial. J Am Coll Cardiol 43(11):1959–1963

Stokes GG (1849) On the theories of the internal friction in fluids in motion, and of the equilibrium and motion of elastic solids. Trans Cambridge Philos Soc 8(22):287–342

Stone GW, Maehara A, Lansky AJ, de Bruyne B, Cristea E, Mintz GS, Mehran R, McPherson J, Farhat N, Marso SP, Parise H, Templin B, White R, Zhang Z, Serruys PW (2011) A prospective natural-history study of coronary atherosclerosis. N Engl J Med 364(3):226–235. https://doi.org/10.1056/NEJMoa1002358

Stone PH, Saito S, Takahashi S, Makita Y, Nakamura S, Kawasaki T, Takahashi A, Katsuki T, Nakamura S, Namiki A (2012) Prediction of progression of coronary artery disease and clinical outcomes using vascular profiling of endothelial shear stress and arterial plaque characteristics: the PREDICTION study. Circulation 126(2):172–181

Suter MJ, Kashiwagi M, Gallagher KA, Nadkarni SK, Asanani N, Tanaka A, Conditt GB, Tellez A, Milewski K, Kaluza GL, Granada JF, Bouma BE, Tearney GJ (2015) Optimizing flushing parameters in intracoronary optical coherence tomography: an in vivo swine study. Int J Cardiovasc Imaging 31(6):1097–1106. https://doi.org/10.1007/s10554-015-0668-0

Tahara S, Bezerra HG, Sirbu V, Kyono H, Musumeci G, Rosenthal N, Guagliumi G, Costa MA (2010) Angiographic, IVUS and OCT evaluation of the long-term impact of coronary disease severity at the site of overlapping drug-eluting and bare metal stents: a substudy of the ODESSA trial. Heart 96(19):1574–1578

Tan KM, Shishkov M, Chee A, Applegate MB, Bouma BE, Suter MJ (2012) Flexible transbronchial optical frequency domain imaging smart needle for biopsy guidance. Biomed Opt Express 3(8):1947–1954

Thim T, Hagensen MK, Wallace-Bradley D, Granada JF, Kaluza GL, Drouet L, Paaske WP, Bøtker HE, Falk E (2010) Unreliable assessment of necrotic core by virtual histology intravascular ultrasound in porcine coronary artery disease. Circulation 3:384–391

Tuchin VV, Xu X, Wang RK (2002) Dynamic optical coherence tomography in studies of optical clearing, sedimentation, and aggregation of immersed blood. Appl Opt 14(1):258–271

Virmani R (2011) Are our tools for the identification of TCFA ready and do we know them? JACC: Cardiovasc Imaging 4(6):656–658

Wang T, Wieser W, Springeling G, Beurskens R, Lancee CT, Pfeiffer T, van der Steen AFW, Huber R, van Soest G (2013) Intravascular optical coherence tomography imaging at 3200 frames per second. Opt Lett 38(10):1715–1717

Waxman S, Ishibashi F, Muller JE (2006) Detection and treatment of vulnerable plaques and vulnerable patients novel approaches to prevention of coronary events. Circulation 114(22):2390–2411

Werkmeister RM, Alex A, Kaya S, Unterhuber A, Hofer B, Riedl J, Bronhagl M, Vietauer M, Schmidl D, Schmoll T (2013) Measurement of tear film thickness using ultrahigh-resolution optical coherence tomography ultrahigh-resolution OCT measurement of tear film. Invest Ophthalmol Vis Sci 54(8):5578–5583

Wykrzykowska JJ, Diletti R, Gutierrez-Chico JL, van Geuns RJ, van der Giessen WJ, Ramcharitar S, Duckers HE, Schultz C, de Feyter P, van der Ent M, Regar E, de Jaegere P, Garcia-Garcia HM, Pawar R, Gonzalo N, Ligthart J, de Schepper J, van den Berg N, Milewski K, Granada JF, Serruys PW (2012) Plaque sealing and passivation with a mechanical self-expanding low outward force nitinol vShield device for the treatment of IVUS and OCT-derived thin cap fibroatheromas

(TCFAs) in native coronary arteries: report of the pilot study vShield evaluated at cardiac hospital in Rotterdam for investigation and treatment of TCFA (SECRITT). EuroIntervention 8:945–954
Yin J, Liu G, Zhang J, Yu L, Mahon S, Mukai D, Brenner M, Chen Z (2009) In vivo early detection of smoke-induced airway injury using three-dimensional swept-source optical coherence tomography. J Biomed Opt 14(6):060503
Yin J, Yang H-C, Li X, Zhang J, Zhou Q, Hu C, Shung KK, Chen Z (2010) Integrated intravascular optical coherence tomography ultrasound imaging system. J Biomed Opt 15(1):010512–010513
Yin J, Li X, Jing J, Li J, Mukai D, Mahon S, Edris A, Hoang K, Shung KK, Brenner M, Narula J, Zhou Q, Chen Z (2011) Novel combined miniature optical coherence tomography ultrasound probe for in vivo intravascular imaging. J Biomed Opt 16(6):060505
Zagaynova E, Gladkova N, Shakhova N, Gelikonov G, Gelikonov V (2008) Endoscopic OCT with forward-looking probe: clinical studies in urology and gastroenterology. J Biophotonics 1(2):114–128
Zhang J, Yang S, Ji X, Zhou Q, Xing D (2014) Characterization of lipid-rich aortic plaques by intravascular photoacoustic tomography: ex vivo and in vivo validation in a rabbit atherosclerosis model with histologic correlation. J Am Coll Cardiol 64(4):385–390
Zivelonghi C, Ghione M, Kilickesmez K, Loureiro RE, Foin N, Lindsay A, de Silva R, Ribichini F, Vassanelli C, Di Mario C (2014) Intracoronary optical coherence tomography: a review of clinical applications. J Cardiovasc Med 15(7):543–553

Chapter 4
Intravascular Photoacoustic Imaging of Lipid-Laden Plaques: From Fundamental Concept Toward Clinical Translation

Jie Hui and Ji-Xin Cheng

Introduction

Coronary artery disease remains the leading cause of morbidity and mortality worldwide. The rupture of vulnerable atherosclerotic plaque, with its vulnerability defined by the propensity for the plaque to rupture, contributes to the majority of acute coronary syndromes and sudden cardiac deaths (Yahagi et al. 2016; Narula et al. 2013). Despite the mechanism and prevalence of plaque vulnerability still under investigation, the thin-capped fibroatheroma has been understood to be the most vulnerable plaque type (Finn et al. 2010; Yahagi et al. 2016). Thin-capped fibroatheromas are grossly defined by hallmarks of a large lipid-rich necrotic core, thin fibrous cap, increased inflammatory infiltrate, and positive vascular remodeling (Narula et al. 2013; Finn et al. 2010; Libby et al. 2010). The plaque rupture occurs where the cap is thinnest and most frequently at cap shoulder region (Bentzon et al. 2014; Falk et al. 1995). Thinning of the fibrous cap is due to either loss of smooth muscle cells or inflammatory infiltrate that secretes matrix metalloproteinases and degrades collagen-rich cap matrix (Bentzon et al. 2014). With plaque rupture, the thrombogenic contents of lipid-rich necrotic core are released into the bloodstream, leading to thrombosis, and acute coronary syndromes. In addition, these vulnerable plaques are often structurally non-obstructive to moderately obstructive, thus clinically unidentifiable by routine angiography and stress testing (Narula et al. 2013; Schoenhagen et al. 2001; Takano et al. 2001). Therefore, accurate identification of vulnerable plaques through advanced imaging or detection technology is necessary for diagnosis and treatment, either by interventional or preventive methods. This eminent need,

J. Hui · J.-X. Cheng (✉)
Department of Electrical and Computer Engineering, Boston University, Boston, MA 02215, USA
e-mail: jxcheng@bu.edu

Photonics Center, Boston University, Boston, MA 02215, USA

J.-X. Cheng
Department of Biomedical Engineering, Boston University, Boston, MA 02215, USA

Q. Zhou and Z. Chen (eds.), *Multimodality Imaging*,
https://doi.org/10.1007/978-981-10-6307-7_4

along with the prevalence of coronary artery diseases, has accelerated the development of various intravascular imaging technologies in the last two decades, including intravascular ultrasound (IVUS), virtual histology IVUS (VH-IVUS), intravascular optical coherent tomography (OCT), and intravascular near-infrared spectroscopy (NIRS). Each of these modalities has their own specific strengths and weaknesses. However, none have shown the ability to detect vulnerable plaques accurately and reliably (Puri et al. 2013; Mintz 2014; Sanidas and Dangas 2013).

IVUS is the current standard for intravascular imaging of coronary atherosclerosis. Pulsed acoustic waves are used to interrogate the vessel wall followed by the detection of echo signals, providing the overall morphology of vessel wall with ultrasonic resolution. This modality has been applied in the clinic to quantify plaque burden, longitudinally monitor disease progression, and guide the stent deployment. Nonetheless, it lacks chemical selectivity to identify plaque composition. As tissue components are found to have different acoustic properties, virtual histology IVUS was then developed to differentiate plaque compositions (e.g., necrotic core, calcium, fibrous, and fibrofatty) based on the radiofrequency analysis of echo signals (Kubo et al. 2010). However, there have been questions raised about its scientific foundation and thorough validation (Nissen 2016; Thim et al. 2010). Intravascular OCT, based on the detection of backscattered light, can accurately detect fibrous cap thickness with micron-scale resolution (Tearney et al. 2012; Jang et al. 2005). However, the optical scattering does not provide chemical contrast. Furthermore, scattering caused by soft tissue limits the penetration depth at a superficial layer (1–2 mm), even with the removal of luminal blood (Tearney et al. 2012). Such imaging depth is insufficient to cover deeper layers, where lipid-rich necrotic cores are typically located. Intravascular NIRS has been shown to reliably detect lipid-rich plaques via the analysis of optical reflection absorption spectrum of the arterial wall (Brugaletta et al. 2011; Caplan et al. 2006). It has been introduced into the clinic as a hybrid-modality product, NIRS/IVUS. However, NIRS lacks the imperative depth resolution to quantify lipid core size and precise location. Instead, it yields a chemogram of lipids with a rough lateral resolution (~1 mm). These limitations highlight an unmet clinical need for the development of a chemically selective imaging modality with sufficient spatial and depth resolution to advance the detection, understanding, and treatment of lipid-laden vulnerable plaques.

As an emerging modality to overcome the abovementioned limitations, catheter-based intravascular photoacoustic (IVPA) imaging provides optical absorption-induced contrast at ultrasonic spatial resolution. Its endogenous lipid-specific mapping is based on the photoacoustic (PA) effect, where overtone absorption-induced thermalized energy is effectively converted into acoustic waves via thermoelastic expansion. Its imaging depth has been shown to be around 5 mm, with the potential for even deeper imaging by fully exploiting the benefits of diffused photons and weak acoustic scattering. Such penetration depth is far beyond the reach of pure optical imaging methods. Furthermore, IVUS is inherently compatible with IVPA to obtain arterial morphology, as they share the same ultrasound transducer. In addition, exogenous contrast agents could be developed and integrated to simultaneously target inflammatory markers of vulnerable plaques. Thus, by providing co-registered,

Table 4.1 Comparison of IVPA/US imaging with major clinically used intravascular imaging modalities in vulnerable plaque identification

Imaging technology	Intravascular ultrasound (IVUS)	Virtual histology-IVUS	Intravascular optical coherence tomography (OCT)	Near-infrared spectroscopy/IVUS (NIRS/IVUS)	Intravascular photoacoustic/ultra sound (IVPA/US)
Contrast mechanism	Acoustic scattering	Acoustic spectrum	Optical scattering	Optical absorption	Photoacousitic effect
Lateral resolution (μm)	200–500	200–500	~10	1000	200–500
Axial resolution (μm)	100–200	150–250	20–40	N/A	100–200
Penetration (mm)	> 5	> 5	1–2	Unknown	~5
Positive remodeling	+++	+++	–	+++	+++
Fibrous cap	–	–	+++	–	–
Lipid core	+	++	+	+++	+++
Inflammation	–	–	+	+	–
Representative image	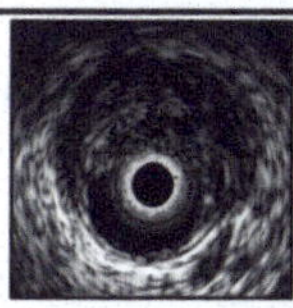	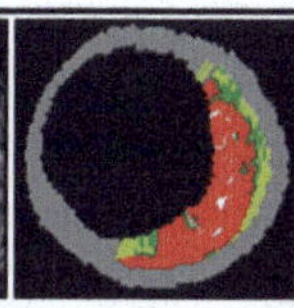	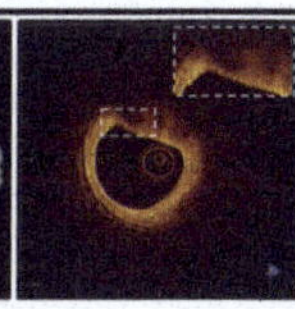	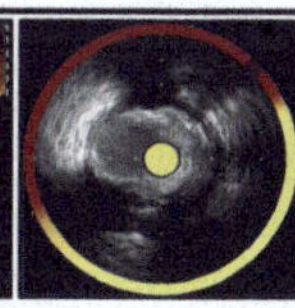	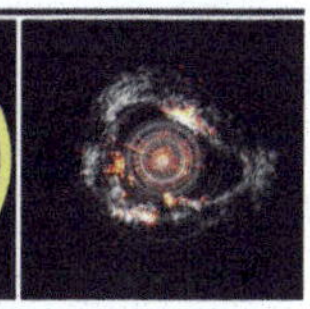

+++, excellent. ++, good. +, poor. –, not currently possible. N/A, not applicable. Data derived from Kilic et al. (2015), Sanidas and Dangas (2013), Suh et al. (2011)

simultaneous and complementary information of the artery wall, IVPA/US imaging can pave the foundation for advanced assessment of lipid-laden vulnerable plaques. The comparison with major clinically used intravascular imaging modalities in vulnerable plaque identification is shown in Table 4.1.

In this chapter, imaging principles and contrast mechanism in IVPA imaging are firstly reviewed. Technical advances in the development of a complete catheter-based IVPA/US imaging system are then summarized with a detailed demonstration of the essential and most up-to-date system components. These include an excitation laser source, hybrid fiber-optic rotary joint, IVPA/US catheter probe, and imaging reconstruction. The current status of preclinical validation and exogenous contrast agent development are further reviewed. Lastly, potential challenges and future clinical applications of IVPA/US imaging are discussed.

Principles of Photoacoustic Imaging

Light-to-Sound Conversion

PA imaging lies in the fundamental concept of light-to-sound conversion, a phenomenon termed as "photoacoustic effect" discovered by Alexander Graham Bell in 1880 (Bell 1880). Imaging technologies utilizing this effect have demonstrated broad biomedical applications (e.g., atherosclerosis diagnosis, breast cancer imaging, and melanoma imaging) (Wang and Hu 2012; Wang and Yao 2016; Beard 2011; Ntziachristos 2010). In PA imaging, a short-pulse laser is typically used to excite targeted chemicals. As photons propagate inside the tissue, part of the photons are significantly absorbed by molecules when the photon wavelength matches the transition frequency between its ground state and excited states. If the absorbed energy is partially or completely thermalized through non-radiative relaxation including vibrational relaxation, it will subsequently induce a local temperature rise and the thermalized energy will be effectively converted into transient waves through thermoelastic expansion. The transient waves propagate through the tissue and can be detected by an ultrasonic transducer or transducer array to reconstruct an image that maps the distribution of targeted molecules in the tissue.

This light-to-sound conversion process is theoretically described and quantified by equation, $p_0 = \xi \Gamma \mu_a F$, where p_0 is the initial pressure rise, ξ is a constant, Γ is the Gruneisen parameter of the absorber quantifying the thermoelastic expansion efficiency, μ_a is the absorption coefficient of the absorber, and F is the local photon fluence. Here, the Gruneisen parameter can be further expressed as $F = \beta v_s^2 / C_p$, where β is the isobaric volume expansion coefficient, v_s is the acoustic speed, and C_p is the specific heat capacity. This equation indicates a 100% relative sensitivity to variations in optical absorption in PA imaging, which means that a small percentage change in optical absorption is reflected by the same percentage change in PA signal amplitude. Therefore, selecting the optical excitation wavelength that matches the absorber's absorption peak can effectively maximize PA signal. This equation is also the foundation for quantitative photoacoustic imaging (Cox et al. 2012), including calculating the contrast from absorber to surrounding tissue components, quantifying the relative concentration of absorber, and estimating parameters of physiological interest derived from absorber concentration.

Figure 4.1a depicts the photon energy transfer on a Jablonski energy diagram, illustrating the difference between PA imaging and one-photon fluorescence imaging and the difference between electronic and vibrational energy transfer in PA imaging. In one-photon fluorescence imaging, the majority of absorbed energy goes through radiative relaxation with photons emitted at a longer wavelength, whereas in PA imaging the majority of absorbed energy is thermalized through non-radiative relaxation. Thus, PA imaging is sensitive for chromophore mapping (e.g., hemoglobin, lipids, melanin, DNA–RNA, and cytochromes). These chromophores can be further separated into two groups: one group absorbs ultraviolet-visible light via electronic transitions (e.g., hemoglobin, melanin, cytochromes, and DNA–RNA), and the other group absorbs light ranging from near-infrared to mid-infrared via vibrational transitions (e.g., lipids and water). More specifically, the vibrational transition from $\nu = 0$ to $\nu = 1$ is called fundamental transition typically located in the mid-infrared range (Weyer and Workman 2007). The transitions from $\nu = 0$ to $\nu = n$ ($n > 1$) are termed "overtone" with its absorption peaks located in the near-infrared range (Weyer and Workman 2007). These vibrational absorption-based mechanisms enrich the endogenous contrasts in PA imaging (Wang et al. 2011; Yakovlev et al. 2010). The hybrid optical excitation and acoustic detection in PA imaging provide far greater penetration depth (up to 7 cm (Mitcham et al. 2015)) than pure optical imaging methods,

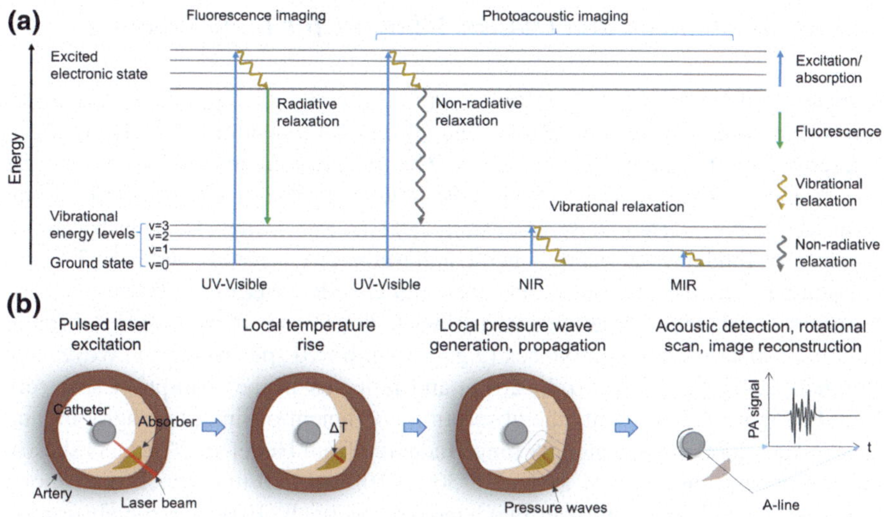

Fig. 4.1 Principles of PA imaging. **a** Jablonski diagram, illustrating the difference in photon energy transfer between one-photon fluorescence imaging and photoacoustic imaging, and the difference between electronic absorption and vibrational absorption in photoacoustic imaging. NIR, near-infrared. MIR, mid-infrared. **b** Principle of IVPA imaging

as acoustic scattering is orders of magnitude weaker than optical scattering and the diffused photons contribute equally to PA signal generation. Through specific implementations (Wang and Yao 2016), these advantages render PA imaging a competitive modality for clinical applications on the abovementioned biomarker-related diseases (e.g., tumor angiogenesis, atherosclerosis, and melanoma cancer).

IVPA imaging has been implemented into an intravascular catheter probe for the clinical detection of atherosclerotic plaques. Figure 4.1b depicts the principle of IVPA imaging. Laser pulses are delivered and side-fired onto the artery wall through an optical fiber. The generated acoustic waves are detected by a miniaturized single-element ultrasound transducer. By using the same transducer, a conventional IVUS signal can be obtained simultaneously. The ultrasound transducer along with the optical fiber is housed in a miniaturized catheter with an ideal diameter of ~1 mm for intravascular access. At each angular position, a depth-resolved PA signal, termed "A-line", is recorded by the catheter probe. This signal contains the information of absorber location, which can be calculated by the time delay. By rotating the catheter at a constant speed, a cross-sectional PA image of the artery wall can be reconstructed. By pulling back the catheter during rotation, a three-dimensional imaging of the artery wall is also obtained.

Contrast Mechanism and Optical Windows for Lipid Imaging

PA imaging of lipids is based on the strong vibrational absorption of C–H bond, abundant in lipids (Wang et al. 2011). The vibrational transitions of C–H bond can be described by the anharmonicity theory. The transition frequency for an overtone band (from $\nu = 0$ to $\nu = n$ $(n > 1)$) has the following relation with the fundamental transition (from $\nu = 0$ to $\nu = 1$), $\Omega = \Omega_0 n - \chi \Omega_0 (n + n^2)$, where Ω_0 is the transition frequency of fundamental band, χ is the anharmonicity constant, and $n = 2, 3, \ldots$ representing the first-, second-, and other higher-order overtones. These overtone transitions and their combinational transitions of C–H bond are well studied through vibrational spectroscopy approaches (e.g., near-infrared spectroscopy) (Weyer and Workman 2007; Sasic and Ozaki 2011) and reflected in the absorption spectrum of lipids. Figure 4.2a presents the absorption coefficient of lipid, hemoglobin, and water in 400–2000 nm wavelength range (adapted from Hui et al. 2016). As shown, lipid has absorption peaks at around 1730, 1210, and 920 nm, which correspond to the first, second, and third overtone bands, respectively. However, the absorption of hemoglobin dominates in the range of 400–1100 nm and overwhelms the third- and higher-order overtone bands of C–H bond. For longer wavelengths in the range of 1100–2000 nm, the absorption of hemoglobin reduces significantly. Thus, two optical windows are identified for PA imaging of lipid (highlighted in blue between 1100 and 1300 nm and 1650–1850 nm), where the absorption of lipid is maximized yet water absorption is locally minimized. In particular, in the first optical window, the hemoglobin absorption is close to one order of magnitude smaller than lipid absorption. The whole blood in the second optical window exhibits almost the same absorption spectrum as pure water, as water is the major content of whole blood (Friebel et al. 2009). Many PA spectroscopic studies have confirmed the two windows for lipid detections within the arterial wall (Wang et al. 2010, 2011, 2012c; Allen et al. 2012; Jansen et al. 2014a).

Notably, the absorption coefficient of lipid is only slightly larger than that of water in both optical windows. Yet, the contrast between the generated PA signals can be reliably observed and separated. This discrepancy between similar absorption coefficients but greatly different generated signals can be realized by the following theoretical calculation (Hui et al. 2016). In equation $p_0 = \xi \Gamma \mu_a F$, only Γ and μ_a have absorber dependence in tissue. Thus, the PA contrast of fat to water can be expressed as $p_{0\_fat}/p_{0\_water} = (\Gamma \mu_a)_{0\_fat}/(\Gamma \mu_a)_{0\_water}$. Based on the Gruneisen parameters and absorption coefficients of fat and water listed in Table 4.2, this contrast is calculated in the range of 9.6–12.4 and 10.9–14.0 at 1210 and 1730 nm, respectively. Such calculation is consistent with experimental observations. As an example, Fig. 4.2b shows the PA spectra of olive oil, water, and oxygenated blood in the range of 1100–1800 nm. The olive oil has around one order of magnitude larger PA signal than water at 1210 and 1730 nm. Collectively, these experimental results and theoretical calculations enable vibration-based PA imaging as a valuable platform for selectively mapping lipids in a complex tissue environment. Compared with 1210 nm, 1730 nm is more favorable for IVPA imaging. The signal amplitude at 1730 nm is significantly

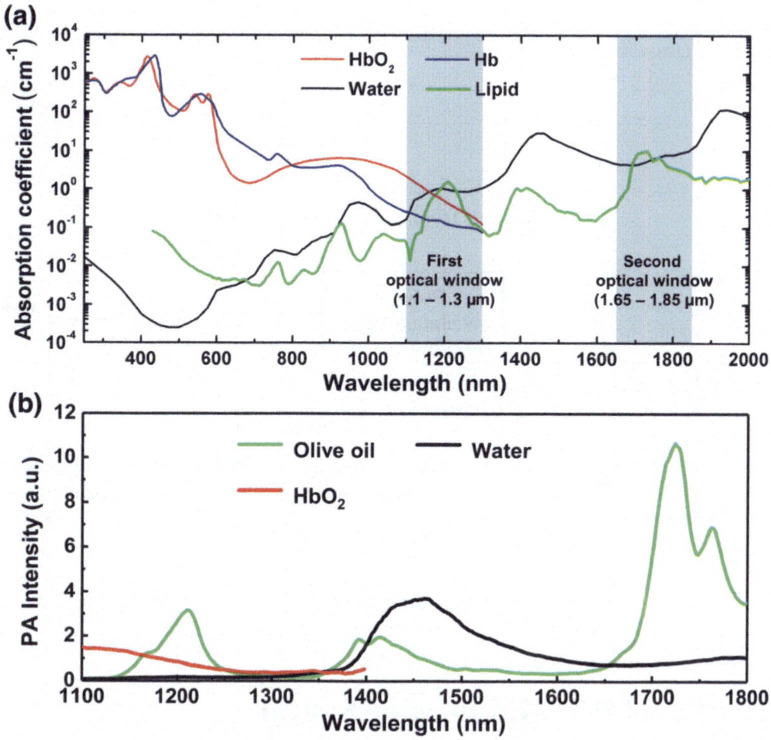

Fig. 4.2 Optical windows for IVPA imaging of lipids. **a** Absorption coefficient profiles of oxygenated hemoglobin (HbO_2), deoxygenated hemoglobin (Hb), lipid, and water in 200–2000 nm wavelength range. **b** PA spectra of oxygenated blood, olive oil, and water in 1100–1800 nm wavelength range. The data set highlights two optical windows (1100–1300 and 1650–1850 nm) for IVPA imaging of lipids in near-infrared region. Adapted from Hui et al. (2016)

larger attributed by stronger absorption at this wavelength. In addition, the optical scattering caused by blood is significantly less at 1730 nm (Friebel et al. 2009), providing the opportunity for IVPA imaging without the need for luminal blood flushing. Notably, due to the heavier mass of deuterium, the prominent overtone and combinational bands of D_2O have their corresponding peaks at wavelengths longer than 1800 nm. Thus, D_2O can be used as acoustic coupling medium for vibration-based PA imaging (Wang et al. 2012c). Particularly, it can be used to flush the catheter head enclosed in catheter sheath to reduce the optical path in water or blood.

Table 4.2 Absorption coefficients and Gruneisen parameters of fat and water at 1210 and 1730 nm

Tissue constituent	Tissue parameter	1210 nm	1730 nm
Fat	μ_a (cm^{-1})	1.65	10.5
	Γ	0.7–0.9	0.7–0.9
Water	μ_a (cm^{-1})	1.00	5.63
	Γ	0.12	0.12

Adapted from Hui et al. (2016)

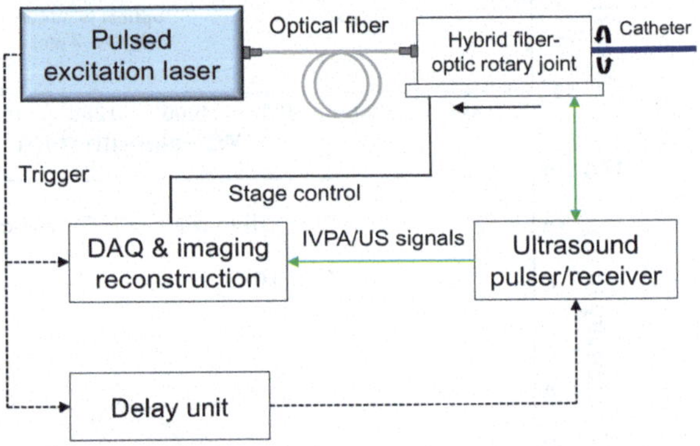

Fig. 4.3 Typical instrumental implementation of IVPA/US imaging. Major components include a pulsed excitation laser, a hybrid fiber-optic rotary joint, an IVPA/US catheter, an ultrasound pulser/receiver, and data acquisition (DAQ) and image reconstruction units

Catheter-Based IVPA/US Imaging System

IVPA/US imaging is a modality that implements PA and US imaging into a miniaturized intravascular catheter probe along with other peripheral equipment. Figure 4.3 shows the major components, detailed connections, and controls within a typical IVPA/US imaging system. A short-pulse laser is used as the optical excitation source. Its output is delivered through a fiber-optic approach and side-fired onto the artery wall for IVPA signal generation. In order to collect IVUS signal at the same angular position, ultrasonic pulse is initiated by an ultrasound pulser, delivered to a single-element transducer through an electric slip ring, and then fired on the artery wall. The timing of optical and ultrasonic pulses is controlled by their trigger signals, with the delay between the optical and ultrasonic pulses precisely set by a delay unit so that the generated IVPA and IVUS signals can be well separated in the time domain. The generated IVPA/US signals are recorded by the same ultrasound transducer, transmitted by the slip ring, processed and amplified by the ultrasound receiver, and digitized by a data acquisition card. By rotating and pulling back the catheter via a controlled stage, their corresponding A-lines are recorded, processed, and reconstructed as co-registered IVPA/US images. Such catheter-based IVPA/US system development includes many detailed technical advancements reported previously. The most up to date and representative system components are shown here. The implementation and key parameters of these components are discussed in details as below.

Excitation Laser Source

There are several key requirements for an excitation laser source, which include wavelength, pulse duration, pulse energy, pulse repetition rate, and pulse-to-pulse stability. In order to maximize the lipid contrast, the laser wavelength should match the overtone absorption bands of C–H bond, particularly at 1.2 and 1.7 μm as mentioned in Section "Contrast Mechanism and Optical Windows for Lipid Imaging". When imaging or differentiating more than one tissue component, multispectral or spectroscopic approaches can be applied. However, the application of these approaches is limited by slow imaging speed. In order to meet the stress and thermal confinements for effective PA signal generation (Jacques 1993), the laser pulse duration is chosen at nanosecond level. As IVPA imaging is generally based on wide-field laser excitation, the required pulse energy is normally at the order of hundred microjoules, administrated under the maximum permissible exposure limits for skin according to ANSI laser safety standards (American National Standard for Safe Use of Lasers, ANSI Z136.1 2014) (100 mJ/cm^2 at 1.2 μm and 1 J/cm^2 at 1.7 μm). With the requirements on pulse duration and pulse energy, solid-state lasers are generally used as the excitation laser source. As each cross-sectional IVPA image is reconstructed from a series of A-line signals, the pulse energy for each A-line signal has to be relatively constant to ensure a uniform contrast in the cross-sectional image.

Each laser pulse yields a depth-resolved A-line signal and a high-quality reconstructed cross-sectional image requires a sufficient number of A-lines. Thus, the laser pulse repetition rate and the A-line requirement determine the final imaging speed. The least number of A-lines for an IVPA image is determined by Nyquist sampling theorem, meaning the A-line number is set by the IVPA lateral resolution (Hui et al. 2017). Therefore, laser pulse repetition rate is the primary determinant of imaging speed. A number of groups have demonstrated the feasibility of IVPA/US imaging of lipid-laden plaques with 10 or 20 Hz lasers, yielding an imaging speed of tens of seconds per frame (Jansen et al. 2011; Wang et al. 2012b). This speed is prohibitive for preclinical or clinical applications. With the advancement in high-repetition rate laser source development, the imaging speed was further improved to several frames per second (fps) at both 1.2 μm (Wang et al. 2014; Li et al. 2015) and 1.7 μm (Hui et al. 2015; Piao et al. 2015). However, the clinical translation requires real-time video-rate imaging speed (video rate, $\geq$15 fps, faster than the frequency response of human vision) to eliminate motion artifacts caused by cardiac pulsation and achieve accurate mapping of lipid-laden plaques. Such speed indicates a laser pulse repetition rate of 1.5 kHz at least. More recently, IVPA imaging speed has been shown achievable at video rates by a 2 kHz laser (Hui et al. 2017). Figure 4.4 shows the detailed design of the laser with output wavelength of 1725 nm and maximum pulse energy of 1 mJ (Hui et al. 2017). With this laser, the motion artifacts induced by cardiac pulsation were successfully suppressed, making IVPA/US imaging translational for in vivo preclinical and clinical applications. Note, the commercial NIRS/IVUS imaging system used in the clinic has a speed of 15 fps (Danek et al. 2016).

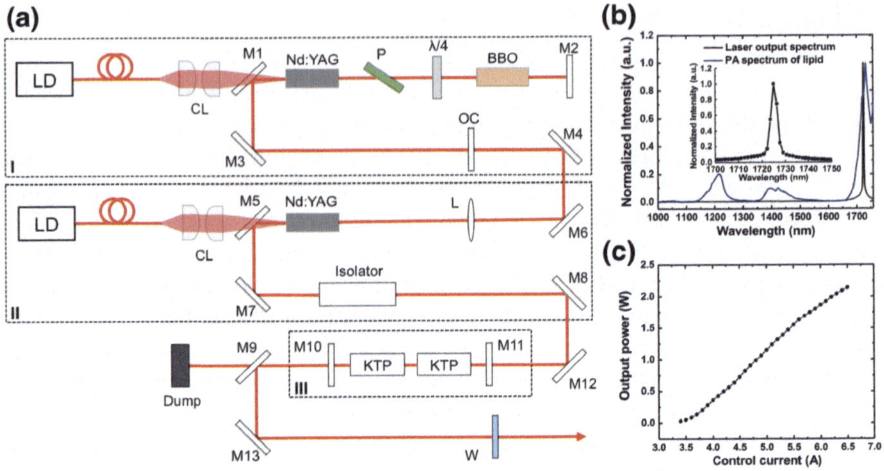

Fig. 4.4 Representative pulsed excitation laser. **a** Schematic of 2 kHz master oscillator power amplifier (MOPA)-pumped OPO, developed for IVPA imaging at video-rate level. Dashed boxes labeled by I, II, and III highlight the master oscillator, optical power amplifier, and OPO, respectively. OPO, optical parametric oscillator. LD, laser diode; CL, coupling lens; M1 and M5, fold mirror; M2, M10, and M11, flat mirror; M3, M4, M6, M7, M8, M12, and M13, reflective mirror; M9, dichroic mirror; P, polarizer; BBO, beta barium borate Pockels cell; OC, output coupler; L, lens; KTP, potassium titanyl phosphate; W, output window. **b** Output wavelength of the laser (black curve), 1725 nm, matching the first overtone absorption of C–H bond in a photoacoustic spectrum of lipid (blue curve). **c** Tunable range of laser power output. Adapted from Hui et al. (2017)

Hybrid Fiber-Optic Rotary Joint

The hybrid fiber-optic rotary joint is an important component, responsible for light delivery and radiofrequency signal transmission under video-rate rotation ($\geq$15 rev/s). As a higher pulse energy is required for IVPA imaging, a multimode fiber is typically used for laser light delivery. Figure 4.5a, b shows a representative design of a hybrid fiber-optic rotary joint that supports high optical coupling efficiency at speed up to 30 rev/s (Hui et al. 2017). In this design, the laser beam is sequentially delivered by a multimode fiber, collimated into another multimode fiber segment through two adjacent collimators, and then coupled into the catheter by a mating sleeve. From left to right, the static components include the first SMA connector, the first SMA collimator, the stator, and the outer ring of slip ring, while the rest rotate with speed driven by a rotator, controlled by an external motor. Under the rotation scheme, the use of adjacent collimators and a mating sleeve enables an overall coupling efficiency of 60% from the initial optical input to the final output at the catheter tip. The design also minimizes the coupling efficiency variation induced by mechanical rotation to 5.4%. In this case, the artery wall is excited with relatively equal pulse energy at each angular position, ensuring uniform contrast in IVPA image. IVPA/US signals are then transmitted back to a host computer by

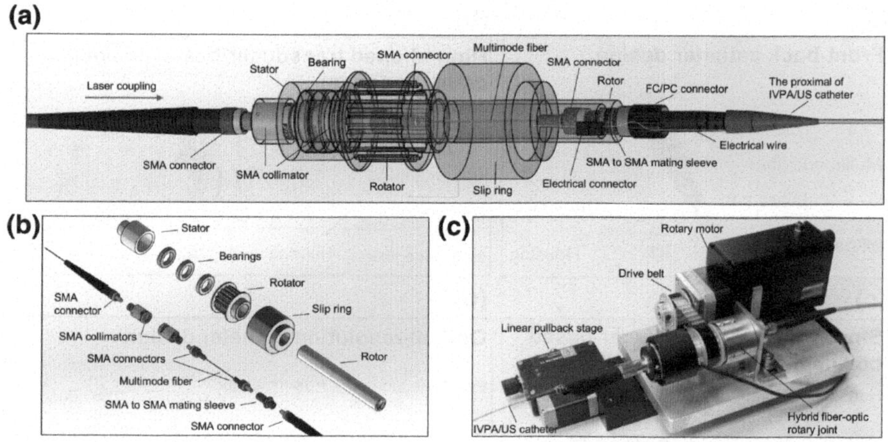

Fig. 4.5 Representative hybrid fiber-optic rotary joint. **a** Schematic of the fully assembled rotary joint design. **b** The exploded assembly view of the hybrid fiber-optic rotary joint. **c** Picture of the hybrid fiber-optic rotary joint module with driving motor and linear pullback stage integrated. Adapted from Hui et al. (2017)

the electrical slip ring for processing and reconstruction. This joint, together with its driving motor and pullback stage, can be integrated together as a compact and portable module for easy operation (Fig. 4.5c).

IVPA/US Catheter

IVPA/US imaging has practical requirements of catheter size, flexibility, sensitivity, imaging depth, protective sheath, and video-rate imaging speed before it can be compatible with clinical use. These requirements collectively define the design and fabrication requirements of IVPA/US catheters. Currently, a number of catheter designs have been reported. Figure 4.6 shows four representative catheter designs. Each of these catheters has its unique advantages and the potential to be further refined for clinical applications.

Figure 4.6a represents a typical front-back catheter design (Jansen et al. 2011). In this design, the optical fiber and the single-element ultrasound transducer are aligned along the catheter axis, offset by a small distance. The fiber end is polished to a precise angle and sealed in an optically transparent glass cap with air, so the output beam can be side-fired onto the artery wall by total internal reflection at the air–silica interface. The overlap between the optical and acoustic fields is maximized by tuning the fiber-polishing angle and positioning the ultrasound transducer at a small angle as well. Two early prototypes in this design were fabricated by combining an optical fiber with an existing commercial IVUS imaging catheter (Karpiouk et al. 2010). Also, in this design, a quasi-focusing optical illumination scheme was used to improve IVPA

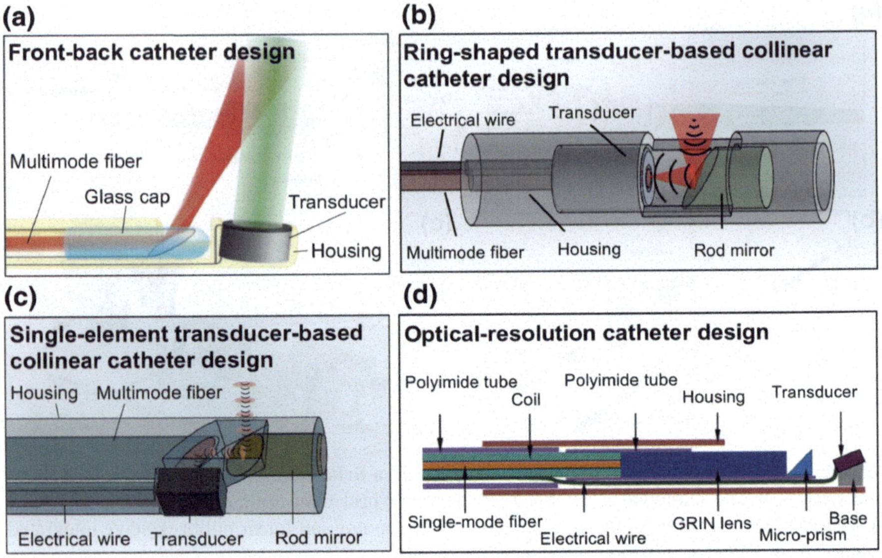

Fig. 4.6 Representative IVPA/US catheter designs. **a** Schematic of front-back catheter design, adapted from Jansen et al. (2011). **b** Schematic of ring-shaped transducer-based collinear catheter design, adapted from Hui et al. (2015). **c** Schematic of single-element transducer-based collinear catheter design, adapted from Cao et al. (2017). **d** Schematic of optical-resolution catheter design, adapted from Bai et al. (2014). Note that the catheter sheath and acoustic coupling medium (e.g. water, D_2O, and saline solution) are not shown

detection sensitivity (Li et al. 2015). Furthermore, rather than front-back alignment of the optical fiber and the single-element transducer along the catheter axis, a side-by-side arrangement of the components was reported (Li et al. 2012). Simplicity and ease of miniaturization are the main advantages of this design. Currently, a 0.9-mm catheter diameter has been achieved in this design, which is comparable to the size of commercially available IVUS catheters.

Figure 4.6b shows a ring-shaped transducer-based collinear catheter design. In this design, a ring-shaped transducer and an optical fiber lie flat distally, where they are collinearly aligned. A 45° rod mirror is used to direct both optical and acoustic excitation pulses perpendicularly to the artery wall and reflect the generated signals back to the ultrasound transducer. Therefore, the optical and acoustic paths in this design are collinearly overlapped, indicating the maximal co-registration along the entirety of the A-line. Currently, the reported catheter size in this design is ~2 mm in diameter, limited by the physical obstacles in miniaturizing a ring-shaped transducer (Hui et al. 2015; Wei et al. 2011).

A third single-element transducer-based collinear catheter also demonstrates collinearly overlapped optical and acoustic paths, in which miniaturization of the total catheter diameter to ~1 mm was recently reported (Fig. 4.6c) (Cao et al. 2016; Hui et al. 2017). In this design, the distal end of the optical fiber is polished to 47°

for acoustic wave reflection, while the optical waves still propagate forward at the water/D_2O-silica interface (detailed optical and acoustic paths and calculation for the polishing angle can be found in Hui et al. (2017)). The single-element transducer is placed in parallel to the optical fiber, with its sensing surface facing the polished fiber plane. The 45° rod mirror is used to direct both optical and acoustic excitations perpendicularly to the artery wall and reflect the generated signals back to the transducer. The collinear nature of the overlapping optical and acoustic paths maximizes the co-registration along the depth direction, up to ~6 mm.

The previously described catheter designs all work in an ultrasonic resolution mode in IVPA imaging as the optical illumination is in a wide-field configuration. In order to achieve optical-resolution IVPA imaging, a tightly focused laser beam has to be achieved. The optical-resolution catheter design shown in Fig. 4.6d is achieved by focusing the laser beam with a gradient-index (GRIN) lens (Bai et al. 2014). The focused beam is then reflected by a micro-prism to the artery wall. The single-element ultrasound transducer is aligned with a ~1 mm offset between the transducer and the micro-prism. In order to maximize the detection sensitivity, the transducer is purposely tilted. In this design, a single-mode fiber is used for better beam shape control, whereas multimode fibers are typically used to deliver more pulse energy for wide-field illumination in the aforementioned catheter designs. Currently, a catheter in this design has been reported with a 1.1 mm diameter and 19.6 μm lateral resolution, significantly greater than that achieved with other designs. However, the imaging depth is sacrificed and the detection sensitivity varies significantly along the depth direction in this design due to the tight optical focusing. Also, the number of A-lines required for IVPA image reconstruction is nearly one order of magnitude higher than that in ultrasonic resolution-mode IVPA imaging.

Most ultrasound transducers used in aforementioned catheter designs have a center frequency in the range of 30–45 MHz. By selecting a low-frequency transducer, IVPA imaging can penetrate deeper, however, at the expense of spatial resolution. In addition, common among all designs is alignment of the optical and acoustic components in a compact, miniaturized housing. The housing is further attached to a flexible torque coil so that the optical fiber and the electrical wire are fully enclosed and the rotation force is transferred to the catheter head. Furthermore, because of the fast mechanical rotation of the catheters, a protective sheath must be between the catheter and the artery. Aside from the traditional method to detect IVPA signals by ultrasound transducers, two IVPA catheter designs based on all-optical detection have been reported. One design uses a Fabry-Perot polymer sensor film (Zhang and Beard 2011), while the other design uses an optically transparent, polymeric microring resonator as the ultrasonic sensor (Dong et al. 2014). These novel detection methods have the potential to further reduce the catheter size. However, they lack the intrinsic capability for complementary IVUS imaging. Further investigations are necessary to explore their potential for clinic applications.

Aside from various permutations of catheter designs, IVPA imaging can be further improved by a number of strategies. First, the IVPA signal, as described in $p_0 = \xi \Gamma \mu_a F$, can be maximized by matching the laser wavelength with the absorption peak or increasing the optical fluence at the absorber. The fluence can be increased by maximization of the excitation pulse energy, while remaining within the safety limits. Second, the electronics noise can be reduced using a number of strategies such as grounding, shielding, filtering, and impedance matching. Lastly, the bandwidth of ultrasound transducer can be further optimized balancing the frequency content and image quality in both IVPA and IVUS imaging.

Image Reconstruction

In IVPA imaging, a single laser pulse generates a depth-resolved A-line signal (Fig. 4.1b). Each A-line signal in time is then numerically reversed back to a one-dimensional image along the radial direction through the known speed of sound in tissue. During the reversing process, digital filters are applied to isolate frequency components of interest and a Hilbert transform is performed to acquire the signal amplitude. As the catheter revolves, a series of adjacent A-lines are recorded, reversed, and then transformed from Cartesian coordinates into polar coordinates as a cross-sectional IVPA image. By pulling back the catheter during rotation, a series of cross-sectional images acquired can be further reconstructed into a three-dimensional image through scaling the pullback distance between two adjacent images (calculated from the pullback speed). Although there are similarities between IVPA and IVUS image reconstruction, several significant differences need to be highlighted. First, the acoustic wave propagation in IVPA imaging is one way, as the light speed is far larger than the acoustic speed propagating in soft tissue. In IVUS imaging, acoustic wave propagation is both ways—to and from the tissue. Additionally, compared with IVUS, IVPA signals lie in lower frequency range, thus the digital filter used in signal processing should be changed accordingly. Considering the limitation caused by laser pulse repetition rate, the number of A-lines used for image reconstruction could also be different (Hui et al. 2017). Lastly, as the optical attenuation in the tissue is significantly larger than acoustic attenuation, the contrast in IVPA and IVUS image should be adjusted through their corresponding time gain compensations along radial direction.

Preclinical Validation

Preclinical validation of IVPA/US imaging in human coronary arteries ex vivo and clinically relevant animal models in vivo is essential, before its translation to the clinic. Through preclinical validation, IVPA/US imaging system parameters—chemical specificity, detection sensitivity, spatial resolution, imaging depth, blood interference, and catheter sheath interference—can be evaluated and optimized. Furthermore, the in vivo imaging procedures can be further developed and validated in these animal models.

Human Coronary Atherosclerosis

In 2011, the first ex vivo IVPA/US imaging of a human coronary artery was demonstrated at 1.2 μm (Jansen et al. 2011). In this work, lipid deposition in the plaque area was mapped using 1210 nm wavelength. Later studies suggest that PA imaging of lipids at 1.7 μm has additional benefits compared to imaging at 1.2 μm (Wang et al. 2012c; Friebel et al. 2009). The optical absorption coefficient of lipids at 1.7 μm is significantly larger than that at 1.2 μm and optical scattering caused by blood at 1.7 μm is smaller. Several works further reported ex vivo IVPA/US imaging results of human coronary arteries at 1.7 μm (Hui et al. 2017; Jansen et al. 2014b). Figure 4.7 shows representative IVPA/US imaging results of an atherosclerotic human coronary artery collected at an imaging speed of 16 fps with optical excitation at 1725 nm (Hui et al. 2017). Two sites of lipid deposition at 2 and 8 o'clock are shown in the IVPA channel (Fig. 4.7a). There is a noticeable lipid-rich core at 2 o'clock, which correlates with luminal encroachment observed in the IVUS channel (Fig. 4.7b). The co-registered IVPA/US image (Fig. 4.7c) further suggests that this lipid-rich core is beneath a fibrous cap, as shown in IVUS channel. The histology slide in Fig. 4.7d–f shows that the lumen and artery structures in the histological section correlate well with the imaged artery morphology. The areas between 5 and 9 o'clock are rich in fibrous tissue, correlating with the strong echogenicity observed in the IVUS channel. Also apparent are two lipid-rich necrotic cores (Fig. 4.7e, f), as identified by the loss of matrix, cholesterol clefts, and macrophage infiltration into the lipid pool with an overlying fibrous cap. These histological hallmarks indicate this plaque as an advanced fibroatheroma, which is consistent with the imaging results.

Animal Models

Validation of IVPA/US imaging in animal models of atherosclerosis is an essential intermediate transition to its clinical applications. It is known that atherosclerosis is a chronic inflammatory disorder. Animal models (e.g. mice, rabbit, pig, and non-human primate) are developed to accelerate the atherosclerotic plaque formation (Getz and Reardon 2012). These accelerating strategies include pro-atherogenic diets, transgenic manipulating of metabolic pathways, or mechanical disruptions of the artery wall. For IVPA imaging, the arteries of mice models are too small for catheter access, while the non-human primates are expensive and highly regulated. Thus, rabbit and pig models are the most frequently used animal models in IVPA/US imaging.

The first in vivo IVPA/US imaging was demonstrated at 1720 nm in a Watanabe heritable hyperlipidemic (WHHL) rabbit (Wang et al. 2012a). Figure 4.8a-c shows the IVPA, IVUS, and merged images collected from the rabbit abdominal aorta. The IVPA image shows the spatially resolved distribution of lipid depositions. Through interpretation with co-registered IVUS image, lipid depositions are mainly found in the intimal layer of the aorta, with some periadventitial fat detected at 5 o'clock.

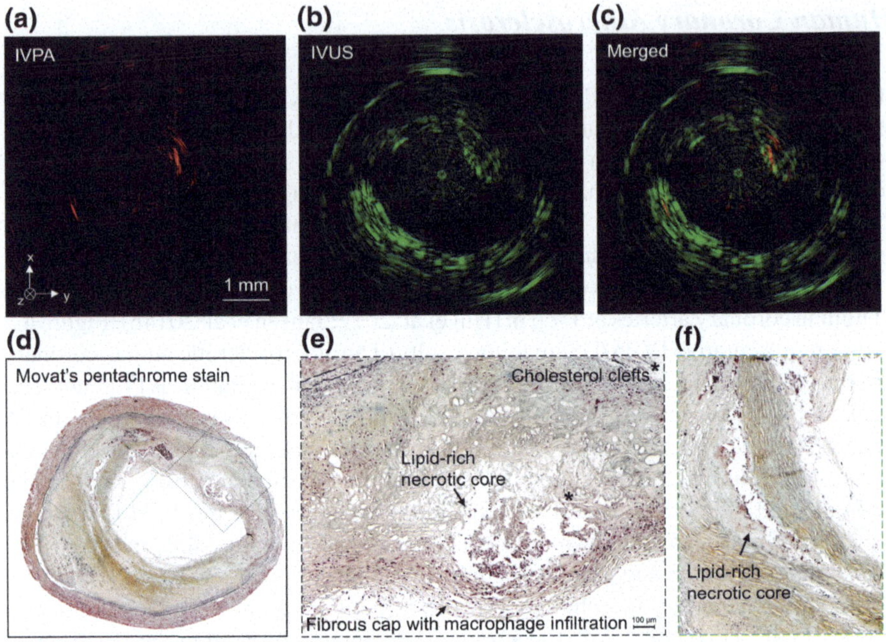

Fig. 4.7 IVPA/US imaging of a high-risk lipid-laden plaque in human coronary artery ex vivo. Cross-sectional **a** IVPA. **b** IVUS. **c** merged images of the human coronary artery at a position of interest, acquired at an imaging speed of 16 fps. **d** Gold-standard histopathology stained with Movat's pentachrome at the region of interest. **g**, **h** Magnified images of lipid deposition sites, corresponding to the dashed boxes in (**d**). The 1 mm scale bar applies to (**a**–**d**). *Indicates the accumulation of cholesterol clefts. Adapted from Hui et al. (2017)

Although this work was performed at a slow imaging speed (25.6 s per frame), it demonstrated for the first time that in vivo IVPA/US imaging without luminal blood flushing was feasible. Similar validation in New Zealand white (NZW) rabbit model was also reported (Zhang et al. 2014). Compared with rabbits, pig models are preferred, as they more closely mimic the human anatomy, physiology, and natural disease progression. PA imaging of lipid-laden plaques in a miniaturized Ossabaw swine model was performed ex vivo under both microscopy and intravascular imaging configurations (Wang et al. 2011, 2014). As an example, Fig. 4.8d-f shows the ex vivo IVPA, merged IVPA/US, and histological images of an iliac artery harvested from an Ossabaw pig with atherosclerosis (Wang et al. 2014). The lipid depositions show clear contrast in the IVPA image, while not visible with IVUS alone. These imaging results are further confirmed on histology, where regions of interest were identified as lipid cores and fatty streaks. Collectively, the aforementioned rabbit and pig models can be used for preclinical validation of IVPA/US imaging.

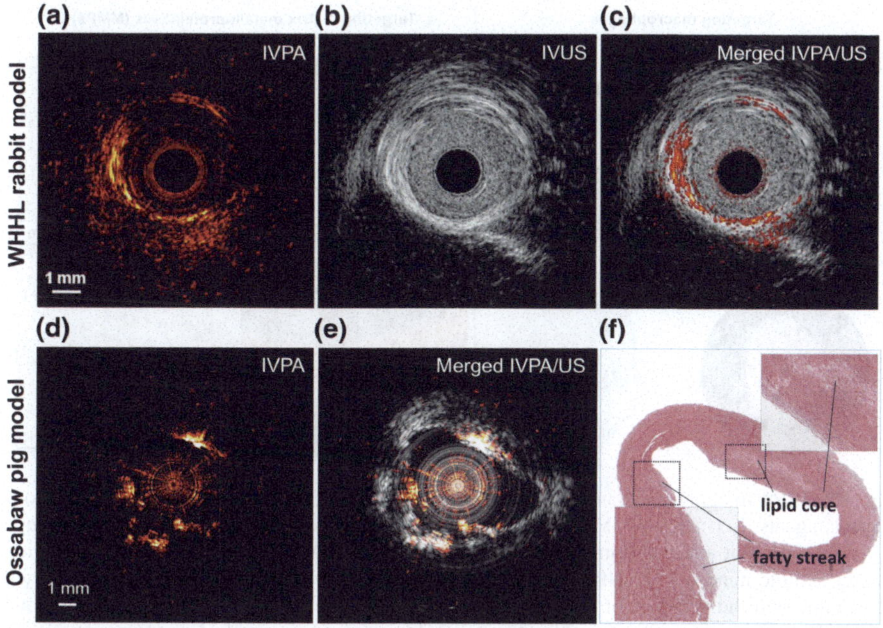

Fig. 4.8 IVPA/US imaging of lipid deposition in rabbit and pig models of atherosclerosis. Top row: cross-sectional **a** IVPA, **b** IVUS, **c** merged images collected in vivo in the abdominal aorta of a Watanabe heritable hyperlipidemic (WHHL) rabbit model. The 1 mm scale bar applies to (**a**–**c**). Adapted from Wang et al. (2012a). Bottom row: cross-sectional **d** IVPA, **e** merged IVPA/US, **f** H&E stain histology images collected ex vivo in the iliac artery of an Ossabaw pig model. The 1 mm scale bar applies to (**d**–**f**). Adapted from Wang et al. (2014)

Exogenous Contrast Agents

Besides endogenous chromophores, exogenous agents could be specifically engineered and targeted in PA imaging to increase sensitivity for specific cellular and molecular biomarkers, as well as the penetration depth. Currently, a variety of contrast agents have been reported in PA imaging, including dyes, nanoparticles, and liposome encapsulations (Weber et al. 2016; Luke et al. 2012). In IVPA imaging, contrast agents are specifically designed to target biomarkers involved in atherogenesis that are unable to be detected without a molecular label. Given that inflammation is involved in nearly every stage of plaque development and a hallmark of plaque vulnerability, several contrast agents targeting the biomarkers of inflammation, including macrophages and matrix metalloproteinases (MMPs), have been demonstrated (Yeager et al. 2012; Qin et al. 2016; Razansky et al. 2012; Ha et al. 2011).

Gold nanoparticles are favored as contrast agent in PA imaging (Li and Chen 2015; Luke et al. 2012). These nanoparticles generate strong PA signals based on surface resonance coupling and their optical spectra tuned by changing the particle size or shape. In addition, the particle surface can be functionalized with different

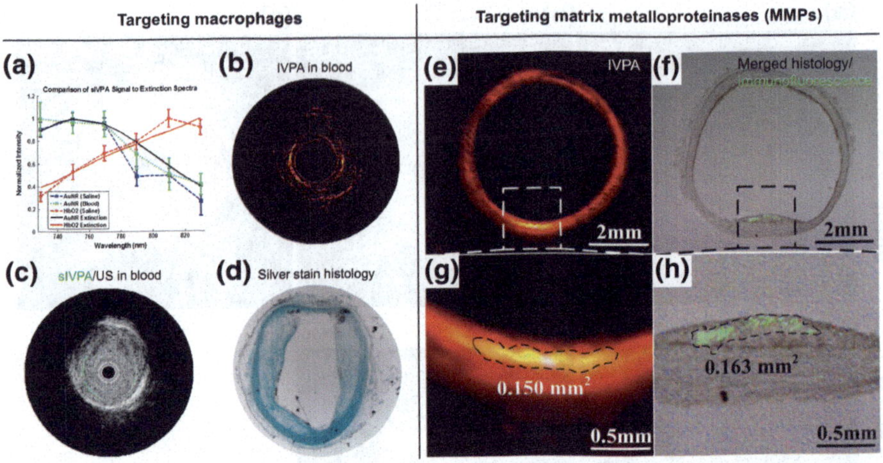

Fig. 4.9 IVPA imaging of inflammation biomarkers in atherosclerotic arteries through exogenous contrast agents. Left: ex vivo IVPA/US imaging of gold nanorods-labeled atherosclerotic plaque in a WHHL rabbit aorta. **a** Comparison of spectroscopic IVPA signals to normalized extinction spectra of gold nanorods (AuNR) and oxygenated hemoglobin (HbO_2). **b** IVPA image collected at 750 nm with the presence of luminal blood. **c** sIVPA/US image with the presence of luminal blood (the regions of spectroscopically detected gold nanorods are overlaid in green). The IVPA and sIVPA/US images displayed represent a total diameter of 22.5 mm. **d** Corresponding liver stain histology. Adapted from Yeager et al. (2012). Right: in situ IVPA imaging of gold nanorods-labeled MMP2 in atherosclerotic plaque in a NZW rabbit aorta. **e** Corresponding IVPA image collected at 695 nm. The yellow area indicates the area of MMP2 expression. **f** The overlay results of microscopic image and immunofluorescence results of MMP2. Images **g** and **h** are the enlarged of the selected areas in (**e**) and (**f**), respectively. Adapted from Qin et al. (2016)

antibodies to specifically target macrophages and MMPs. The feasibility of targeting macrophages by IVPA imaging was first shown in a rabbit aorta harvested and then injected with macrophage-loaded gold nanoparticles (Wang et al. 2009). In a follow-up study (Yeager et al. 2012), the authors systematically injected polyethylene glycol stabilized gold nanorods into a balloon-injured WHHL rabbit and performed the ex vivo IVPA/US imaging on a freshly excised aorta (Fig. 4.9a-d). They reported the detected signal from localized nanorods was significantly larger than the signal from blood at 750 nm (based on PA spectra in Fig. 4.9a). The extravasation of these nanorods at sites of dysfunctional endothelium within atherosclerotic regions at 3 and 9 o'clock was reliably imaged even with the presence of luminal blood (Fig. 4.9b, c). The imaging results were further confirmed by silver stain histology in Fig. 4.9d, as silver stain, used to identify the presence of gold nanorods near the luminal boundary of plaque region, correlates with the distribution of macrophages.

MMPs are a family of enzymes that degrade extracellular matrix components in atherosclerotic plaques and weaken the fibrous cap, making them prone to rupture. Currently, few contrast agents have been developed to quantify and map the expression of MMPs in atherosclerotic plaques via PA imaging (Razansky et al. 2012; Qin et al. 2016). MMPSence, a MMP-sensitive activated fluorescent probe, was reported as one of such agents (Razansky et al. 2012). Under a PA tomography, the MMP activity in atherosclerotic human carotid arteries was mapped and monitored through the MMPSence response. Recently, given the advantage of significantly higher absorption of gold nanoparticles, gold nanorods conjugated with MMP_2 antibody were demonstrated as another PA imaging probe for MMP_2 in atherosclerotic plaques (Qin et al. 2016). This probe was validated by in situ IVPA imaging of atherosclerotic aorta excised from a NZW rabbit with ear vein injection. Figure 4.9e, g shows the corresponding IVPA images at 695 nm with MMP_2 expression highlighted and quantified by area. The results were further confirmed by histology and immunofluorescence (Fig. 4.9f, h). Moving forward, more elaborate work is needed to address systemic delivery, toxicity, stability, and biocompatibility for clinical introduction of these exogenous contrast agents.

Future Development and Potential Clinical Applications

The past decade has witnessed the rapid development of IVPA/US imaging from a fundamental concept to a miniaturized device toward clinical translation. These achievements span endogenous contrast mechanisms, laser excitation sources, various catheter designs, preclinical validation, and exogenous contrast agents. Although encouraging, there are still specific challenges to overcome before the clinical translation is feasible. Below, we summarize and discuss these challenges in IVPA/US imaging development along with the potential clinical applications it would enable.

In order to perform in vivo IVPA/US imaging, a thin catheter sheath has to be integrated surrounding the catheter. A sheath protects both the catheter components and the artery endothelium from rotating probe. The sheath material should have minimum attenuation of both optical photons and acoustic waves. Furthermore, the sheath should be flexible and biocompatible. Lastly, with sheath enclosed, the final diameter of the catheter probe should be 1 mm, the current standard for safe human coronary access. Due to these critical requirements, it is challenging to find an optimal sheath material.

Currently, in vivo validation of IVPA/US imaging in animal models is lacking. While ex vivo imaging results are valuable, they lack the real challenges of in vivo imaging, such as the presence of luminal blood. A few studies have demonstrated the feasibility of IVPA/US imaging in the presence of luminal blood, however with attenuated signals (Wang et al. 2012a, b). Strategies that can reduce blood interference or enhance IVPA detection sensitivity need to be developed and tested in vivo. Also, the IVPA imaging sensitivity and specificity for lipid-rich necrotic cores must be systematically validated through comparison with gold-standard histopathology. Longitudinal studies of lipid-rich plaque formation can be further performed in these animal models. Compared with small animal models, large animal models, e.g. pigs, are preferred for IVPA/US imaging.

Novel excitation laser sources and ultrasound transducers are expected to be developed to improve the performance of current IVPA/US imaging systems. For example, laser sources with a higher repetition rate would enable even faster imaging speed (≥30 fps). By reducing the transducer size, the final catheter diameter can be miniaturized by less than 1 mm. Ultrasound transducer with improved detection sensitivity and extended bandwidth can be used to improve IVPA/US imaging sensitivity and resolution. Furthermore, IVPA imaging is complementary to other intravascular imaging modalities, such as OCT, IVUS, and fluorescence imaging, in terms of contrast mechanism, penetration depth, and spatial resolution. Therefore, other hybrid or multimodal approaches can be developed into a single catheter probe for more advanced identification of vulnerable atherosclerotic plaques.

After preclinical and clinical validations, IVPA/US imaging can provide the unmet clinical need for the detection of atherosclerotic plaques vulnerable to rupture. Based on the imaging results, proper treatment or prevention strategy can be utilized to a patient. Furthermore, IVPA imaging can also be a powerful research tool for a number of clinical applications. For example, it can be used to monitor the plaque formation process so that the role of lipid composition in atherosclerotic plaque progression or plaque rupture can be further elucidated. Currently, no methods exist to evaluate the use of lipid-lowering therapies in reducing true lipid core size due to the lack of in vivo methods to longitudinal monitor lipid content inside the coronary arterial wall (Puri et al. 2013). IVPA/US imaging can potentially fill this gap by monitoring therapy-induced lipid content changes, which can be performed in living animals and eventually human patients.

References

Allen TJ, Hall A, Dhillon AP, Owen JS, Beard PC (2012) Spectroscopic photoacoustic imaging of lipid-rich plaques in the human aorta in the 740 to 1400 nm wavelength range. J Biomed Opt 17(6):061209. https://doi.org/10.1117/1.JBO.17.6.061209

American National Standard for Safe Use of Lasers, ANSI Z136.1 (2014) Laser Institute of America

Bai X, Gong X, Hau W, Lin R, Zheng J, Liu C, Zeng C, Zou X, Zheng H, Song L (2014) Intravascular optical-resolution photoacoustic tomography with a 1.1 mm diameter catheter. PLoS One 9(3):e92463. https://doi.org/10.1371/journal.pone.0092463

Beard P (2011) Biomedical photoacoustic imaging. Interface Focus 1(4):602–631. https://doi.org/10.1098/rsfs.2011.0028

Bell AG (1880) Upon the production and reproduction of sound by light. J Soc Telegraph Eng 9(34):404–426. https://doi.org/10.1049/jste-1.1880.0046

Bentzon JF, Otsuka F, Virmani R, Falk E (2014) Mechanisms of plaque formation and rupture. Circ Res 114(12):1852–1866. https://doi.org/10.1161/CIRCRESAHA.114.302721

Brugaletta S, Garcia-Garcia HM, Serruys PW, de Boer S, Ligthart J, Gomez-Lara J, Witberg K, Diletti R, Wykrzykowska J, van Geuns RJ, Schultz C, Regar E, Duckers HJ, van Mieghem N, de Jaegere P, Madden SP, Muller JE, van der Steen AF, van der Giessen WJ, Boersma E (2011) NIRS and IVUS for characterization of atherosclerosis in patients undergoing coronary angiography. JACC Cardiovasc Imaging 4(6):647–655. https://doi.org/10.1016/j.jcmg.2011.03.013

Cao Y, Hui J, Kole A, Wang P, Yu Q, Chen W, Sturek M, Cheng JX (2016) High-sensitivity intravascular photoacoustic imaging of lipid-laden plaque with a collinear catheter design. Sci Rep 6:25236. https://doi.org/10.1038/srep25236

Cao Y, Kole A, Lan L, Wang P, Hui J, Sturek M, Cheng JX (2017) Spectral analysis assisted photoacoustic imaging for lipid composition differentiation. Photoacoustics 7:12–19. https://doi.org/10.1016/j.pacs.2017.05.002

Caplan JD, Waxman S, Nesto RW, Muller JE (2006) Near-infrared spectroscopy for the detection of vulnerable coronary artery plaques. J Am Coll Cardiol 47(8 Suppl):C92–C96. https://doi.org/10.1016/j.jacc.2005.12.045

Cox B, Laufer JG, Arridge SR, Beard PC (2012) Quantitative spectroscopic photoacoustic imaging: a review. J Biomed Opt 17(6):061202. https://doi.org/10.1117/1.JBO.17.6.061202

Danek BA, Karatasakis A, Madder RD, Muller JE, Madden S, Banerjee S, Brilakis ES (2016) Experience with the multimodality near-infrared spectroscopy/intravascular ultrasound coronary imaging system: principles, clinical experience, and ongoing studies. Curr Cardiovasc Imaging 9(2):7

Dong B, Chen S, Zhang Z, Sun C, Zhang HF (2014) Photoacoustic probe using a microring resonator ultrasonic sensor for endoscopic applications. Opt Lett 39(15):4372–4375

Falk E, Shah PK, Fuster V (1995) Coronary plaque disruption. Circulation 92(3):657–671

Finn AV, Nakano M, Narula J, Kolodgie FD, Virmani R (2010) Concept of vulnerable/unstable plaque. Arterioscler Thromb Vasc Biol 30(7):1282–1292. https://doi.org/10.1161/ATVBAHA.108.179739

Friebel M, Helfmann J, Netz U, Meinke M (2009) Influence of oxygen saturation on the optical scattering properties of human red blood cells in the spectral range 250 to 2,000 nm. J Biomed Opt 14(3):034001. https://doi.org/10.1117/1.3127200

Getz GS, Reardon CA (2012) Animal models of atherosclerosis. Arterioscler Thromb Vasc Biol 32(5):1104–1115. https://doi.org/10.1161/ATVBAHA.111.237693

Ha SH, Carson A, Agarwal A, Kotov NA, Kim K (2011) Detection and monitoring of the multiple inflammatory responses by photoacoustic molecular imaging using selectively targeted gold nanorods. Biomed Opt Express 2(3):645–657. https://doi.org/10.1364/Boe.2.000645

Hui J, Yu Q, Ma T, Wang P, Cao Y, Bruning RS, Qu Y, Chen Z, Zhou Q, Sturek M, Cheng J-X, Chen W (2015) High-speed intravascular photoacoustic imaging at 1.7 μm with a KTP-based OPO. Biomed Opt Express 6(11):4557–4566. https://doi.org/10.1364/boe.6.004557

Hui J, Li R, Phillips EH, Goergen CJ, Sturek M, Cheng JX (2016) Bond-selective photoacoustic imaging by converting molecular vibration into acoustic waves. Photoacoustics 4(1):11–21. https://doi.org/10.1016/j.pacs.2016.01.002

Hui J, Cao Y, Zhang Y, Kole A, Wang P, Yu G, Eakins G, Sturek M, Chen W, Cheng JX (2017) Real-time intravascular photoacoustic-ultrasound imaging of lipid-laden plaque in human coronary artery at 16 frames per second. Sci Rep 7(1):1417. https://doi.org/10.1038/s41598-017-01649-9

Jacques SL (1993) Role of tissue optics and pulse duration on tissue effects during high-power laser irradiation. Appl Opt 32(13):2447–2454. https://doi.org/10.1364/AO.32.002447

Jang IK, Tearney GJ, MacNeill B, Takano M, Moselewski F, Iftima N, Shishkov M, Houser S, Aretz HT, Halpern EF, Bouma BE (2005) In vivo characterization of coronary atherosclerotic plaque by use of optical coherence tomography. Circulation 111(12):1551–1555. https://doi.org/10.1161/01.CIR.0000159354.43778.69

Jansen K, van der Steen AF, van Beusekom HM, Oosterhuis JW, van Soest G (2011) Intravascular photoacoustic imaging of human coronary atherosclerosis. Opt Lett 36(5):597–599. https://doi.org/10.1364/OL.36.000597

Jansen K, van der Steen AF, Wu M, van Beusekom HM, Springeling G, Li X, Zhou Q, Shung KK, de Kleijn DP, van Soest G (2014a) Spectroscopic intravascular photoacoustic imaging of lipids in atherosclerosis. J Biomed Opt 19(2):026006. https://doi.org/10.1117/1.JBO.19.2.026006

Jansen K, Wu M, van der Steen AF, van Soest G (2014b) Photoacoustic imaging of human coronary atherosclerosis in two spectral bands. Photoacoustics 2(1):12–20. https://doi.org/10.1016/j.pacs.2013.11.003

Karpiouk AB, Wang B, Emelianov SY (2010) Development of a catheter for combined intravascular ultrasound and photoacoustic imaging. Rev Sci Instrum 81(1):014901. https://doi.org/10.1063/1.3274197

Kilic ID, Caiazzo G, Fabris E, Serdoz R, Abou-Sherif S, Madden S, Moreno PR, Goldstein J, Di Mario C (2015) Near-infrared spectroscopy-intravascular ultrasound: scientific basis and clinical applications. Eur Heart J Cardiovasc Imaging 16(12):1299–1306. https://doi.org/10.1093/ehjci/jev208

Kubo T, Maehara A, Mintz GS, Doi H, Tsujita K, Choi SY, Katoh O, Nasu K, Koenig A, Pieper M, Rogers JH, Wijns W, Bose D, Margolis MP, Moses JW, Stone GW, Leon MB (2010) The dynamic nature of coronary artery lesion morphology assessed by serial virtual histology intravascular ultrasound tissue characterization. J Am Coll Cardiol 55(15):1590–1597. https://doi.org/10.1016/j.jacc.2009.07.078

Li W, Chen X (2015) Gold nanoparticles for photoacoustic imaging. Nanomedicine (Lond) 10(2):299–320. https://doi.org/10.2217/nnm.14.169

Li X, Wei W, Zhou Q, Shung KK, Chen Z (2012) Intravascular photoacoustic imaging at 35 and 80 MHz. J Biomed Opt 17(10):106005. https://doi.org/10.1117/1.JBO.17.10.106005

Li Y, Gong X, Liu C, Lin R, Hau W, Bai X, Song L (2015) High-speed intravascular spectroscopic photoacoustic imaging at 1000 A-lines per second with a 0.9-mm diameter catheter. J Biomed Opt 20(6):065006. https://doi.org/10.1117/1.jbo.20.6.065006

Libby P, DiCarli M, Weissleder R (2010) The vascular biology of atherosclerosis and imaging targets. J Nucl Med 51(Suppl 1):33S–37S. https://doi.org/10.2967/jnumed.109.069633

Luke GP, Yeager D, Emelianov SY (2012) Biomedical applications of photoacoustic imaging with exogenous contrast agents. Ann Biomed Eng 40(2):422–437. https://doi.org/10.1007/s10439-011-0449-4

Mintz GS (2014) Clinical utility of intravascular imaging and physiology in coronary artery disease. J Am Coll Cardiol 64(2):207–222. https://doi.org/10.1016/j.jacc.2014.01.015

Mitcham T, Dextraze K, Taghavi H, Melancon M, Bouchard R (2015) Photoacoustic imaging driven by an interstitial irradiation source. Photoacoustics 3(2):45–54. https://doi.org/10.1016/j.pacs.2015.02.002

Narula J, Nakano M, Virmani R, Kolodgie FD, Petersen R, Newcomb R, Malik S, Fuster V, Finn AV (2013) Histopathologic characteristics of atherosclerotic coronary disease and implications of the findings for the invasive and noninvasive detection of vulnerable plaques. J Am Coll Cardiol 61(10):1041–1051. https://doi.org/10.1016/j.jacc.2012.10.054

Nissen SE (2016) IVUS virtual histology: unvalidated gimmick or useful technique? J Am Coll Cardiol 67(15):1784–1785. https://doi.org/10.1016/j.jacc.2016.02.037

Ntziachristos V (2010) Going deeper than microscopy: the optical imaging frontier in biology. Nat Methods 7(8):603–614. https://doi.org/10.1038/nmeth.1483

Piao Z, Ma T, Li J, Wiedmann MT, Huang S, Yu M, Kirk Shung K, Zhou Q, Kim CS, Chen Z (2015) High speed intravascular photoacoustic imaging with fast optical parametric oscillator laser at 1.7 m. Appl Phys Lett 107(8):083701. https://doi.org/10.1063/1.4929584

Puri R, Tuzcu EM, Nissen SE, Nicholls SJ (2013) Exploring coronary atherosclerosis with intravascular imaging. Int J Cardiol 168(2):670–679. https://doi.org/10.1016/j.ijcard.2013.03.024

Qin H, Zhao Y, Zhang J, Pan X, Yang S, Xing D (2016) Inflammation-targeted gold nanorods for intravascular photoacoustic imaging detection of matrix metalloproteinase-2 (MMP2) in atherosclerotic plaques. Nanomedicine 12(7):1765–1774. https://doi.org/10.1016/j.nano.2016.02.016

Razansky D, Harlaar NJ, Hillebrands JL, Taruttis A, Herzog E, Zeebregts CJ, van Dam GM, Ntziachristos V (2012) Multispectral optoacoustic tomography of matrix metalloproteinase activity in vulnerable human carotid plaques. Mol Imaging Biol 14(3):277–285. https://doi.org/10.1007/s11307-011-0502-6

Sanidas E, Dangas G (2013) Evolution of intravascular assessment of coronary anatomy and physiology: from ultrasound imaging to optical and flow assessment. Eur J Clin Invest 43(9):996–1008. https://doi.org/10.1111/eci.12119

Sasic S, Ozaki Y (2011) Raman, infrared, and near-infrared chemical imaging. Wiley, New Jersey

Schoenhagen P, Ziada KM, Vince DG, Nissen SE, Tuzcu EM (2001) Arterial remodeling and coronary artery disease: the concept of "dilated" versus "obstructive" coronary atherosclerosis. J Am Coll Cardiol 38(2):297–306

Suh WM, Seto AH, Margey RJ, Cruz-Gonzalez I, Jang IK (2011) Intravascular detection of the vulnerable plaque. Circ Cardiovasc Imaging 4(2):169–178. https://doi.org/10.1161/CIRCIMAGING.110.958777

Takano M, Mizuno K, Okamatsu K, Yokoyama S, Ohba T, Sakai S (2001) Mechanical and structural characteristics of vulnerable plaques: analysis by coronary angioscopy and intravascular ultrasound. J Am Coll Cardiol 38(1):99–104

Tearney GJ, Regar E, Akasaka T, Adriaenssens T, Barlis P, Bezerra HG, Bouma B, Bruining N, Cho JM, Chowdhary S, Costa MA, de Silva R, Dijkstra J, Di Mario C, Dudeck D, Falk E, Feldman MD, Fitzgerald P, Garcia H, Gonzalo N, Granada JF, Guagliumi G, Holm NR, Honda Y, Ikeno F, Kawasaki M, Kochman J, Koltowski L, Kubo T, Kume T, Kyono H, Lam CCS, Lamouche G, Lee DP, Leon MB, Maehara A, Manfrini O, Mintz GS, Mizuno K, Morel MA, Nadkarni S, Okura H, Otake H, Pietrasik A, Prati F, Raber L, Radu MD, Rieber J, Riga M, Rollins A, Rosenberg M, Sirbu V, Serruys PWJC, Shimada K, Shinke T, Shite J, Siegel E, Sonada S, Suter M, Takarada S, Tanaka A, Terashima M, Troels T, Uemura S, Ughi GJ, van Beusekom HMM, van der Steen AFW, van Es GA, van Soest G, Virmani R, Waxman S, Weissman NJ, Weisz G (2012) Consensus standards for acquisition, measurement, and reporting of intravascular optical coherence tomography studies a report from the international working group for intravascular optical coherence tomography standardization and validation. J Am Coll Cardiol 59(12):1058–1072. https://doi.org/10.1016/j.jacc.2011.09.079

Thim T, Hagensen MK, Wallace-Bradley D, Granada JF, Kaluza GL, Drouet L, Paaske WP, Botker HE, Falk E (2010) Unreliable assessment of necrotic core by virtual histology intravascular ultrasound in porcine coronary artery disease. Circ Cardiovasc Imaging 3(4):384–391. https://doi.org/10.1161/CIRCIMAGING.109.919357

Wang LV, Hu S (2012) Photoacoustic tomography: in vivo imaging from organelles to organs. Science 335(6075):1458–1462. https://doi.org/10.1126/science.1216210

Wang LV, Yao J (2016) A practical guide to photoacoustic tomography in the life sciences. Nat Methods 13(8):627–638. https://doi.org/10.1038/nmeth.3925

Wang B, Yantsen E, Larson T, Karpiouk AB, Sethuraman S, Su JL, Sokolov K, Emelianov SY (2009) Plasmonic intravascular photoacoustic imaging for detection of macrophages in atherosclerotic plaques. Nano Lett 9(6):2212–2217. https://doi.org/10.1021/nl801852e

Wang B, Su JL, Amirian J, Litovsky SH, Smalling R, Emelianov S (2010) Detection of lipid in atherosclerotic vessels using ultrasound-guided spectroscopic intravascular photoacoustic imaging. Opt Express 18(5):4889–4897. https://doi.org/10.1364/OE.18.004889

Wang H-W, Chai N, Wang P, Hu S, Dou W, Umulis D, Wang LV, Sturek M, Lucht R, Cheng J-X (2011) Label-free bond-selective imaging by listening to vibrationally excited molecules. Phys Rev Lett 106(23). https://doi.org/10.1103/physrevlett.106.238106

Wang B, Karpiouk A, Yeager D, Amirian J, Litovsky S, Smalling R, Emelianov S (2012a) In vivo intravascular ultrasound-guided photoacoustic imaging of lipid in plaques using an animal model of atherosclerosis. Ultrasound Med Biol 38(12):2098–2103. https://doi.org/10.1016/j.ultrasmedbio.2012.08.006

Wang B, Karpiouk A, Yeager D, Amirian J, Litovsky S, Smalling R, Emelianov S (2012b) Intravascular photoacoustic imaging of lipid in atherosclerotic plaques in the presence of luminal blood. Opt Lett 37(7):1244–1246. https://doi.org/10.1364/OL.37.001244

Wang P, Wang HW, Sturek M, Cheng JX (2012c) Bond-selective imaging of deep tissue through the optical window between 1600 and 1850 nm. J Biophotonics 5(1):25–32. https://doi.org/10.1002/jbio.201100102

Wang P, Ma T, Slipchenko MN, Liang S, Hui J, Shung KK, Roy S, Sturek M, Zhou Q, Chen Z, Cheng JX (2014) High-speed intravascular photoacoustic imaging of lipid-laden atherosclerotic plaque enabled by a 2-kHz barium nitrite Raman laser. Sci Rep 4(6889). https://doi.org/10.1038/srep06889

Weber J, Beard PC, Bohndiek SE (2016) Contrast agents for molecular photoacoustic imaging. Nat Methods 13(8):639–650. https://doi.org/10.1038/nmeth.3929

Wei W, Li X, Zhou Q, Shung KK, Chen Z (2011) Integrated ultrasound and photoacoustic probe for co-registered intravascular imaging. J Biomed Opt 16(10):106001. https://doi.org/10.1117/1.3631798

Weyer L, Workman J Jr (2007) Practical guide to interpretive near-infrared spectroscopy. CRC Press, Boca Raton

Yahagi K, Kolodgie FD, Otsuka F, Finn AV, Davis HR, Joner M, Virmani R (2016) Pathophysiology of native coronary, vein graft, and in-stent atherosclerosis. Nat Rev Cardiol 13(2):79–98. https://doi.org/10.1038/nrcardio.2015.164

Yakovlev VV, Zhang HF, Noojin GD, Denton ML, Thomas RJ, Scully MO (2010) Stimulated Raman photoacoustic imaging. Proc Natl Acad Sci U S A 107(47):20335–20339. https://doi.org/10.1073/pnas.1012432107

Yeager D, Karpiouk A, Wang B, Amirian J, Sokolov K, Smalling R, Emelianov S (2012) Intravascular photoacoustic imaging of exogenously labeled atherosclerotic plaque through luminal blood. J Biomed Opt 17(10). doi: Artn 106016. https://doi.org/10.1117/1.Jbo.17.10.106016

Zhang EZ, Beard PC (2011) A miniature all-optical photoacoustic imaging probe. Proc Spie 7899. doi: Artn 78991f. https://doi.org/10.1117/12.874883

Zhang J, Yang S, Ji X, Zhou Q, Xing D (2014) Characterization of lipid-rich aortic plaques by intravascular photoacoustic tomography: ex vivo and in vivo validation in a rabbit atherosclerosis model with histologic correlation. J Am Coll Cardiol 64(4):385–390. https://doi.org/10.1016/j.jacc.2014.04.053

Chapter 5
Contrast-Enhanced Dual-Frequency Super-Harmonic Intravascular Ultrasound (IVUS) Imaging

Jianguo Ma and Xiaoning Jiang

Background

Atherosclerotic cardiovascular disease is a leading cause of death worldwide, and one which often manifests without warning (Naghavi 2003). According to the 2014 update of Heart Disease and Stroke Statistics by the American Heart Association (Go et al. 2014), there are more than 2000 deaths every day in the USA on average, which is 1 death every 40 s. For up to 75% of acute coronary syndromes, the underlying pathological mechanism is hypothesized to be atherosclerotic plaque rupture (Naghavi 2003). Unfortunately, a high percentage of vulnerable plaques are also angiographically occult, and these are responsible for a high proportion of ensuing cardiac events resulting in either fatalities or requiring further interventional treatment (Glaser 2005; Goertz et al. 2007). For this reason, detection and characterization of plaques which are rupture prone is one of the most active areas of research in cardiology and biomedical imaging (Constantinides 1990). The vasa vasorum is a network of microvessels which supports larger vessels such as the aorta, and increased density of the vasa vasorum has been associated with a plaque advancing from a stable state to a rupture-prone state (Naghavi 2010; Moulton et al. 2003). Additionally, intraplaque hemorrhage occurring from thin-walled, immature microvessels has been present in plaques in many cases of sudden coronary death (Virmani 2005). Evidence suggests that vasa vasorum proliferation and associated angiogenesis and inflammation is

J. Ma (✉)
School of Instrumentation and Optoelectronic Engineering, Beihang University, Beijing 100191, China
e-mail: majianguo@buaa.edu.cn

Beijing Advanced Innovation Center for Big Data-Based Precision Medicine, Beihang University, Beijing 100191, China

X. Jiang
Department of Mechanical and Aerospace Engineering, North Carolina State University, Raleigh, NC 27695, USA

Q. Zhou and Z. Chen (eds.), *Multimodality Imaging*,
https://doi.org/10.1007/978-981-10-6307-7_5

associated with plaque instability and rupture (Virmani 2005; Kolodgie et al. 2003; Milei et al. 1998; Moreno and Fuster 2004). As our ability to predict the instability of atherosclerotic lesions remains a substantial challenge, there is an unmet need for new imaging methods to identify, detect, and differentiate these pathologies (Jaffer et al. 2006).

The new technology of ultrasound molecular imaging utilizes contrast agents displaying targeting ligands to identify areas of inflammation and angiogenesis associated with disease progression (targets that cannot be identified by B-mode ultrasound) (Lindner 2004; Choudhury et al. 2004; Gessner and Dayton 2010). Prior data suggest that ultrasound molecular imaging will provide a unique opportunity for plaque biomarker evaluation (such as inflammatory or angiogenic markers) and for identification of vulnerable plaques (ten Kate et al. 2010). Additionally, a new high-frequency contrast imaging technique, acoustic angiography (Gessner et al. 2013c), takes advantage of exciting microbubbles near resonance and detecting their high-frequency, broadband harmonics with sufficient bandwidth separation to achieve both high resolution and high contrast-to-noise ratio (CNR). Data have shown that acoustic angiography enables detailed visualization and analysis of microvascular structure (Gessner et al. 2012, 2013c), and will likely be applicable to vasa vasorum imaging. Thus, we hypothesize that there is a role for contrast-enhanced ultrasound imaging in the assessment of atherosclerosis at high-order harmonic frequencies.

Feinstein has illustrated the potential of contrast-enhanced transcutaneous ultrasound imaging on the carotid artery (Feinstein 2006), but the potential of transcutaneous ultrasound has limitations with resolution and motion artifacts (Staub et al. 2010), especially if the target is the deeper coronary arteries. This may present an opportunity for intravascular ultrasound (IVUS) (Slager et al. 2000), which has been widely utilized for the characterization of coronary vessel walls (Tobis et al. 1991), morphology of plaques (Jang et al. 2002), and so on. However, conventional IVUS transducers are not optimized for contrast imaging (Nissen et al. 1991), and, therefore, are ineffective for vasa vasorum imaging. This absence of technology may be due to the fact that nonlinear detection strategies for contrast imaging are most effective near the resonant frequency of microbubble contrast agents, which is typically between 1 and 10 MHz (Kasprzak et al. 1999). Thus, conventional contrast imaging strategies are not very effective with high-frequency ultrasound (35–50 MHz) that is typically used with IVUS. To overcome this challenge, Goertz and collaborators have been evaluating both subharmonic and harmonic contrast IVUS imaging, with the goal of vasa vasorum imaging (Goertz et al. 2006, 2007). Their research showed a contrast-to-tissue ratio (CTR) of 28 dB in subharmonic imaging with a fundamental frequency of 30 MHz (Goertz et al. 2007) and 25 dB in second-harmonic imaging with a fundamental frequency of 20 MHz (Goertz et al. 2006).

We hypothesize that resolution and contrast-to-tissue ratios can be further improved over subharmonic or harmonic contrast IVUS imaging by excitation of microbubbles near resonance and detecting their backscatter at a bandwidth substantially higher than that of the transmission, previously called "super-harmonic," "ultra-broadband," or "transient" imaging. In prior work, de Jong et al. (2002), Bouakaz et al. (2003), and Kruse and Ferrara (2005) demonstrated that substantial improvements in CTR could be achieved by detecting the high-frequency energy produced by microbubbles excited at lower frequencies. Bouakaz et al. utilized a dual-frequency transducer to illustrate that the CTR of the fourth and fifth harmonic could be 15 and 7 dB higher than that of the second harmonic, respectively (Bouakaz et al. 2003). Kruse et al. also utilized a dual-frequency transducer arrangement, and demonstrated that substantial scattered energy from microbubbles excited near 2 MHz could be detected at a frequency as high as 45 MHz. More recently, Gessner et al. (2010, 2012, 2013c) utilized mechanically scanned dual-frequency transducers (transmit at 2 or 4 MHz, and receive at 30 MHz) on the VisualSonics Vevo770 to perform high-frequency 3-D contrast imaging of in vivo microvasculature and achieved resolution on the order of 100 microns with a CTR high enough so that microvessels could be readily segmented from the images and their morphology analyzed.

Despite the promising CTR and vessel imaging capability of this imaging approach, there is a substantial challenge for "ultra-broadband" contrast-enhanced intravascular ultrasound (CE-IVUS), which is likely why it is yet relatively undeveloped. The primary limitation is the large frequency span, which is outside of the current bandwidth of commercially available single-frequency transducers. Such difficulty could be surmounted by using multiple confocal transducers as described by Gessner et al. (2010), however, such transcutaneous exposure method is almost impossible to be used for coronary vasa vasorum imaging due to penetration depth limitation and existence of ribs at the chest. Furthermore, decreasing the frequency to <20 MHz is not a solution for a larger penetration depth because the high resolution is needed to image the vasa vasorum with an average diameter of 161 μm for the first order and 68 μm for the second order (Kwon et al. 1998). Intravascular ultrasound (IVUS) imaging may be a solution to access the vasa vasorum deep inside the body, and has been widely utilized for the characterization of coronary vessel walls (Tobis et al. 1991), the morphology of plaques (Jang et al. 2002), and so on. For the CE-IVUS super-harmonic vasa vasorum imaging, a specifically designed dual-frequency IVUS transducer is essential.

In this chapter, small aperture, dual frequency, and IVUS transducers are introduced (Ma et al. 2013a, b), which provide sufficient bandwidth separation for high CTR, high-resolution CE-IVUS imaging. Design, fabrication, characterization, and super-harmonic imaging capabilities are demonstrated. The imaging performance was compared among transducers with different materials and excitation frequencies. High-resolution (70 μm), high CTR (23 dB) images of microbubbles could be generated by the dual-frequency transducers, indicating the potential capability of imaging the second-order vasa vasorum.

Dual-Frequency IVUS Transducer

Dual-Frequency Transducer Structure Design

There are several alternative structures of the dual-frequency transducer design (Fig. 5.1a–d). One structure is to put the high-frequency element and the low-frequency element side by side, which is named as interleaved structure (Fig. 5.1a) (Van Neer et al. 2010; Martin et al. 2014). Advantage of this structure is the low cross talk between the high-frequency and the low-frequency elements. However, this structure has misaligned beam profile between the two frequencies and the transmitting pressure was too low in the sensitive region of the high-frequency beam ($x = -0.2$, $y = 0$ in Fig. 5.2a). If the elements are stacked in layers (Fig. 5.1b), then the transmitting and receiving beam would have the best overlap because of the spatial alignment. However, if the two active layers were directly bonded together, then the high-frequency receiving signal would continue propagation after being detected by the receiving element, the reflection of which at the back side of the transducer would cause aliasing echoes that show up after the real signal. Such aliasing echoes could either cause fake imaging signal (if each echo is considered as a signal) or reduce the resolution (if the series of echoes are considered as one signal). An acoustic filter (Fig. 5.1c) with a quarter wavelength for high frequency was placed between the two piezoelectric layers to enhance the low-frequency transmitting wave penetration, while reflecting the high-frequency receiving wave to suppress the aliasing echo (Azuma et al. 2010). Detailed analysis of the acoustic filter will be described in the next subsection. Furthermore, it is impossible to make the electrical impedance match for both elements if the two elements are same in the aperture. The low-frequency element is thicker than the high-frequency element, so that the capacitance is lower. Consequently, the impedance ($Z = \frac{1}{j\omega C}$) of the low-frequency element is significantly higher than the high-frequency element due to low ω and low C. Furthermore, the natural focus points (near-field and far-field transition) of the two beams are also significantly different if the apertures of the two elements are the same. Based on these concerns, the structure of the transducer was designed as stacked layers with different apertures between the two active elements (Fig. 5.1d).

With the above consideration, the dual-frequency transducer was designed as a dual layer transduction structure, composed of a low-frequency transmitting layer and a high-frequency receiving layer (Fig. 5.1e) (Ma et al. 2013a). The transmission layer was placed behind the receiving layer (with respect to a forward traveling sound wave) since low-frequency transmitted acoustic waves could propagate through the smaller, high-frequency element. The placement of the receiving layer counted in the matching network of the low-frequency transmission element, which showed positive effect that amplifies the low-frequency propagation wave. The selection of 6.5 MHz as the transmitting layer's center frequency was because it was both close to the contrast agents' resonant frequency and since the piezoelectric material for this element was readily available. The high-frequency receiving layer was positioned in the front of the transducer to achieve an optimal receiving performance of the

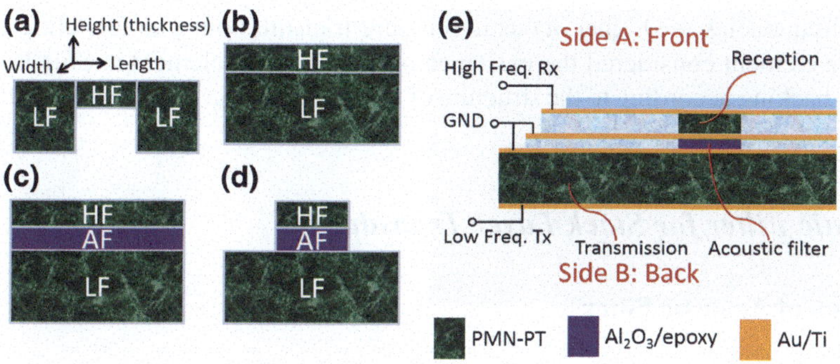

Fig. 5.1 Structural design of dual frequency transducer design alternatives. **a** Interleaved structure. **b** Stack layer structure. **c** Stack layer structure with acoustic filter sandwiched between the high–low-frequency piezoelectric layers. **d** Stack layer structure with acoustic filter and different apertures of the high–low-frequency elements. **e** Cross-sectional view of the final dual-frequency ultrasonic transducer design

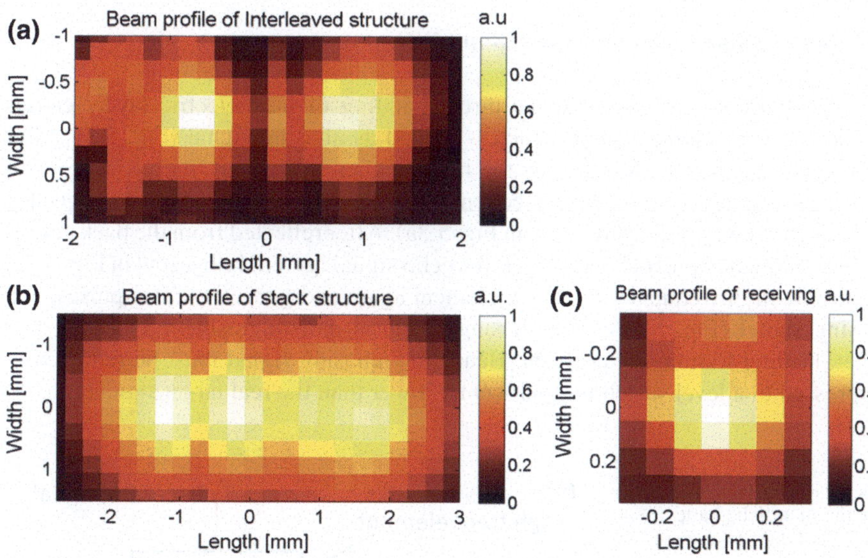

Fig. 5.2 Beam profile of **a** the low-frequency transmitter in the interleaved structure transducer, **b** the low-frequency transmitter in the stack layer transducer, and **c** the high-frequency receiver

high-frequency element. Transducer dimensions were optimized using a KLM model (Krimholtz et al. 1970) to validate the thickness of layers, length, and width of each component for ideal thickness mode excitation. The aperture of the high-frequency (0.5 mm × 0.6 mm) receiving layer was designed to be similar to that of commercial IVUS transducers (Garcia-Garcia et al. 2010), and thus significantly smaller than that of the transmission layer (3×0.6 mm^2). The relatively large aperture of the low-frequency component was designed to obtain reasonably low electrical impedance at low frequencies for higher acoustic pressure transmission. The modeling of the 30 MHz element considered the existence of the 6.5 MHz element at Side B (back side) as backing according to the structure of the design in Fig. 5.1e (Ma et al. 2014b).

Acoustic Filter for Stack Layer Transducers

Function of Acoustic Filter

If the dual-frequency transducer is designed with stacked layers as illustrated in Fig. 5.1b, then one major concern would be the acoustic interference between the two elements. Such interference might cause advantages or disadvantages. Unfortunately, the direct bonding of the two active layers leads to interferences between the two layers which are disadvantages for the dual-frequency intravascular ultrasound transducer.

Isolation of High-Frequency Ultrasound Wave

For high-frequency ultrasound, the direct bonding of piezoelectric layers causes an additional receiving signal, which is named as aliasing signal (Fig. 5.3). After the receiving signal is detected by the high-frequency element, the acoustic wave further propagates from the high-frequency piezoelectric layer to the low-frequency piezoelectric layer (red solid arrow in Fig. 5.3a). After reflected from the back side of the low-frequency piezoelectric layer, the echo signal (green dash arrow in Fig. 5.3a) excites the high-frequency receiving element again and generates extra pulses, i.e., aliasing signal (Fig. 5.3b). The aliasing signal is indistinguishable from the real target, leading to imaging artifacts. Hence, the aliasing signal must be removed or suppressed to a level which is significantly lower than the real target signal.

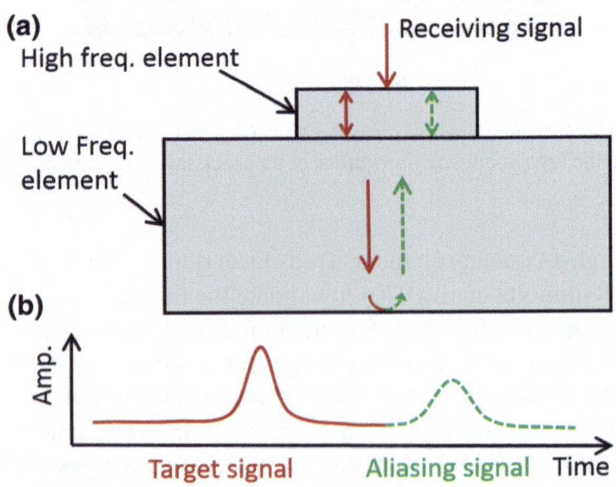

Fig. 5.3 Aliasing echo generation. **a** Schematic diagram of dual-frequency ultrasonic transducer with high- and low-frequency piezoelectric layers directly bonded; **b** signal detected by the receiver

The traditional method of removing the aliasing signal was to use high absorptive material to attenuate the signal transmitted backward, which does not work well in intravascular dual-frequency transducer application except that the two frequencies are significantly different ($f_H > 14 f_L$). First, there are barely any high absorptive materials that can eliminate the signal within a couple of wavelengths according to the space allowed in the intravascular ultrasound transducer. Second, high absorptive material with large thickness attenuates the low-frequency transmission ultrasound as well. High attenuation on the low-frequency transmission is not acceptable for the non-focused small-aperture transducer, which needs to generate high enough pressure for microbubble nonlinear excitations. Taking these concerns into account, the aliasing echo should not be removed using the absorption method if the two frequencies are not significantly (>10 times) different.

Reflection can be used to suppress the backward wave by changing the acoustic impedance matching conditions at the boundary. With two active layers bonded together (Fig. 5.3a), the impedance of them are well matched (PZT, PMN-PT …) or perfectly matched (for the same material of the two layers). Most ultrasound energy is transmitted backward and hence, causing the aliasing echo (Fig. 5.3b). If an intermediate layer is inserted between the two active layers, then the boundary condition can be changed, which could possibly reflect most of the energy directly without propagating to the low-frequency element. Impedance of the intermediate layer should be different from the active layers as much as possible. Low impedance material, with impedance much lower than piezoelectric materials, was chosen in this application because such materials are much more easily available. With a thickness of a quarter wavelength, the resultant impedance at the interface can be much lower than that of piezoelectric materials, so that most energy is reflected and little energy propagates. The suppression further happens when the tiny back-reflected wave interface with the boundary again. As a result, there is very little aliasing echo from the back side of the transducer. Such an intermediate layer changes the equivalent impedance at the boundary from well matched to mismatched. As a result, this particular intermediate layer is denominated as anti-matching layer.

Promotion of Low-Frequency Ultrasound Wave Propagation

For low-frequency ultrasonic wave, the direct bonding of the piezoelectric layers usually causes low wave transmission efficiency. In super-harmonic imaging, the frequency of receiving ultrasonic waves is at least three times of the transmitting wave. Therefore, the frequency-specific matching mechanism (like quarter-wavelength matching layer) does not match both frequencies. Matching layer in front of the high-frequency piezoelectric layer is usually designed to match the high-frequency wave because of its short wavelength and it is more sensitive to the thickness. In this case, the transmission efficiency is low without proper matching layer. Actually, both layers in front of the low-frequency piezoelectric layer, the high-frequency piezoelectric layer and the high-frequency matching layer, are thin and almost negligible for the low-frequency wave propagation. Consequently, the low-frequency ultrasonic wave suffers from low transmission efficiency.

A properly designed low impedance layer sandwiched between the high impedance piezoelectric layers promotes the low-frequency ultrasonic wave transmission. This low impedance layer can be the same physical layer as the anti-matching layer for high-frequency ultrasonic wave. It functions oppositely for high–low ultrasonic waves, and is denominated as acoustic filter.

Microwave Analysis of Piezoelectric Transducers

Both microwave and mechanical wave (ultrasound) share the same wave propagation properties although they are in different field. General wave equation $u(x,t) = Ae^{j(\omega t-kx)} + Be^{j(\omega t+kx)}$ applies for the propagating wave in both domains. Reflection and refraction properties are also the same for both types of waves. The two waves could be studied together with shared theories and methods (Mason 1930).

Mass-spring-damper model in mechanical vibration is equivalent to the lumped element resistor–inductor–capacitor (RLC) circuit in electrical vibration (Fig. 5.4). Equivalences of the parameters are shown in Table 5.1. Lumped element analysis is usually used for conceptual design of transducers, but is not widely used for detailed parameters of the transducers.

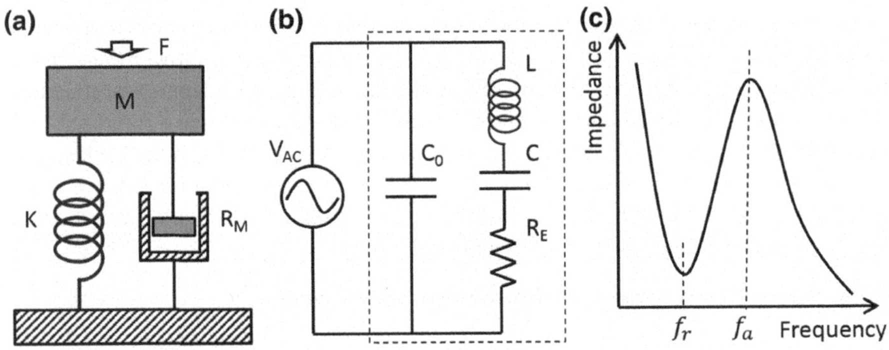

Fig. 5.4 **a** Lumped element equivalence of the dynamic behavior of the piezoelectric material, **b** the electrical equivalent circuit, and **c** impedance response of the vibrating system

Table 5.1 Equivalence matching table

Mechanical properties	Electromagnetic properties
Force	Voltage
Velocity	Current
Mass	Inductance
Spring constant	Inversion of capacitance (1/C)
Damping factor	Resistance

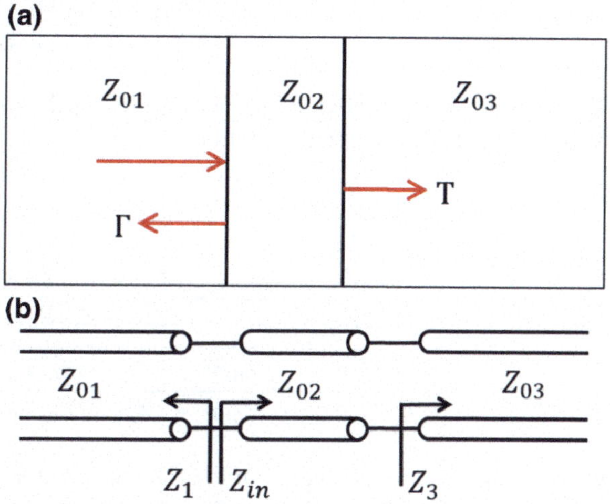

Fig. 5.5 Schematic diagram of transmission and reflection in **a** a three-layer structure and **b** its equivalent microwave transmission line

Continuous variations of strain and stress are the principal features of piezoelectric transducers, which is typically equivalent to a distributed circuit like transmission line for electromechanical wave. Electromechanical wave equations are defined as

$$\left(\nabla^2 - \frac{1}{c^2}\frac{\partial^2}{\partial t^2}\right)E = 0, \tag{5.1}$$

$$\left(\nabla^2 - \frac{1}{c^2}\frac{\partial^2}{\partial t^2}\right)B = 0, \tag{5.2}$$

where E and B are the electric field and magnetic field, respectively, and c is the wave speed in the medium defined as $c = 1/\sqrt{\mu\varepsilon}$ (μ and ε are the magnetic permeability and dielectric constant in the medium.). Such wave equations are mathematically identical to a mechanical wave

$$\frac{\partial^2 u}{\partial x^2} - \frac{1}{c^2}\frac{\partial^2 u}{\partial t^2} = 0, \tag{5.3}$$

where u is the particle displacement from still state and c is the sound speed defined as $c = \sqrt{E/\rho}$. Electromagnetic wave reflection coefficient is also identical as that for mechanical wave. The cascade transmission line theories could be applied to the ultrasound wave propagation in multiple layers.

A typical multi-layer wave propagation problem is a thin intermediate layer placed between two infinitely large media (Fig. 5.5a), and the equivalent circuit is a short section of transmission line connecting two infinitely long lines (Fig. 5.5b). If Z_{01} and Z_{03} are long, then $Z_1 = Z_{01}$ and $Z_3 = Z_{03}$. Neglecting the loss, the input impedance Z_{in} is calculated as

$$Z_{\text{in}} = Z_{02}\frac{Z_{03} + jZ_{02}\tan kl}{Z_{02} + jZ_{03}\tan kl}, \tag{5.4}$$

where k is the wave number in Z_{02}, and l is the length of Z_{02}. With the input impedance available, the reflection coefficient is

$$\begin{aligned}\Gamma &= \frac{Z_{\text{in}} - Z_{01}^*}{Z_{\text{in}} + Z_{01}^*} \\ &= \frac{Z_{02}(Z_{03} - Z_{01}) + j(Z_{02}^2 - Z_{01}Z_{03})\tan kl}{Z_{02}(Z_{03} + Z_{01}) + j(Z_{02}^2 + Z_{01}Z_{03})\tan kl} \\ &= \sqrt{\frac{Z_{02}^2(Z_{03} - Z_{01})^2 + (Z_{02}^2 - Z_{01}Z_{03})^2\tan^2 kl}{Z_{02}^2(Z_{03} + Z_{01})^2 + (Z_{02}^2 + Z_{01}Z_{03})^2\tan^2 kl}}\, e^{j(\phi_1 - \phi_2)}\end{aligned}, \tag{5.5}$$

where ϕ_1 and ϕ_2 are the phase angle of the numerator and denominator of Γ. Transmission coefficient T (amplitude) through the intermediate transmission line is calculated as

$$\begin{aligned}T^2 &= 1 - |\Gamma|^2 \\ &= \frac{4Z_{01}Z_{03}}{(Z_{01} + Z_{03})^2\cos^2 kl + \left(Z_{02} + \frac{Z_{01}Z_{03}}{Z_{02}}\right)^2 \sin^2 kl}\end{aligned}, \tag{5.6}$$

which is mathematically identical to the acoustic transmission coefficient (intensity) through an intermediate layer (Kinsler et al. 2000)

$$T_I = \frac{4}{2 + (Z_{03}/Z_{01} + Z_{01}/Z_{03})\cos^2 kl + (Z_{02}^2/Z_{01}Z_{03} + Z_{01}Z_{03}/Z_{02}^2)\sin^2 kl}. \tag{5.7}$$

For a multi-layer vibration system, the cascade calculation from microwave (Steer 2013) provides a simple and detailed method to estimate the input impedance and reflection coefficient. If the loss of the multi-layer system is negligible, then the transmission coefficient could be calculated from Eq. 5.7. If the loss of the system needs to be taken into account, then there is no analytical calculation of the transmission coefficient, in which case, a software simulation (AWR Microwave Office, AWR corp., EI Segundo, CA) could be used instead.

Smith Chart

The Smith chart (Steer 2013; Smith 1995) illustrates the equivalent impedance and the reflection coefficient, demonstrating the explicit matching effect. The phasor from the origin (center) to a position in the Smith Chart represents the reflection coefficient, length of the phasor indicating the amplitude of the reflection, and the direction indicating the phase. With each layer added in sequence, the loci shift

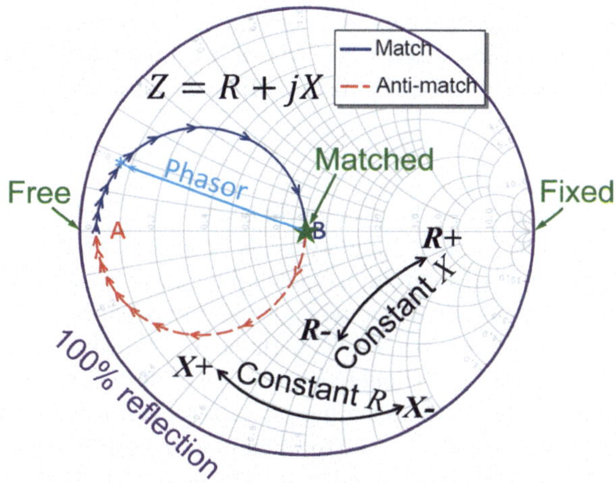

Fig. 5.6 Illustration of an impedance Smith chart. The phasor from the center to a certain point in the chart means the reflection coefficient. The purple circle at the edge of the chart denotes the 100% reflection with free boundary condition at the leftmost point and the fixed boundary condition at the rightmost point (magnitude of the phasor is 1). The center point if the matching point with phasor magnitude of 0. Real and imaginary values of the magnitude are plotted according to the orthogonal black curves. Loci of matching (blue solid line) and anti-matching (red dash line) effect of the quarter wavelength impedance inverter present the equivalent impedance if the matching or anti-matching layer increase the thickness from 0 to quarter wavelength

clockwise as a circle with centers on the horizontal middle line ($X = 0$) and the position dependent on the impedance of the material (the center of the circle is not the material impedance value on the chart). Meanwhile, the impedance can be directly read from the chart according to the markers (constant resistance or reactance arcs, black curves in Fig. 5.6). As labeled in the chart, the center is the matched point and the edge means 100% reflection because the magnitude of the phasor is 1. On the curve marked with "Constant *R*," together with those locally parallel with it, the imaginary value reactance of the impedance changes while the real value resistance keeps constant. Contrarily, on the curves marked as "Constant *X*" which are locally perpendicular with the "Constant *R*" curves, the real value resistance varies while the imaginary value reactance remains constant. As both reflection information and the impedance information are plotted in the same chart, the reflection information can be directly read from the impedance value and vice versa.

Quarter Wavelength Impedance Inverter

Quarter wavelength layer in ultrasound system also acts as an impedance inverter with the equivalent input impedance $Z_{in} = Z_0^2/Z_L$. Characteristic acoustic impedances of materials are real values, so that the impedance inverter transforms one real impedance to another. Such impedance inverter could be applied for different purposes, either making the impedance better matched or mismatched. A matching layer for an ultrasonic transducer has been widely applied for enhancing the wave propagation efficiency from piezoelectric material to medium such as water. Detailed design and applications anti-matching layer will be discussed in the next sections.

Matching Layer

A matching layer shifts the mismatched impedance to a matched impedance. In ultrasound transducer design, the matching layer can be either considered as promoting the equivalent impedance of medium to match the characteristic impedance of the piezoelectric materials, or decreasing the equivalent impedance of the piezoelectric material to match the medium. In a Smith chart, the matching procedure is to shorten the phasor of the reflection coefficient from the edge of the Smith chart to the center. A demonstrative example of the matching effect is shown from A to B in Fig. 5.1 with normalized impedance $Z_0 = 0.2$ (for matching material) and $Z_L = 0.04$ (for load, i.e., medium). As the input impedance changes from 0.04 to 1, the reflection coefficient changes from -0.923 to 0, with the minus sign denoting the phase of the reflection. It can be observed that the matching layer can be used to decrease the reflection and to enhance the transmission efficiency.

Anti-matching Layer

The impedance inverter could be used in the reversed way as well. If a wave propagation is desired to be prohibited, the impedance inverter could be utilized as anti-matching layer, shifting the matched impedance to a mismatch. In the Smith chart, the anti-matching procedure is to elongate the phasor of the reflection coefficient from the center of the chart to the edge. A demonstration of the anti-matching effect is shown from B to A in Fig. 5.6 with normalized impedance $Z_0 = 0.2$ (for matching material) and $Z_L = 1$ (for load, i.e., medium). As the input impedance changes from 1 to 0.04, the reflection coefficient changes from 0 to -0.923. Anti-matching layer increased the reflection and suppressed the transmission efficiency.

Anti-matching Layer for High-Frequency Receiving Wave

Mechanism Analysis

The mechanism of the anti-matching layer is rooted in the acoustic wave propagation theory. The acoustic intensity transmission coefficient T_I is given by

$$T_I = \frac{4Z_H Z_L}{(Z_H + Z_L)^2 \cos^2 kl + \left(Z_{\mathrm{AM}} + \frac{Z_H Z_L}{Z_{\mathrm{AM}}}\right)^2 \sin^2 kl}, \tag{5.8}$$

where Z_H, Z_L, and Z_{AM} are the characteristic acoustic impedance of the high-frequency element, the low-frequency element, and the anti-matching material, respectively; k is the wave number and l is the thickness of the anti-matching layer. Minimum transmission happens when l is equal to the quarter wavelength $\left(\frac{2n+1}{4}\lambda, n = 0,\ 1,\ 2\ldots\right)$ and the equation reduces to

$$T_I = \frac{4Z_H Z_L Z_{\mathrm{AM}}^2}{\left(Z_{\mathrm{AM}}^2 + Z_H Z_L\right)^2}. \tag{5.9}$$

Usually the two active elements are made of the same materials. In that case, the equation is further reduced to

$$T_I = \frac{4Z_H^2 Z_{\mathrm{AM}}^2}{\left(Z_{AM}^2 + Z_L^2\right)^2}. \tag{5.10}$$

This transmission coefficient is very low for a low impedance of anti-matching Z_{AM}. The equivalent impedance Z_e at the interface is

$$Z_e = \frac{Z_{\mathrm{AM}}^2}{Z_L}. \tag{5.11}$$

This equation indicates that this layer acts as an impedance inverter, which is the same form as that for a matching layer. However, in this case, the high impedance of Z_L and relatively low impedance Z_{AM} resulted in a very low Z_e and, therefore, the very low acoustic intensity transmission coefficient T_I.

Compared to the matching layer design, the mechanism and equations of anti-matching layer design are similar. However, very different values lead to significantly different effects. In matching layer design, the impedance of the medium (or load) Z_L is much lower than Z_H, the active material. With a moderate impedance Z_s, the equivalent impedance is changed to very high value to match Z_H. On the other hand, for anti-matching design, Z_L is very high, which makes the Z_e very low so that impedance is mismatched and wave cannot propagate.

The effect of the matching and anti-matching could also be understood from the boundary conditions. For matching layers between a high impedance material and a low impedance material, boundary condition of the matching layer is almost fixed free, so that resonance occurs with a quarter wavelength thickness to ensure good wave propagation. Contrarily for the anti-matching layer, it is almost fixed–fixed boundary condition and no vibration mode exists within a quarter wavelength layer. In other words, the anti-matching layer could hardly vibrate, so that ultrasonic wave hardly propagates through this layer.

Microwave Analysis

Anti-matching layers could be analyzed as transmission line networks because the mechanical wave and the electromagnetic wave share identical mathematic equations. Specifically, the dual-frequency ultrasound transducer (Fig. 5.7a) could be considered as the equivalent circuit (Fig. 5.7b).

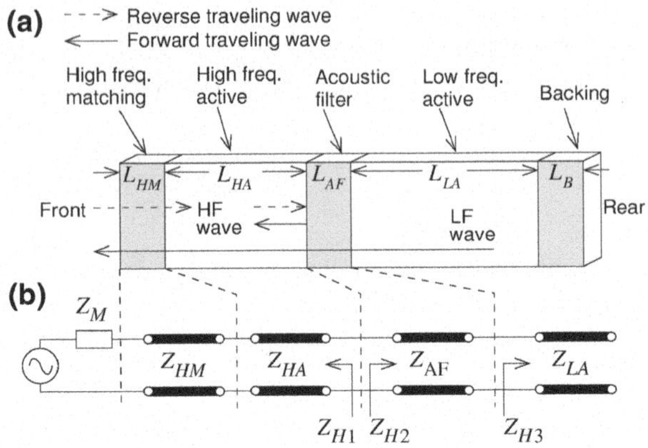

Fig. 5.7 Anti-matching (acoustic filter) design for the high-frequency receiving wave. **a** Structure of the dual-frequency transducer; **b** equivalent circuit for the high-frequency wave propagation

The anti-matching layer for the high-frequency element could be designed based on wave reflection (Type I) or wave absorption (Type II) in the stop band. For Type I transducer design, an anti-matching layer shifts the perfectly matched impedance (if $Z_{HA} = Z_{LA}$) into a great mismatch between Z_{H1} and Z_{H2}. As shown in Fig. 5.7b, the incident high-frequency wave through the matching layer of the high-frequency element is equivalent to a voltage (stress) source with Z_{HA} Thevenin impedance. As a result, the output impedance at high-frequency element is $Z_{H1} = Z_{HA}$. Pulse length of the high-frequency wave was spatially shorter than the twice of the low-frequency element thickness. If the short pulse entered the low-frequency element, the pulse would terminate before the reflected wave arrived at the input interface again. Then the reflected wave from the back side of the low-frequency element could not interfere with the incident wave at the input boundary of the low-frequency element. In this case, $Z_{H3} = Z_{LA}$ and the input impedance Z_{H2} at the front surface of the anti-matching layer is

$$Z_{H2} = Z_{AF}\frac{Z_{LA}\cosh(\gamma_{HF}l) + Z_{AF}sinh(\gamma_{HF}l)}{Z_{AF}\cosh(\gamma_{HF}l) + Z_{LA}sinh(\gamma_{HF}l)}, \tag{5.12}$$

where γ is the propagation constant and l is the thickness of the anti-matching layer. The propagation constant is defined as $\gamma_{HF} = \alpha_{HF} + j\beta_{HF}$ where α_{HF} is the attenuation coefficient and β_{HF} is the phase constant. The anti-matching layer is thin compared to the low-frequency wavelength, as a result, the loss is negligible $\alpha_{HF}l \approx 0$ if the attenuation is not very high, and $\gamma_{HF}l$ reduces to $j\beta_{HF}l$ so that the input impedance at Z_{H2} reduces Eq. (5.4)

$$Z_{H2} = Z_{AF}\frac{Z_{LA} + jZ_{AF}\tan(\beta l)}{Z_{AF} + jZ_{LA}\tan(\beta l)}, \tag{5.13}$$

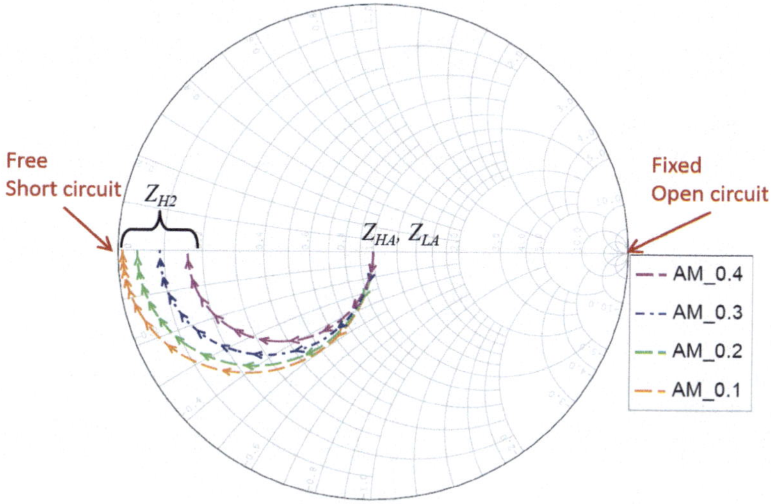

Fig. 5.8 Loci of the quarter wavelength anti-matching layer. The numbers in the legend followed by "AM" indicate the relative impedance of the anti-matching layer

where the phase constant β is equal to the wave number k in this lossless case. The intensity transmission coefficients T_I are identical in both mechanical and electromagnetic calculations.

Insertion loss (IL) is also a parameter to indicate the performance of the anti-matching layer, which mainly depends on Z_{AF} and $\gamma_{HF}l$. Starting with $Z_{P1} = Z_{P2}$, the lossless anti-matching layer, acting as a section of transmission line connected between Z_{P1} and Z_{P2} (Fig. 5.7b), shifts from ideal match condition Z_{H3} (center) to free-moving boundary condition Z_{H2} (left side close to the edge). The lower Z_{AF} is accompanied by the lower Z_{H2} (see loci of the Smith chart in Fig. 5.8). At $\lambda_{HF}/4$, reflection coefficient becomes maximum in amplitude and 0 in phase, so that insertion loss becomes maximum (Fig. 5.9). An insertion loss of 10 dB and 20 dB are achievable using a lossless anti-matching layer with a relative impedance $Z_{\text{rel}} = Z_{AF}/\sqrt{Z_{P1}Z_{P2}}$ of 0.15 and 0.05, respectively. The AWR software simulation result proved that the loss is negligible (<1 dB) on insertion loss at $\lambda_{HF}/4$ if the attenuation is within 30 dB/cm/MHz (Type I in Fig. 5.10). On the contrary, as the thickness of acoustic filter layer increases to $>1.2\lambda_{HF}$ with 30 dB/cm/MHz attenuation, the insertion loss is dominated by the loss and converges to a backing layer (Type II in Fig. 5.10).

The anti-matching layer design was validated by comparing the experimental results of piezoelectric transducer prototypes with the theoretical calculation. Anti-matching performance of the Type II transducer is very straightforward and widely accepted (Wang et al. 2001; Kossoff 1966). The performance in Type I was verified by transducer prototypes. In the verification, $f_{HF} = 30$ MHz and $f_{LF} = 3.5$ MHz.

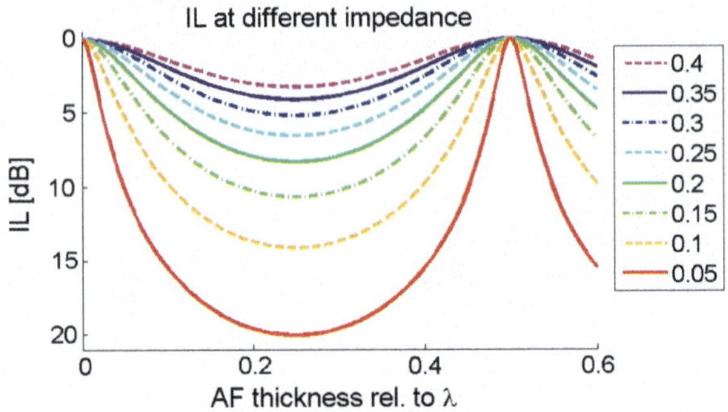

Fig. 5.9 Insertion loss of the anti-matching layer dependent on the relative impedance ZAF/ZLA (values labeled in legend) and thickness LAF (x-axis)

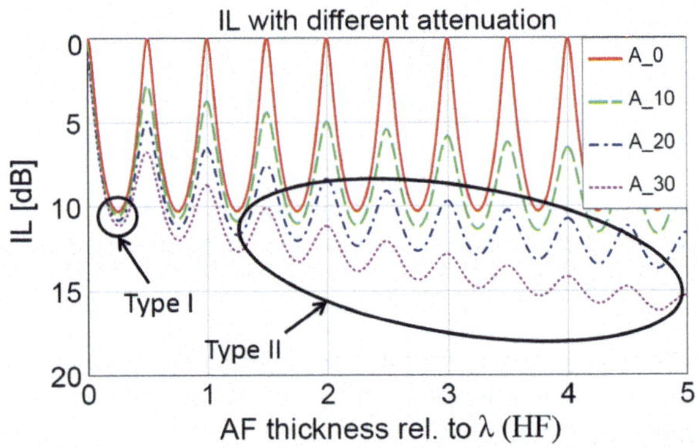

Fig. 5.10 Insertion loss of the anti-matching layer with attenuation (in dB/cm/MHz) labeled as legend numbers followed by “A”. Relative impedance of the passive amplifier is 0.157 (from an actual case: silver epoxy vs. PMN-PT)

The impedances are $Z_{P1} = Z_{P2} = 35.2$ MRayl and $Z_{AF} = 5.53$ MRayl. The thicknesses for anti-matching verification are $L_{AF} = 0.25\lambda_{HF}$ and $L_{HF} = 0.5\lambda_{HF}$.

The anti-matching effect is revealed by the amplitude of aliasing echo reflected from rear side of the low-frequency element. Without the anti-matching layer, the received ultrasound wave would continue propagating from the high-frequency element to the low-frequency element, and the reflected wave from the rear surface of low-frequency element would excite the high-frequency receiver again, appearing as an aliasing echo. Such echo may be reflected back and forth which result in a serial of wave packages after the main pulse from the real target (Fig. 5.11a). On the contrary, such aliasing echoes were suppressed to <−20 dB in a round loop (Fig. 5.11b) with the acoustic filter layer sandwiched between the high-frequency and low-frequency elements. Pulse length (−20 dB) was significantly reduced from 555 to 152 ns, leading to an axial imaging resolution enhancement from 0.416 mm to 0.114 mm in water.

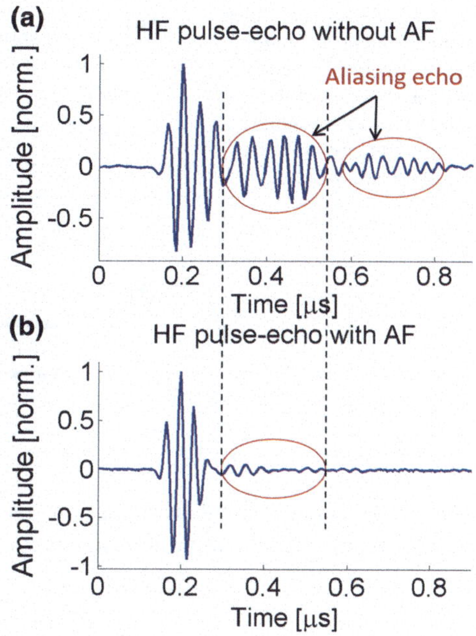

Fig. 5.11 High-frequency pulse-echo response on transducer prototypes, **a** without anti-matching layer and **b** with anti-matching layer

Passive Amplifier for the Low-Frequency Transmitting Wave

The passive amplifier effect of the acoustic filter for low-frequency transmitting wave was not straightforward in mechanical analysis. However, applying the multi-section of transmission lines (Eq. 5.14) and Smith chart, the analysis is still applicable and intuitive.

For the low-frequency transmitting wave, the transducer structure (Fig. 5.12a) is equivalent to the transmission line network shown in Fig. 5.12b. The amplification effect is studied by comparing the transmission efficiency with and without the passive amplifier. In the analysis of the low-frequency transmission, reflection from the backing layer is neglected because the backing layer dissipates most of the energy backwards. In the actual case without perfect backing, the reflected wave from the rear side would inevitably lead to the energy accumulation in the low-frequency element and compensate the pressure output enhancement. Such compensation will be shown in the experiment. As for analysis, the active low-frequency element is considered as a voltage (stress) source with Thevenin impedance of Z_{P2}. The medium is large in space with no phase delay between strain and stress, and hence it is considered as a pure resistive load Z_M. Because the thickness of the front layers (high-frequency matching, high-frequency active and the passive amplifier layers) are all very short compared to the low-frequency transmission wavelength, there are strong inference between the incident wave and the reflected wave. In this case, these layers are

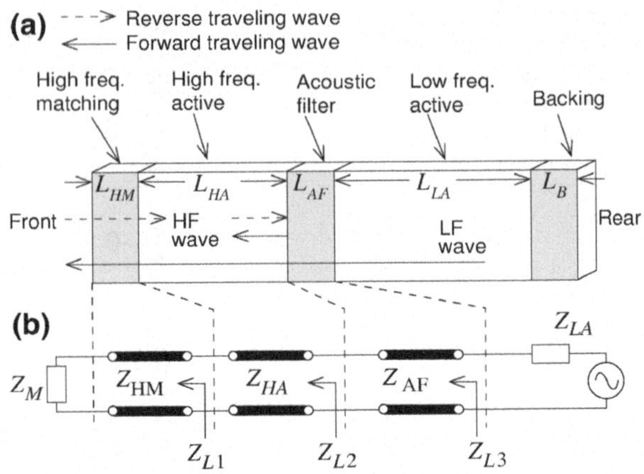

Fig. 5.12 Passive amplifier (acoustic filter) design for the low-frequency receiving wave. **a** Structure of the dual-frequency transducer; **b** equivalent circuit for the low-frequency wave propagation

considered as sections of transmission lines in cascade connection. As shown in Eqs. 5.14 and 5.15, the input impedance Z_{L3} could be calculated from

$$Z_{L3} = \frac{P_{L3}}{v_{L3}} \quad (5.14)$$

and

$$\begin{bmatrix} p_{L3} \\ v_{L3} \end{bmatrix} = \begin{bmatrix} A & B \\ C & D \end{bmatrix}_{AF} \begin{bmatrix} A & B \\ C & D \end{bmatrix}_{HA} \begin{bmatrix} A & B \\ C & D \end{bmatrix}_{HM} \begin{bmatrix} p_M \\ v_M \end{bmatrix}, \quad (5.15)$$

where P_{L3} and P_M are the pressure and v_{L3} and v_M are the volume velocity at the specific material, which satisfy $p_{L3}/v_{L3} = Z_{L3}$ and $p_M/v_M = Z_M$. The subscripts $[\,]_{AF}$, $[\,]_{HA}$, and $[\,]_{HM}$ at each *ABCD* matrix indicate the transmission line sections of the passive amplifier (acoustic filter) layer, high-frequency active layer, and the high-frequency matching layer, respectively. The *ABCD* parameter of each section of the transmission line is defined as

$$\begin{bmatrix} A & B \\ C & D \end{bmatrix} = \begin{bmatrix} \cosh(\gamma_{LF} l) & Z_0 \sinh(\gamma_{LF} l) \\ Y_0 \sinh(\gamma_{LF} l) & \cosh(\gamma_{LF} l) \end{bmatrix}, \quad (5.16)$$

where Z_0 denotes the characteristic impedance of the specific transmission line and Y_0 is the conductance defined as $1/Z_0$. Propagation constant γ_{LF} and the thickness l are also referred to the specific transmission line section. If the attenuation is neglected, then the transmission coefficient is $T_p = \frac{2Z_{L3}}{Z_{L3}+Z_{P2}^*}$, where $Z_{L3} = \frac{p_{L3}}{v_{L3}}$ and Z_{P2}^* denotes the conjugate of Z_{P2}. Nevertheless, when the attenuation is non-trivial, no analytical calculation exists and the performance was estimated as a scattering parameter of a two-port transmission line network simulated by AWR software. Due to the existence of the high-frequency element in front of the low-frequency element, the traditional quarter wavelength matching layer is always disturbed by the high-frequency element. According to wave theory, the insertion of the passive amplifier layer enhances the transmission efficiency and results in a negative insertion loss, suggesting this passive layer functions as an amplifier, and consequently, denominated as a passive amplifier for the low-frequency wave.

The AWR software was used to simulate the performance of the passive amplifier as well on the reflection coefficient and the transmission coefficient. Different characteristic impedances were assigned to the passive amplifier layer to shown the corresponding reflection coefficient. Gain of the passive amplifier was evaluated by the thickness combinations of the high-frequency active layer and the passive amplifier layer. The transmission efficiency with these multilayers is compared to that with $L_{AF} = 0$ (Fig. 5.13) because it is the case without the acoustic filter layer and the TE is almost constant (<0.005 dB variance) at high-frequency active layer $L_{HA} \leq 0.1\lambda_{LF}$ (see leftmost edge of Fig. 5.13). If $L_{HA} = 0$ and $L_{AF} = 0.25\lambda_{LF}$, then it is a traditional matching layer as shown in Point *A*. As L_{HA} increases from 0, L_{AF} with maximum gain decreases as shown on point $A \Rightarrow E$. The magnitude of the maximum gain remains almost constant up to $L_{HA} = 0.03\lambda_{LF}$ (Point *C*) and decreases as L_{HA} further increasing (Point *D* and *E*). This result suggests the feasibility of designing dual-frequency transducers with high-frequency element in front of the low-frequency element without sacrificing the transmission efficiency of the low-frequency element. At the green dash line contour marked as Type II in Fig. 5.13, $L_{AF} > 2.5L_{HA}$, also $L_{AF} > 1.25\lambda_{HF}$, at least 10 dB insertion loss could be achieved if the attenuation of the acoustic filter layer is 30 dB/cm/MHz Fig. 5.10. For the low-frequency ultrasound, however, as the frequency $f_{LF} < f_{HF}/14$, the attenuation is still sufficiently low so that the transmission efficiency is still good. This is Type II dual-frequency transducer (*A*, *B*, and *C* in Fig. 5.13), in which the acoustic filter function as an absorber for high-frequency element and passive amplifier for low-frequency element. If $f_{HF} \leq 10f_{LF}$, and the difference of loss is not sufficient for high-frequency absorption and low-frequency amplification, then the anti-matching effect could be used based on reflection from impedance mismatch. $L_{HF} \approx 0.5\lambda_{HF}$ and $L_{AF} = 0.25\lambda_{HF}$, so $L_{AF} \approx 0.5L_{HF}$, as shown on the green dash line marked as Type I in Fig. 5.13. Although amplification gain is not as good as Type II, it still reaches up to 5 dB gain at certain frequency ratio compared to that without such an acoustic filter layer. This is Type I dual-frequency transducer (*G* in Fig. 5.13) which uses the acoustic as an anti-matching layer for high-frequency element and passive amplifier for low-frequency element. For both Type I and Type II (green dash lines in Fig. 5.13), this acoustic filter resulted in positive gain for low-frequency ultrasound. The existence of the high-frequency active layer is not negligible as assumed in traditional analysis although it is very thin $L_{HF} \approx 0.02\lambda_{LF}$. This happens because impedance shifting efficiency is not uniform along the locus. If the starting point of the transmission line is almost a short circuit (leftmost point in Smith chart), $Z_{L3} \approx jZ_{AF}\tan(\beta_{LF}l)$, and $\Delta Z_{L3} \approx jZ_{AF}\tan(\beta_{LF}\Delta l)$ is a small value if $\beta_{LF}l$ is small. This effect is illustrated at the high-frequency matching layer (the gray solid line at the starting point of the HF active black solid line), where $L_{HM} = 0.024\lambda_{LF}$ with an impedance of $Z_{\text{rel}} = 0.157$. The low impedance thin

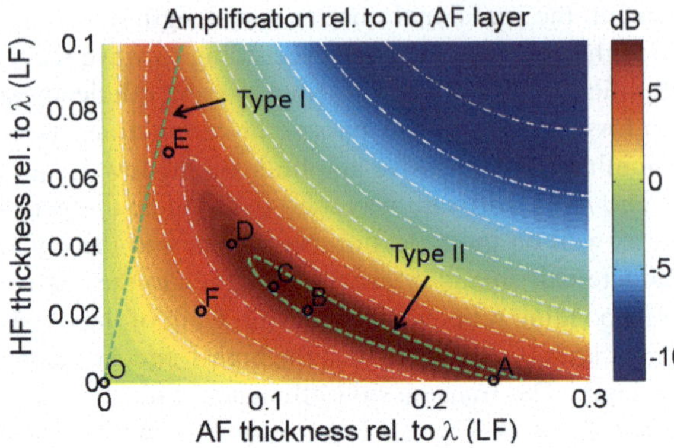

Fig. 5.13 Gain of the passive amplifier with different thicknesses of the high-frequency active layer and passive amplifier layer. Relative impedance of the passive amplifier is 0.157 (from an actual case: silver epoxy vs. PMN-PT)

layer is negligible if its load is almost short circuit (free to move with no phase). However, if the locus goes to the almost open circuit (rightmost point of Smith chart, then $Z_{L3} \approx -jZ_{AF}/\tan(\beta_{LF}l)$ and $\Delta Z_{L3} \approx jZ_{AF}\tan(\beta_{LF}\Delta l)/\tan^2(\beta_{LF}l)$, indicating ΔZ_{L3} can be large value if $\beta_{LF}l$ is small. In Loci of $L_{HF} = 0.02 \sim 0.1\lambda_{LF}$, the existence of the high-impedance, high-frequency active layer shifted Z_{L2} away from the short circuit condition so that the acoustic filter layer shifts Z_{L3} very efficiently. It is not quarter wavelength for L_{AF} to make zero phase on Z_{L3}, but is much shorter (0.03–0.125), as shown in the numbers followed the underscores in the legend of Fig. 5.14.

Performance of the passive amplifier was also validated by comparing the pressure output from transducer prototypes with and without such layer. Thicknesses of the layers for such verification are marked in Fig. 5.13 as (*A*) ideal traditional matching, (*F*) non-ideal condition with high-frequency active and acoustic layer, and ($B \Rightarrow E$) ideal condition with high-frequency piezoelectric layer and acoustic filter layer. Transmitting pressure was measured by a hydrophone (HGL-0085, Onda Corp., Sunnyvale, CA) at a distance of 1 cm away from the transducers (aperture: 4 × 4 mm, natural focus at about 9.3 mm). The results for transducer *A* and *C* are shown in Fig. 5.15 and compared with that without matching (Point *O*). Pressures from both *A* and *C* are approximately 1.7 times of that from non-matching transducers. The measured gain was weakened by the reflection from the rear side of the low-frequency element. Actual gains of the transducers (*O* and *A* through *F*) with the rear-reflection compensated were compared in Fig. 5.16. The measured results were in good agreement with the calculations.

Overall Performance of the Acoustic Filter

In order to solve wave propagation problems in the multi-layer dual-frequency ultrasound transducer, an acoustic filter design was designed and applied, sandwiching an intermediate layer between the low-frequency and the high-frequency active

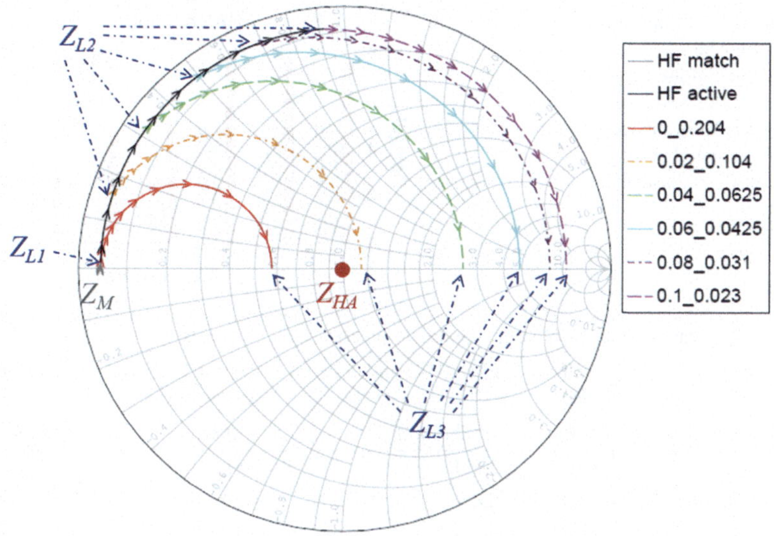

Fig. 5.14 Loci of the input impedance of the front layers (high-frequency matching, high-frequency active and the passive amplifier layers). The numbers ahead of the underscores are the thickness of the high-frequency active layer relative to low-frequency wavelength. Numbers followed the underscores are the corresponding passive amplifier thickness that make ZL3 zero phase. Relative impedance of the passive amplifier is 0.157 (from an actual case: silver epoxy vs. PMN-PT)

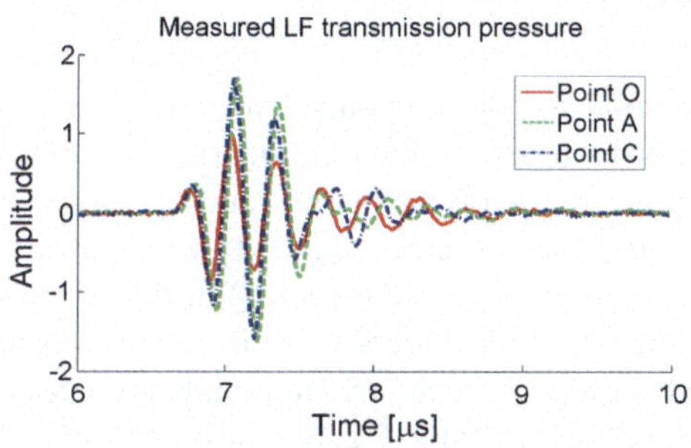

Fig. 5.15 Pressure amplitude measured from the transducer with traditional matching (transducer A, quarter wavelength matching) and transducer with high-frequency active and acoustic layer (transducer C, P1 and AF) normalized to the one without matching layers (transducer O, no matching)

elements. For the high-frequency receiving wave, the acoustic filter functions as an anti-matching layer, reflecting the majority of the high-frequency incoming energy of the wave due to impedance mismatch between the high-frequency active layer and the anti-matching layer. For the low-frequency transmitting wave, the same acoustic filter acted as a passive amplifier that enhances the low-frequency wave propagation. The anti-matching layer and passive amplifier can be achieved with the same acoustic filter layer that performs differently for different wave frequencies.

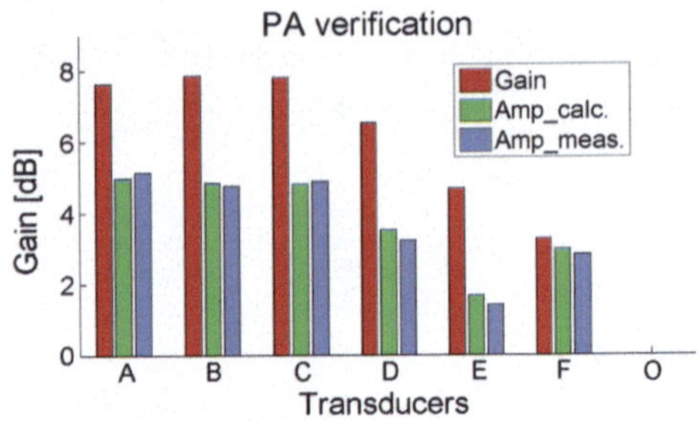

Fig. 5.16 Actual gain of the acoustic filter on transmission and the output amplitude gain considering the reflection from the rear surface (calculated and measured). PA = passive amplifier

Table 5.2 Typical parameters of some candidate piezoelectric materials

Parameters	PZT-5H	PMN-PT single crystal	PMN-PT 1-3 composite	PVDF
Dielectric ε_r	1800	4000–6000	~3000	9
d_{33} (pC/N)	640	2000	2000	−30
k_t	0.5	0.6	>0.7	0.15
Sound speed (m/s)	4400	4400	3950	1400
Impedance (MRayl)	34	35	18	2.5
Curie temperature (°C)	350	160	160	135

Materials Selection

High efficiency is essential for the transmission in the space limited intravascular environment. First, in order to excite the nonlinearities of microbubbles, the super-harmonic imaging requires relatively higher peak negative pressure (PNP, >0.6 MPa) than B-mode imaging. High-performance (d_{33}, k_t) piezoelectric materials are critical. Second, high dielectric constant material is preferable due to the small aperture of the IVUS transducers. Third, the transducers work in low-power and low-temperature conditions so that high-power piezoelectric materials are not necessary. According to the material properties listed in Table 5.2, magnesium niobate-lead titanate (PMN-PT) was selected to meet the requirements. The design of the transducer used lead PMN-PT instead of polyvinylidene fluoride (PVDF) for the receiving layer since the sensitivity could be higher due to the following (Shung et al. 2007): (1) Compared to PVDF, PMN-PT has a much higher piezoelectric and electromechanical coupling coefficient and (2) the dielectric constant of PMN-PT is 400–700 times higher than that of PVDF, making it easier to match the electrical impedance to the receiving system. Material properties and final optimized parameters for modeling and fabrication are summarized in Table 5.3.

Table 5.3 Fabrication parameters of the dual-frequency transducer

Parameters	6.5 MHz layer	30 MHz layer
Active material	PMN-PT	PMN-PT
Thickness (μm)	300	65
Width (mm)	0.6	0.6
Length (mm)	3	0.5
Sound speed (m/s)	4400	4400
Impedance (MRayl)	35	35
Matching material	Al_2O_3/epoxy	Parylene
Thickness (μm)	80	15
Sound speed (m/s)	2800	2770
Impedance (MRayl)	5.5	3.16
Attenuation (dB/cm/MHz)	4.3	0.1
Backing material	Ag/epoxy	Ag/epoxy (anti-matching)
Thickness (μm)	200	15
Sound speed (m/s)	1900	1900
Impedance (MRayl)	5.15	5.15
Attenuation (dB/cm/MHz)	8	8

Fabrication

In the fabrication of the dual-frequency transducer, a 5 × 5 mm piezoelectric acoustic stack was first assembled and then diced into 0.6 mm wide slices as individual transducers. The assembly process started with a 5 × 5 mm PMN-PT plate which was lapped to 300 μm ($f = 6.5$ MHz) in thickness and then coated with Ti/Au (Ti: 10 nm and Au: 100 nm, E-Beam, Jefferson Hills, Pennsylvania, USA) on both surfaces (Fig. 5.17(1)). The second piece of PMN-PT (0.5 × 5 × 0.3 mm) was then bonded on the first PMN-PT layer using conductive silver epoxy to form the high-frequency receiving element and anti-matching layer (Ma et al. 2014c, in press). Polystyrene microspheres (Polysciences Inc., Warrington, Pennsylvania, USA) having a nominal diameter of 10 μm was added (about 1% in volume) to the silver epoxy so that the thickness of bonding layer was controlled to be 13–15 μm in order to function as a anti-matching layer previously mentioned (Fig. 5.17(2)). After the silver epoxy cured, a composite layer of Al_2O_3 powder (1 μm grain size, Logitech Limited, Glasgow, UK) and Epo-tek 301 (Epoxy Technology Inc., Billerica, Massachusetts, USA) was mixed 1:1 by weight and centrifuged at 10,000 RPM (5590 g) for 10 min (Microfuge Lite, Beckman Coulter Inc., Brea, California, USA). The prepared composite was then cast onto the front of the 5 × 5 mm PMN-PT layer beside the 0.5 mm width slice. A small margin (0.5 mm) at one edge was left as an electrical connection site

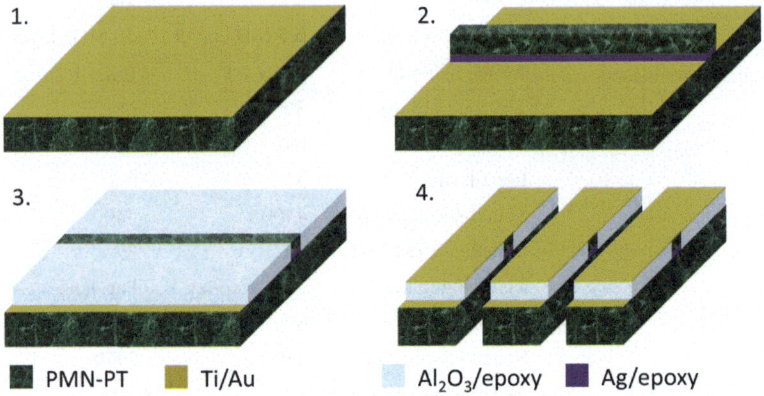

Fig. 5.17 Fabrication process diagram of the dual-frequency transducer

for later wiring. After the Al_2O_3/epoxy composite cured, it was then lapped until the bonded composite layer was 80 μm thick, making the thickness of the top PMN-PT layer 65 μm ($f = 30$ MHz) on top of the 15 μm anti-matching layer (Fig. 5.17(3)). Another Ti/Au layer (Ti: 10 nm and Au: 100 nm) was then deposited onto the top surface to form the top electrodes. The final stack was then diced into 0.6 mm wide slices to form individual dual-frequency transducers (Fig. 5.17(4)). Each slice was bonded to the tip of a 20-gauge hypodermic needle (Fisher Scientific International Inc., Hampton, New Hampshire, USA) and then coated with a layer of parylene film (15 μm). The parylene film served two purposes: (1) to act as a matching layer of the 30 MHz, high-frequency element and (2) to provide electrical isolation in the form of a passivation layer for the entire transducer. Finally, the transducer was poled with a DC electrical field of 10 kV/cm for 15 min in silicone oil at room temperature.

After the transducers were polarized, electrical characteristics such as capacitance, loss, and input electrical impedance were characterized using an Agilent 4294A Precision Impedance Analyzer (Agilent Technologies, Inc., Santa Clara, CA, USA). Capacitance and loss were measured at 1 kHz and the input electrical impedance was measured near the resonant frequency of each element, individually.

Acoustic Characterization

The main design consideration for the low-frequency (6.5 MHz) transmission element was peak negative pressure (PNP) and the pulse length. Large PNP (e.g., >1 MPa) is desirable to produce detectable nonlinear oscillations of microbubble contrast agents. Less pulse length with a fewer number of negative peaks leads to higher resolution because each high-pressure negative induces nonlinearities in the bubble, generating a super-harmonic signal and decreasing the axial resolution. The pressure output of the small aperture transducer was measured using a calibrated needle hydrophone (HNA-0400, Onda Co., Sunnyvale, California, USA) positioned axially at 3 mm away from the transducer. This distance was kept constant (3 mm) between pressure measurements and subsequent microbubble tests in order to estimate pressure levels applied to the contrast agents. The excitation pulse used was a sinusoidal burst (1–5 cycles) at 6.5 MHz generated by the arbitrary function generator (AFG3101, Tektronix Inc., Beaverton, Oregon, USA) and amplified by 55 dB with a radio frequency amplifier (Model 3200L, Electronic Navigation Industries Inc., Rochester, New York, USA). The amplitude of the transmission signal was adjusted from 50 to 350 mV (peak to peak) prior to amplification. Pressure output of the transducer was recorded using an in-house LabVIEW (National Instruments Co., Austin, Texas, USA) data acquisition system. Pulse length was defined as the −6 dB amplitude relative to the largest negative peak pressure, because pressure with amplitude less than −6 dB was considered too low to generate super-harmonic signals on microbubbles.

Sensitivity and pulse length are important characteristics of the receiving element for high CTR and high-resolution imaging. Receiving elements were characterized using the pulse-echo method. The transducer was excited by a customized pulser/receiver system (Li et al. 2014) using a 20 V 1-cycle impulse. A steel block was placed in front of the transducer as the reflection target. Envelope of the echo was calculated as the absolute value of the Hilbert transformation on the time domain signal. Sensitivity of the transducer was defined as the amplitude of the envelope divided by the 20 V amplitude input. Axial resolution of the transducer was also calculated from the envelope with amplitude higher than a threshold (−6 dB or −20 dB).

Imaging Process

Microbubble Response Test

To validate the contrast response of the transducer, microbubbles were excited by the low-frequency element and the nonlinear responses from microbubbles were detected using the high-frequency element. Relative positions of the transducer and the tube were carefully adjusted in a water bath using a three-axis precision rectilinear stage. In the alignment process, an acoustically transparent 200 μm diameter microtube was filled with air to provide a strong echo to indicate alignment in the lateral dimension of the 30 MHz element. Time of arrival of the echo was used to calculate the distance between the transducer and the tube in order to position it axially 3 mm away from the transducer. Poly-disperse lipid shelled microbubbles (Fig. 5.18) were formulated as described previously (Borden et al. 2006) and pumped through the aligned microtube at a concentration of 4.8×10^8 MBs/mL at a velocity between 1.8 and 4.4 cm/s to maximize the signal response. The tube was slightly angled (~10°) with respect to

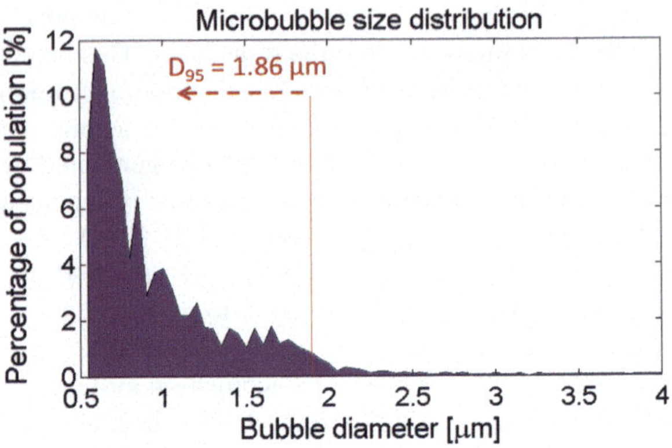

Fig. 5.18 Microbubble size distribution. 95% of the microbubbles are within 1.86 μm

the front surface of the transducer to reduce specular reflections from the wall of the tube.

Once the transducer was aligned and positioned for the contrast imaging, the excitation pulse sequence was adjusted for optimal imaging quality. A sinusoidal burst excited the transmitting element with pulse lengths between 1 to 5 cycles and voltage variations from 50 to 350 mV_{pp} with 50 mV_{pp} increments. For each combination of testing parameters, 100 A-lines of microbubble echoes were received and recorded by the high-frequency element for offline analysis. The contrast imaging data was evaluated using both time domain amplitude analysis and short-time Fourier transform.

Fundamental Frequency Imaging

Fundamental imaging at 30 MHz and dual-frequency super-harmonic contrast imaging were tested with the transducer in tissue-mimicking phantoms immersed in water (Fig. 5.19). Typical phantoms had a speed of sound similar to tissue (1496 m/s), relatively high attenuation (0.9 dB/cm at 3 MHz), and had fully developed speckle. A hole was drilled through the phantom by a thin wall steel tube (5.5 mm OD, 0.4 mm wall thickness) to simulate a vessel. After drilling, the dual-frequency probe was placed in the center of the lumen and rotated to scan for an IVUS image. The rotation was controlled using a microcontroller that stepped the transducer at 0.9° angular increments for one revolution and provided a trigger for the excitation pulse. An acoustically transparent microtube with 200 μm diameter was placed through the phantom, running parallel to the channel to mimic the vasa vasorum. Diluted microbubbles were pumped through the tube (4.4 cm/s) while imaging. A 0.6 mm diameter steel rod was placed in the phantom opposite of the microbubble tube to provide a strong reflection target for comparison. Both the microbubble tube and the

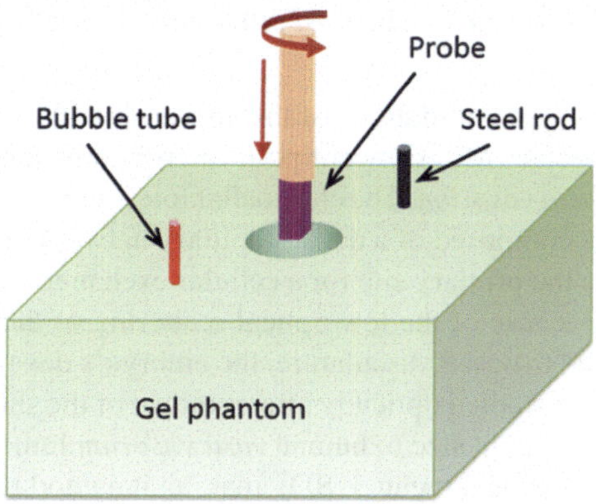

Fig. 5.19 Experimental setup for the ex vivo imaging in a gel phantom

steel rod were embedded at ~4 mm radially in the phantom to test if the system could detect contrast in a scattering and attenuating medium, which is typical of tissue.

The 30 MHz high-frequency element was used for the fundamental imaging. In other words, both acoustic excitation and signal reception were performed using the 30 MHz element. A pulser/receiver supplied the transducer with a 1 μJ excitation pulse at every step and the reflected signal was recorded by the LabVIEW data acquisition system described previously.

Super-Harmonic Imaging

In super-harmonic imaging, the setup is similar to that for fundamental imaging except that the existence of the steel rod. For super-harmonic imaging, a 2-cycle sinusoidal burst generated by AFG3101 was amplified by the 3200L RF amplifier to excite the 6.5 MHz transducer while the 30 MHz element was used for receiving. A synchronized three-dimensional (3D) motion stage and a stepper motor (400 steps/rev) produced the mechanical scan for 3D imaging. The 6.5 MHz element was excited by 1-cycle or 2-cycle burst using the imaging system, and the 30 MHz element received the high-frequency super-harmonic signals. The signal was digitally band-pass filtered with a frequency window of 25–35 MHz corresponding to the receive element's bandwidth. Envelope of the wave was defined as the absolute value of Hilbert transform from the filtered signal. Contrast-to-tissue ratio (CRT) was defined as the amplitude of the envelope for contrast signal to the amplitude of the noise.

Chorioallantoic Membrane Imaging In Vivo

In vivo, validations of the approach with selective imaging of microbubble contrast agents (MCAs) were performed in developing chicken embryos as a surrogate for *vasa vasorum*. The chorioallantoic membrane (CAM) of developing chicken embryos is composed of a dense capillary network cradled in a sheet of connective tissue and is the primary site for a cellular exchange of respiratory gases and metabolic wastes. Because of the low optical scattering of the albumen and the direct visualization of the exposed vasculature, the embryo's developing circulatory system can be imaged and studied optically after removal of the shell. Vessel diameters within the CAM are similar in size to human *vasa vasorum* lumens reported previously (43.4 ± 47.4 μm diameters, mean ± SD), making it a good model for evaluating the sensitivity of the imaging system to detect small vasculature (Sluimer and Daemen 2009).

Fertilized chicken eggs (broiler line, Ross 708) were collected from a local poultry farm (North Carolina State Chicken Educational Unit, Raleigh, NC, USA) and refrigerated at 6 °C on arrival for 3–7 d until incubation. Eggs were first incubated in vivo at 37.5 °C with 70% relative humidity for 3 d, being turned every 4 h with an automated egg rocker (Model 4200/3200, Farm Innovators, Plymouth, IN, USA). Eggs were then cracked and explanted into disposable holders, as described by Schomann et al. (2013), and incubated for 14 d in a humidified incubator at 37.5 °C, 70% humidity, and 2.0% CO_2 (NAPCO 8000 Series, Thermo Scientific, Waltham, MA, USA). Chicken embryo morphology was classified at the time of imaging according to the Hamburger and Hamilton criteria, with the majority of embryos being classified as HH39–40. The vitelline vein was cannulated to allow the injection of MCAs at a concentration of 1010 MCAs/mL and a flow rate of 1.0 mL/h using a syringe pump. A continuous infusion of contrast agents was administered during the entire imaging session. The CAM was prepared for imaging by coupling the transducer to the structure with 37 °C PBS.

The intravascular ultrasound probe was positioned adjacent to the CAM, and a volumetric acquisition was performed 2 min after starting the flow of contrast (Fig. 5.20). Image slices orthogonal to the direction of transducer pullback were acquired at 200 μm intervals while operating in super-harmonic mode for contrast-specific imaging. Conventional B-mode and super-harmonic mode volumetric scans were acquired both before and after administration of MCAs for comparison. Additionally, contrast infusion was monitored by acquiring volume scans in super-harmonic mode at 2 min intervals for a total of 12 min. Photographs of the CAM corresponding to the region that was scanned were taken to provide an optical reference to measure vessel diameters; 11 embryos were imaged. Photographs were analyzed in ImageJ software to measure the width of the vessels within the imaging region.

Signals were acquired from the dual-frequency probe using a custom imaging system capable of volumetric acquisitions. A programmable microcontroller was used to mechanically rotate the probe using a stepper motor having 400 discrete angular positions per revolution (angular step size, $\Delta\theta = 0.9°$), and images were acquired at a pulse repetition rate of 100 Hz. The motor and transducer assembly was mounted

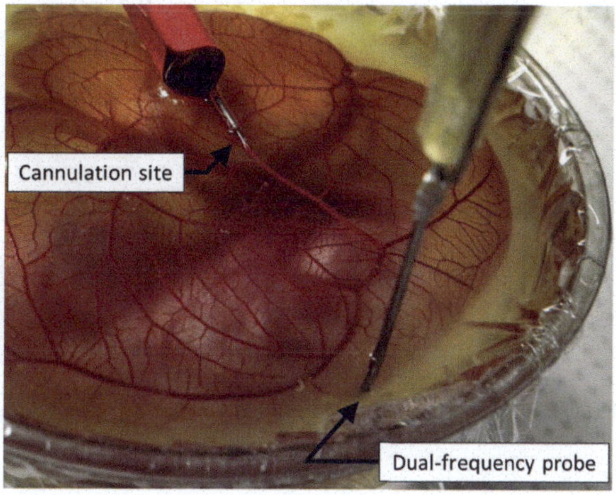

Fig. 5.20 Positioning used to acquire images from the chorioallantoic membrane

to a three-axis computer-controlled motion stage (Newport XPS, Irvine, CA, USA), which controlled transducer pullback to collect images of the entire volume for 3D rendering. The transducer was operated in either super-harmonic mode for contrast detection or B-mode for conventional pulse-echo IVUS imaging. In super-harmonic mode, the low-frequency element was excited using a 5.5 MHz, 50% bandwidth Gaussian enveloped pulse from an arbitrary function generator (AFG3101, Tektronix, Beaverton, OR, USA). The pulse was amplified to 275 Vpp using a 60 dB radio-frequency amplifier (A-500, Electronic Navigation Industries, Rochester, NY, USA) to generate 1.2 MPa of peak rarefractional pressure at a depth of 2 mm in water, which was measured using a needle hydrophone (HNA-0400, Onda, Sunnyvale, CA, USA). The transmission pressure was selected to produce detectable nonlinear responses above the fourth harmonic and being received by the high-frequency element.

Preliminary Results

Prototype and Housing

The dual-frequency transducer prototypes were housed on the tip of a 20 gauge hypodermic needles (Fig. 5.21a) for demonstration. The back surface (Side B in Fig. 5.1e) of the transducer was bonded to the needle with conductive epoxy, allowing the needle to function as an electrode leading to the back side of the 6.5 MHz transducer. Coaxial wires (25-gauge) were attached to the top electrode and common electrode between the two PMN-PT layers and then threaded through the needle.

Eventually, the transducers would be housed inside catheters used for IVUS imaging. The viability was demonstrated by housing the transducer into a commercial

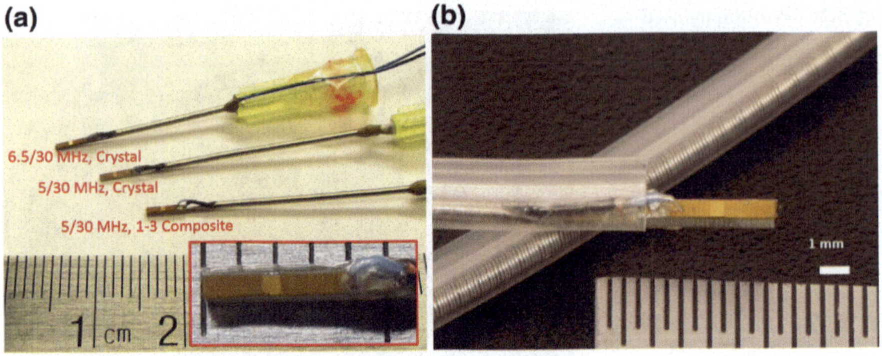

Fig. 5.21 Prototype transducers **a** housed on the tip of a 20-gauge hypodermic needle or **b** integrated in a commercial catheter

catheter (Fig. 5.21b) from Boston Scientific (Natick, Massachusetts, USA). Obviously, the transducers are small enough to fit inside the IVUS catheters.

Electrical Characterization

The capacitance and loss at 1 kHz (measured with an Agilent 4294A Precision Impedance Analyzer) showed good agreement to predicted values. Capacitance values of the transducers were 344 pF for the 6.5 MHz transmission element and 131 pF (including a 2 pF parasitic capacitance from Al_2O_3/epoxy layer) for the 30 MHz receiving element. These data were in agreement with theoretical calculations using a relative dielectric constant of 4000 (HC Materials, Inc.). Loss of the transducer was 1.1% for transmission and 2.7% for reception, similar to the properties of PMN-32%PT.

The measured input electrical impedance at the resonant frequency (Fig. 5.22b, d) matched well with the results of the KLM modeling (Fig. 5.22a, c). In order to obtain a strong resonance for high pressure output in the transmission element, the backing layer of the transmission element was designed with reasonable absorption for compromise of achieving sufficient pressure output while minimizing ringing. In both the modeling and measured results, a strong resonance occurred at 6.5 MHz (Fig. 5.22a, b). The center frequency of the receiving element was designed to be 30 MHz (Fig. 5.22c, d) and the measured results agreed very well with simulations.

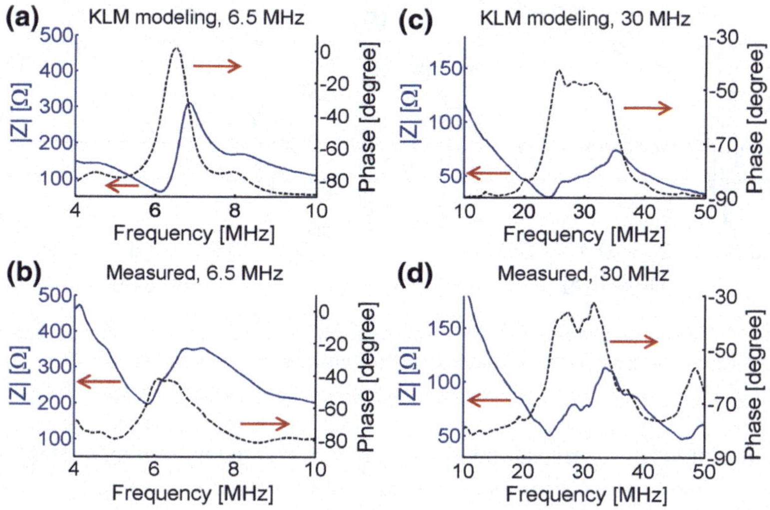

Fig. 5.22 Electrical impedance of the **a**, **b** 6.5 MHz element and **c**, **d** 30 MHz from **a**, **c** KLM modeling, and **b**, **d** measurement from impedance analyzer

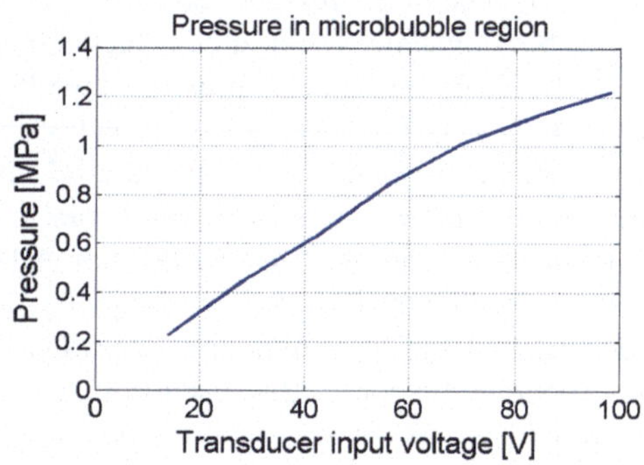

Fig. 5.23 Peak rarefractional pressure at the contrast imaging area, 3 mm away from the center of the transducer, using a 5-cycle burst excitation on the low-frequency (6.5 MHz) element at varied voltages

Acoustic Characterization

The peak negative pressure of the 6.5 MHz element was recorded with different excitation voltages. Measurements were recorded at the contrast imaging area in order to verify that this would be the pressure applied in the region of the microbubbles. As shown in Fig. 5.23, the response of the low-frequency transducer was nearly linear at excitation voltages lower than 70 V, having an average transmitting sensitivity of 14.5 kPa/V. Nonlinearities showed up when the input was higher than 70 V. At about 100 V, more than 1.2 MPa rarefractional pressure (MI: 0.48) was generated, which was sufficient to produce a high-frequency, broadband response from microbubbles imaged in tissue in prior studies (Gessner et al. 2012, 2013a).

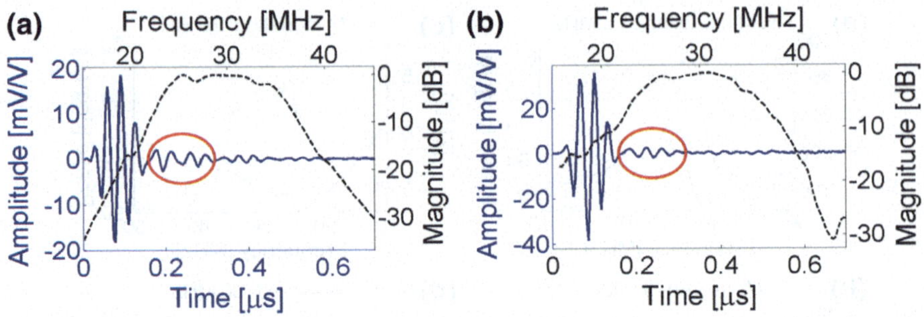

Fig. 5.24 Pulse-echo response and its FFT spectrum of 30 MHz element in **a** modeling and **b** measurement. The aliasing echoes are marked by the red ellipses

Beam profiles of the transmitter were scanned with the hydrophone (HNA-0400, Onda Co., Sunnyvale, California, USA) and motion stage controlled by LabVIEW (National Instruments Co., Austin, Texas, USA). With the stacked layers design with different transmitter/receiver apertures, the output pressure (Fig. 5.2b) at the center of the beam ($x = -1$, $y = 0$) was almost as high as that at the side ($x = -2$ to $x = 1$, $y = 0$). The high pressure beam covers the center region of the transducer so that super-harmonic signal could be detected by the receiver.

Pulse-echo experiments illustrated the broad bandwidth of the receiving element (Fig. 5.24b). The pulse length of the echo signal was about 119 ns (at −20 dB for the envelope), corresponding to a spatial axial length of 89 μm in tissue, which means that the high-frequency transducer can be used in pulse-echo mode for high-resolution fundamental imaging. The −6 dB fractional bandwidth of the receiving element was measured to be 46%, covering a frequency span of 22.9–36.6 MHz, providing good reception of the high-frequency, broadband (fourth- to seventh-order harmonics) microbubble response. Because of the anti-matching layer, the high-frequency element behaves as if there was little backing for it. As such, the bandwidth predicted by the modeling result was broad (Fig. 5.24a) and was validated in the measured bandwidth (Fig. 5.24b). According the measurement result, the loop sensitivity of the high-frequency receiving element is −27 dB.

The transmit frequency spectrum was measured by hydrophone using a 2-cycle burst excitation, and the receive bandwidth was measured from the pulse-echo experiment driven with a 1 μJ impulse. The −20 dB frequency response of the transmit element was 4.0–8.9 MHz. The frequency response of the transmitting and receiving elements were well separated (5.6 MHz stop band at −20 dB) (Fig. 5.25), which is ideal for detecting microbubble broadband frequency content and achieving high contrast-to-tissue ratios.

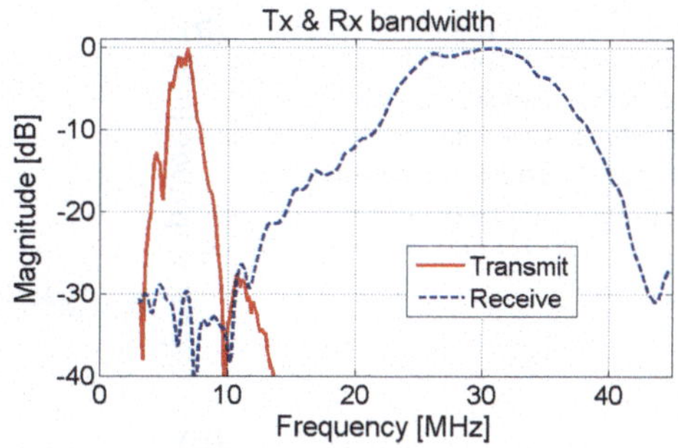

Fig. 5.25 Transmission (2-cycle burst) and receiving (pulse-echo) bandwidth separation

Microbubble Response

Microbubble response was clearly detected with the dual-frequency transducer. In the time domain amplitude analysis, a root-mean-square (RMS) value through the 100 lines of data was taken as the amplitude of the microbubble response. The received data was first high-pass filtered at 10 MHz to thoroughly remove residual tissue/phantom responses at the fundamental or low harmonic frequencies. Residual stationary signals from the tube wall were filtered using a clutter filter, leaving only the transient high-frequency signal from microbubbles. RMS values of the microbubble response with different excitations were shown in Fig. 5.26. As shown in the figure, the background noise was about 0.5 mV, the source of which primarily comes from the noise of the amplifier. With 1-cycle burst excitation, the peak negative pressure was low (PNP < 0.65 MPa) at all voltage inputs (14–98 V), resulting in low (6 dB) SNR (Fig. 5.26a). With 2 or more cycles in each burst, microbubble response could be more than 1.5 mV at 70 V excitation (PNP ≥ 0.8 MPa), and the SNR was larger than 10 dB (Fig. 5.26b, c). Because of the small Q-factor of the receiving element, each negative peak of the transmission wave was clearly discernible temporally (Fig. 5.26c). At 2 or more cycles in each burst, 42 V input excitation was high enough to excite the nonlinear microbubble response (however, at a low SNR of 4 dB), corresponding to a rarefractional pressure of 0.65 MPa. The microbubble response was just slightly higher at 98 V input compared to 70 V input, and we hypothesize that increasing the pressure too high may decrease the CTR due to relatively high nonlinear response in tissue. No signal was detected at low-level voltage excitation (14 V) and the response was always within the noise. In summary, the SNR increased rapidly with increasing driving voltage and then leveled off (e.g. <70 V). If the input voltage was higher than 70 V, the signal increase of the microbubble response was less significant. Gessner et al. also observed a similar plateau for nonlinear microbubble response as a function of pressure when excited at 2 MHz (Gessner et al. 2013b). The excitation voltage may be further optimized for in vivo conditions.

The spectra of the received signals were analyzed to investigate the microbubble response in frequency space. Spectra of all the 100 measurements were calculated using the short-time Fourier transform and the spectrum for a given test condition was

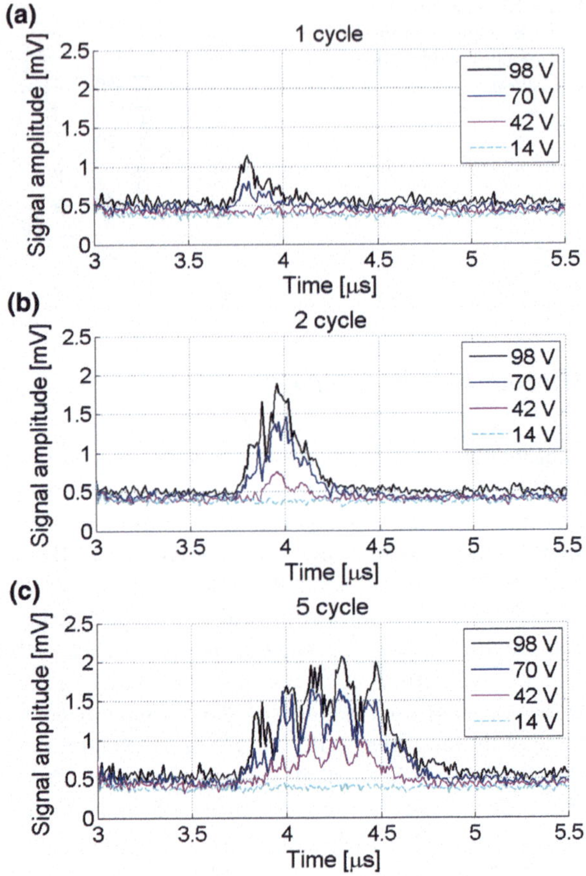

Fig. 5.26 Amplitude of nonlinear microbubble response with sinusoidal bursts of **a** 1-cycle, **b** 2-cycle, and **c** 5-cycles at different voltage excitations

analyzed. As shown in Fig. 5.27, frequencies lower than 20 MHz were intrinsically suppressed by the transducer's receiving sensitivity so low-frequency harmonics of the fundamental do not greatly contribute to the image formation in dual-frequency imaging. Input voltage was kept at 70 V while number of cycles in the burst varied (Fig. 5.27). Fewer number of cycles produced higher axial resolution, however, a 1-cycle burst suffered from low SNR because the peak negative pressure was too low to induce sufficient microbubble harmonics. With a 2-cycle burst at 70 V input, microbubble response was detected while minimizing loss of axial resolution (about 200 μm). The frequency content of the noise was less than −12 dB relative to the peak frequency content of the microbubble signal (Fig. 5.27b).

The presence of the microbubble signal was verified by rinsing the tube with pure water. The high-frequency response observed when imaging microbubbles vanished instantly when the microbubbles were cleared (Fig. 5.27f).

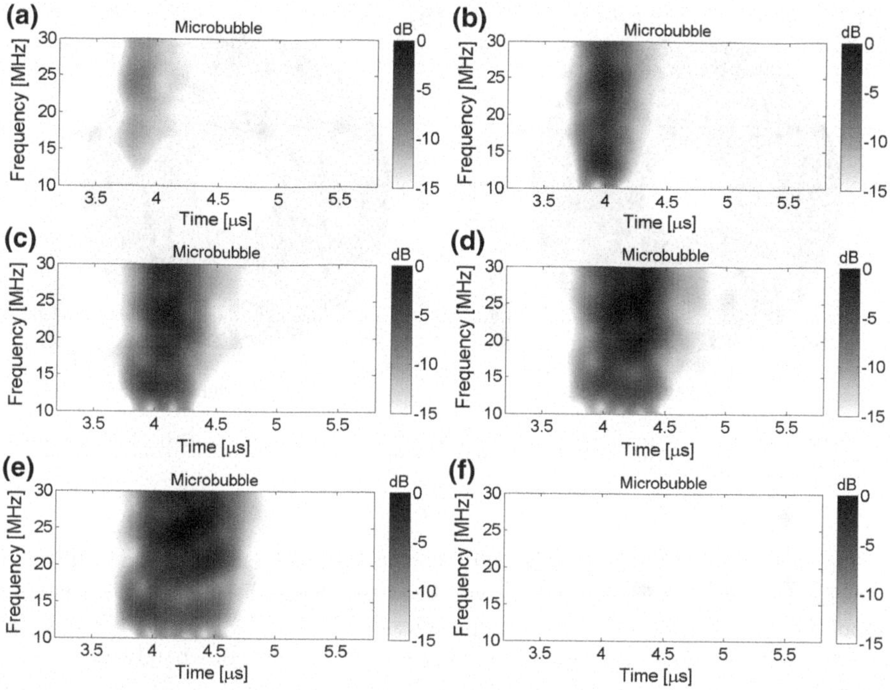

Fig. 5.27 Spectrum of microbubble responses at 70 V with a burst of **a** 1-cycle, **b** 2-cycles, **c** 3-cycles, **d** 4-cycles, and **e** 5-cycles. The spectrum of a water filled tube at 70 V with a burst of **f** 5-cycles shown for comparison

Fundamental Imaging

Fundamental imaging mode was performed using the 30 MHz element in pulse-echo mode. An impulse with 1 μJ energy was used to produce acoustic waves and the reflected signal was recorded by a LabVIEW data acquisition system at 100 MHz sampling rate. The envelope was detected and the magnitude was scan converted to polar system as shown in Fig. 5.28a. In fundamental imaging mode, the reflection from the 0.6 mm steel rod was nearly 20 dB stronger than that from the 200 μm tube containing microbubbles.

Super-Harmonic Imaging

In the super-harmonic imaging mode, microbubbles were excited using the low-frequency element and backscatter was received using the 30 MHz element. The transmission input was a 6.5 MHz 2-cycle sinusoidal burst at 84 V. The received signal was high-pass filtered at 10 MHz before being digitized.

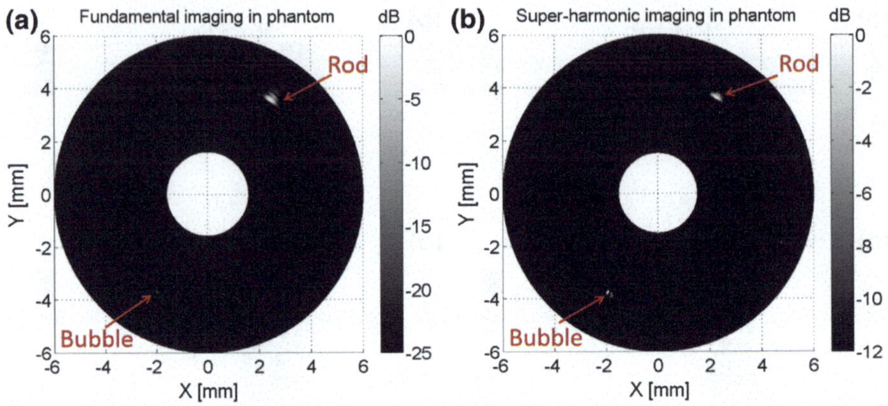

Fig. 5.28 **a** Fundamental imaging at 30 MHz. **b** Super-harmonic imaging with 6.5 MHz transmission and 30 MHz reception

The data were filtered digitally and reconstructed to create an image offline in MATLAB. First, each line in the scan was band-pass filtered with a 60th order finite impulse response filter with corner frequencies of 25 and 35 MHz, which is also the sensitivity range of the transducer. The frequency content in this bandwidth covers the majority of the fourth and fifth harmonics of the transmission frequency. After filtering, the wave package was detected from the signal and the magnitude was scan converted to polar coordinate as shown in Fig. 5.28b.

The microbubble filled tube was clearly detected with a CTR of 12 dB. There was some nonlinear response on the rod signified by its presence in the image even after filtering, but the signal magnitude was strongly decreased in comparison to the fundamental mode image. In dual-frequency imaging, the microbubble response was almost the same magnitude as the reflection from the steel rod. Both axial and lateral resolutions were high enough to clearly resolve the 200 μm microbubble tube. The diameter of the steel rod can also be clearly visualized to be 0.6 mm in the lateral direction but not easily seen in the axial direction. Because the sound speed in steel is about four times of that in water, the reflection from the back surface of the steel rod seems very close to the reflection from the front surface, introducing a high distortion in the axial resolution on the rod.

In Vivo *Contrast Detection of Microvascular Flow*

Translation of this technique into the in vivo environment was performed after establishing feasibility from in vitro experiments. Photographs were taken of the chicken embryo to provide optical verification of vascular structures after embryos were cannulated for vascular perfusion of contrast agents (Fig. 5.29a). Three-dimensional renderings of vessel networks were visualized by performing maximum intensity

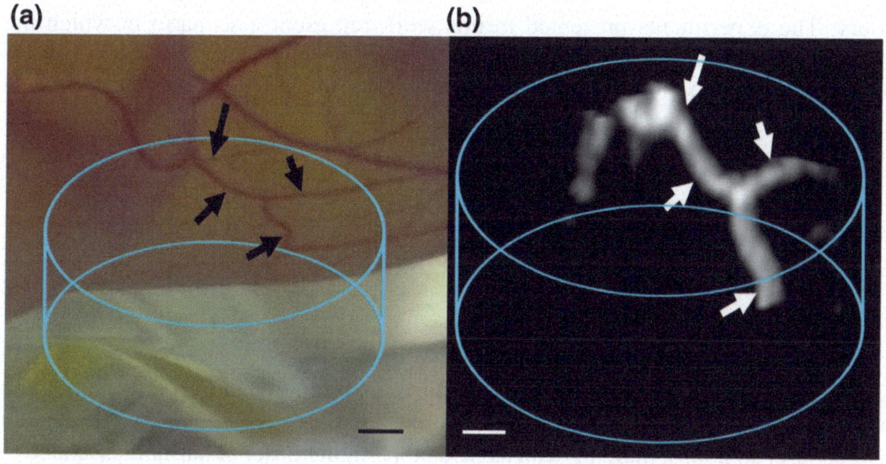

Fig. 5.29 **a** Chorioallantoic membrane at 14 d, revealing the microvasculature. The intravascular ultrasound transducer was positioned relative to the vessels of interest (arrows), where a 4 mm pullback was performed (cylindrical outline). **b** Three-dimensional contrast-specific rendering of the same region photographed using the dual-frequency intravascular ultrasound transducer. Fiduciary marks (arrows) are used to aid the reader in co-registration of the optical and ultrasound images. Bars = 1 mm

projections on volumetric data sets (Fig. 5.29b). Contrast flow was monitored over time as a continuous infusion of contrast agents was administered while acquiring super-harmonic volumetric pullbacks.

During the MCAs injection, super-harmonic mode images started with a statistically zero CNR ($p = 0.28$) and improved to a maximum value of 3.49 dB after 6 min of infusion. Both axial and lateral resolutions were computed in optical and super-harmonic contrast images of the CAM vasculature. Optical measurements of vessels were taken such that they would correspond to lateral measurements in super-harmonic images, and it was assumed that the diameter of the vessel in this plane would also be the diameter in the axial direction. The smallest vessel detected in vivo was measured to have a diameter of 140 ± 8.1 μm (mean ± SE) optically, whereas the super-harmonic image measured 626.5 ± 16.9 μm axially and 603.2 ± 22.6 μm laterally. The relatively lower resolution was achieved because the long burst excitation of the low-frequency ultrasonic wave in this case (50% bandwidth Gaussian enveloped). A higher resolution is achievable with further optimizations, which will be discussed in the next section for the details.

At frequencies relevant for IVUS imaging, blood scattering effects become significant and attenuation becomes more severe (Lockwood et al. 1991). Additionally, the presence of MCAs in the lumen of the vessel where the IVUS catheter is placed results in attenuation and nonlinear propagation effects, which can produce pseudo-enhancement artifacts, or contrast signal where no MCAs are present (Tang and Eckersley 2006; Ten Kate et al. 2012). This artifact may mask the presence of true contrast in IVUS if vessels supplying the lesion lie close to the lumen of the parent artery. The experiments presented in this work represent a scenario in which the majority of microbubbles have been cleared by either temporary balloon occlusion or a saline flush to clear blood flow in the main artery while collecting an image (Jang et al. 2002).

Performance Optimization

Optimization Approaches

Low-frequency transmitter is subjected to the main limitation of the performance and has more rooms for performance optimization. First, as shown in previous research (Bouakaz et al. 2003), imaging with a reasonably higher order of harmonics tended to generate higher CTR. However, the receiving frequency is the major limitation of the penetration depth, which could not be increased without sacrificing the penetration depth. The way to increase the CTR would be realized by decreasing the transmitting frequency (Ma et al. 2014a). Second, for this dual-frequency super-harmonic imaging transducer, the wavelength of the low-frequency wave is several (4–6) times longer than the high-frequency wave. As a result, decreasing the pulse length of the low-frequency element is much more efficient to increase the axial resolution. Based on this consideration, the first step of performance optimization was focused on the transmitting frequency and the pulse length of the low-frequency transmitters. Of cause, the high-frequency sensitivity and pulse length are also important features of the dual-frequency transducers. These minor optimizations are left for future work on this project.

First, the frequency of the transmitter was decreased from 6.5 to 5 MHz. Because the transmitter was much larger than the receiver, the intravascular environment dominantly limited the transmitter. As a result, the aperture size of the transducer remained constant as when decreasing the transmitter frequency. In this case, the width-to-thickness ratio would be less than 2 in the 5 MHz transmitter, so that the lateral mode would be coupled in the thickness mode vibration, leading to a decreased pressure output. The lateral mode vibration could be suppressed by microstructures like 1-3 composites. So in this frequency optimization, constant pressure was used to compare the CTR between 5 and 6.5 MHz transmitters.

Second, pulse length was decreased by replacing the PMN-PT single crystal by PMN-PT 1-3 composite. As an imaging transducer, a short pulse length is preferable, which determines the axial imaging resolution. PMN-PT 1-3 composite possesses high coupling factor (k_t is close to $k_{33} \sim 0.9$) and relatively low acoustic impedance (~18 MRayl), which is preferred for a short impulse response and high-resolution imaging. However, the pressure output efficiency of PMN-PT 1-3 composite is affected by two controversial effects: the ultrasonic wave generation and propagation efficiencies. Low fraction (~50%) of active material leads to low wave generation efficiency while low impedance of the composite materials causes high

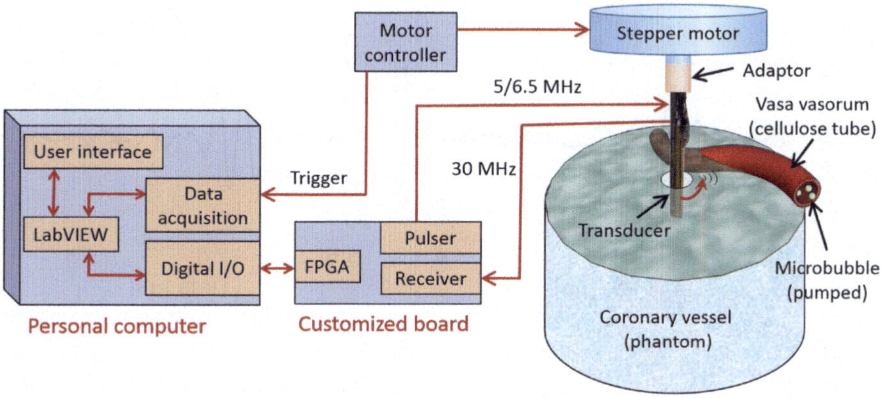

Fig. 5.30 Setup of vasa vasorum imaging in vitro. Phantom was used to mimic the coronary vessel wall and an acoustically transparent microtube was used to mimic the vasa vasorum

wave propagation efficiency. As stated above, the lateral mode could be suppressed in the 1-3 composite. The resonant frequency of these 1-3 composite transmitters was 5 MHz.

Fabrication and characterization process of these 5 MHz transmitter transducers were very similar to that of the 6.5 MHz transmitter transducers. These transducers were characterized to obtain the transmitting pressure output and the contrast imaging result. In the imaging, the setup (Fig. 5.30) was similar to that in the preliminary experiment, except that the steel rod was not included because the receiver was designed to be identical and it was not necessary to compare the B-mode imaging.

Transmission Output

Transmission characteristics of the transducers were verified by hydrophone with 1- or 2-cycle bursts excitations. Pressure measurement results were normalized to unit voltage (1 V) applied on the samples, which are shown in Fig. 5.31.

Normalized pressure value of each transducer in each test condition is marked by the dash horizontal line in Fig. 5.31. Because the width-to-thickness ratio (W/T) of the 5 MHz single crystal material is lower than 2 (0.6 mm over 0.39 mm), it vibrated in a coupled mode and hence the pressure output is relatively low. The 6.5 MHz transducer, with W/T of 2 (0.6 mm over 0.3 mm), vibrated mainly in the thickness mode (Lerch 1990) and generated the highest pressure in these three groups. The 5 MHz 1-3 composite transducer is also mainly in the thickness mode because the lateral mode is highly suppressed by the periodic structure within this frequency range. However, the pressure of the 5 MHz 1-3 composite transducer was still relatively low compared to the 6.5 MHz single crystal transducer, mainly due

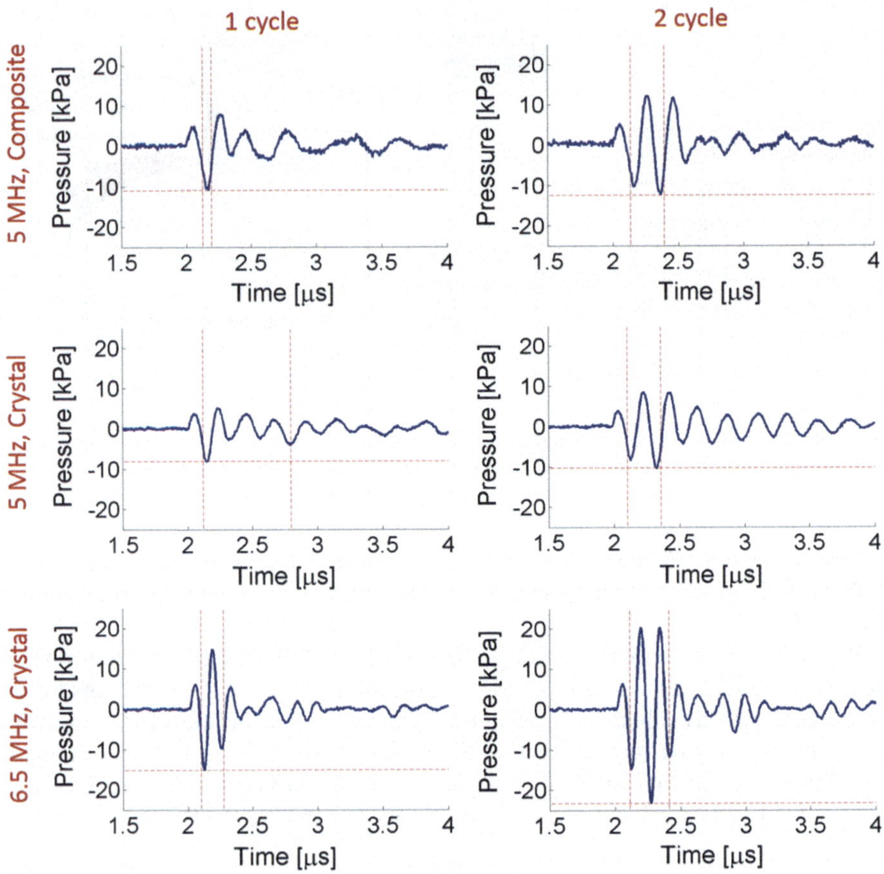

Fig. 5.31 Transmission sensitivity of the prototype transducers with the transducers shown at the left and the excitations shown at the top. The horizontal red dash lines indicate the peak negative pressure and the vertical ones indicate the pulse length at −6 dB. All the values are normalized to unit voltage as kPa/V

to the low volume ratio of active material. As a result, the pressure output of the 6.5 MHz single crystal transducers is the highest among the groups.

Relatively high acoustic pressures induced detectable super-harmonics from microbubbles while low pressures did not. As a result, the pulse length (marked as a vertical dashed line in Fig. 5.31) was defined as a real-time signal with −6 dB negative pressure value. Because of mode coupling, very long pulse showed up from the 5 MHz single crystal transducers when driven by broadband, 1-cycle excitation. Such low-frequency components were insignificant when excited by a 2-cycle burst. On the contrary, the 5 MHz 1-3 composite transducer, with little low-frequency components coupled, generated clean 5 MHz ultrasound wave at both 1-cycle and 2-cycle excitations. Because of the relatively low impedance of the 1-3 composite material, a very short pulse (1 peak) was achieved with the 1-cycle excitation. With an aspect ratio of 2, the 6.5 MHz single crystal transducers had almost no low-frequency components either, which vibrate in the 6.5 MHz thickness mode dominantly. However, the pulse length of this group of transducers was still long because of the high acoustic impedance mismatch (2 peaks in 1-cycle excitation and 3 peaks in 2-cycle). Therefore, according to the pulse length in the pressure output, the 5 MHz 1-3 composite transducers were preferable.

The tradeoff between the high pressure from the 6.5 MHz crystal transducers and the short pulse from the 5 MHz composite transducers may be determined for imaging. First, with 1-cycle excitation, the pressure output from the composite material was about 70% of the 6.5 MHz crystal, which was not dramatically low. Within 100 V excitation, the composite could generate about 1 MPa pressure, which was sufficient for the contrast super-harmonic imaging. Second, a short pulse length (single negative peak) is highly desirable for a high-resolution imaging, which is, however, hardly achievable by the 6.5 MHz single crystal transmitter. Taking the two considerations into account, the 5 MHz composite transmission transducers were believed to be suitable for a high-contrast, high-resolution imaging.

Imaging Results

Transmitter Frequency Comparison

Contrast imaging using the prototyped transducers were validated and compared for their CTR and resolution. The comparison of CTR was made between the two frequencies with 2-cycle excitations. With the similar PNP (~1 MPa), the 5 MHz transducer generated higher CTR (23 dB, Fig. 5.32b) image than the 6.5 MHz transducer (15 dB, Fig. 5.32a). There are three possible reasons for this phenomenon. The primary reason is that the high-order harmonics from tissue are much lower than that from microbubbles. The second reason is that lower-frequency ultrasonic wave relates to higher mechanical index (MI) that would induce more significant nonlinear vibrations on microbubbles. Another possible reason is that the 5 MHz ultrasound is closer to microbubble resonance (<4 MHz) (Doinikov et al. 2009) than the 6.5 MHz one, so that the 5 MHz ultrasound is likely to induce stronger vibration on the microbubbles. Because of these reasons, the 5 MHz transmitter is more effective than 6.5 MHz one in super-harmonic contrast imaging for a 30 MHz receiver.

Transmitter Material Comparison

With 2-cycle excitations, the pulse lengths of the contrast signals were long (Fig. 5.32). Pulse length of the 6.5 MHz single crystal transmitter was ~ 200 μm because of multiple negative peaks existed in the transmission. The 5 MHz single crystal ultrasound has longer pulse length than the 6.5 MHz one because of longer wavelength and more cycles of reverberation due to lateral mode coupling. As a

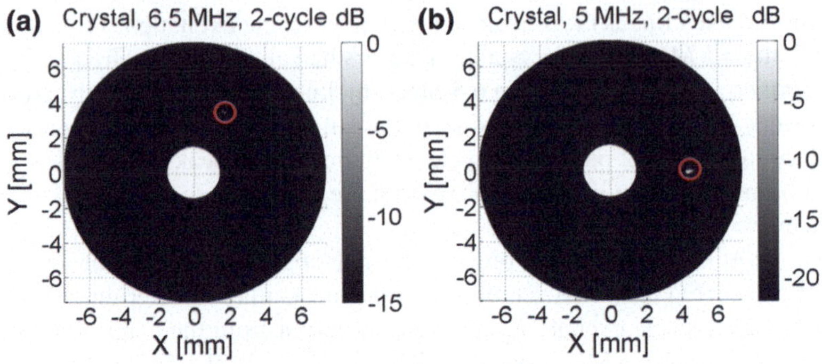

Fig. 5.32 Comparison of contrast imaging with the **a** 6.5 MHz and **b** 5 MHz transmitters, and both of which were made of PMN-PT single crystals are excited by 2-cycle bursts

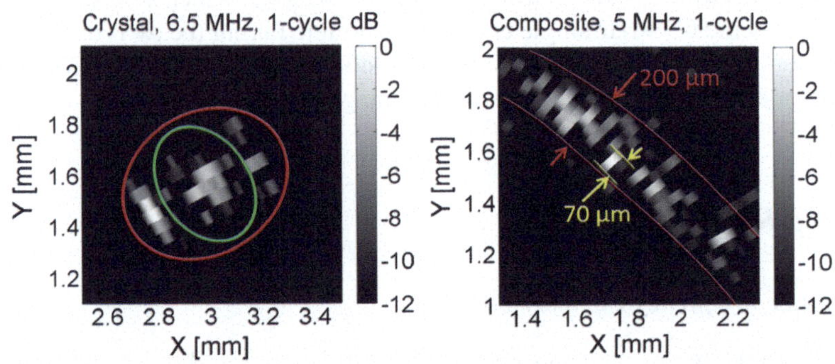

Fig. 5.33 Comparison of contrast imaging with the 5 MHz transmitters made of PMN-PT single crystal and PMN-PT 1-3 composite. Both were excited by 1-cycle burst

result, the 5 MHz single crystal transducer generated the lowest resolution imaging among the three groups.

A fewer number of cycles (1-cycle) excitation generated shorter pulse length with a compromise on the CTR. However, because of intrinsic high quality factor (*Q*-factor) of the single crystal material, the two peaks in the negative pressure generated by the 6.5 MHz transmitter with 1-cycle excitation (Fig. 5.31) resulted in the resolution of >50 μm (Fig. 5.33a). In comparison, the 1-3 composite transmission transducer under 1-cycle excitation generated a very short pulse (single peak). The image generated with the 5 MHz 1-3 composite transmitter with 1-cycle excitation is shown in Fig. 5.33b. The 200 μm tube was clearly detected with a CTR of 12 dB. Pulse length of each bubble response is equivalent to a 70 μm spatial distance, indicating an axial resolving capability of 70 μm. It is worth being pointed out that in this contrast imaging, the microtube was randomly distributed in the phantom with arbitrary angle relative to the transducer orientation. As a result, the lateral dimension of the imaging results did not indicate the lateral resolution.

Summary

In this research, a dual frequency IVUS transducer was designed, fabricated, and characterized, and its contrast imaging and fundamental imaging capability were evaluated. The measured transducer performance matched the modeling results very well for both transmission and receiving components.

In the transducer design, two key features were proposed. First, the aperture of the receiving element was significantly smaller than the transmission element, which yields both mechanical and electrical benefits. In mechanics, the large aperture transmitter delivered more acoustic excitation energy, while the small aperture of the receiver reduced the distance of the near-field within 1.5 mm so that the valid imaging depth could as close as 1.5 mm. In electrical consideration, the capacitance per unit area of the low-frequency element is lower than that of the high-frequency element due to larger thickness; the different apertures made electrical impedance the two elements closer to 50 Ω to match the input/output impedance of commercial machines. Second, the acoustic filter layer placed intermediately between the two active layers enabled the low-frequency transmit wave to pass through, but the high-frequency receive wave to be reflected which aided in decoupling signal content. Thickness of the anti-matching layer was controlled by mixing a 1% microsphere (10 μm diameter) solution into the silver epoxy. The addition of the microspheres had little effect on the acoustic or electrical properties of the silver epoxy. Finally, a layer of parylene coating acted both as shielding and as the matching layer of the high-frequency receiver. Such design and fabrication processes were demonstrated successfully in prototyping dual-frequency transducers for both fundamental mode imaging and contrast-specific imaging.

The feasibility of ultra-broadband contrast-enhanced intravascular ultrasound imaging was evaluated in vitro, using a prototype of dual-frequency small aperture transducer (T6.5/R30). Peak negative pressures higher than 1.2 MPa were generated at 3 mm away axially from the low-frequency element (0.6 × 3 mm) of the transducer, which was proved to be sufficient for contrast imaging. Sensitivity of the receiving element (0.6 × 0.5 mm) provided the capability of detecting the broadband high-frequency response of microbubbles. Initial imaging of a 200 μm tube filled with microbubbles showed reasonably high SNR (>12 dB). The small aperture dual-frequency transducer design presented demonstrates the first of its kind for contrast-enhanced high-frequency ultra-broadband intravascular imaging.

Both fundamental mode and dual frequency super-harmonic imaging were tested in vitro using a tissue-mimicking gelatin based phantom. The 30 MHz pulse-echo fundamental imaging showed a very high SNR (>25 dB) with a reasonable resolution (200 μm). While microbubble backscatter was very weak in the fundamental mode, dual-frequency super-harmonic imaging generated reasonable CTR (12 dB) and good resolution (200 μm) in resolving the microbubble filled tube. The steel rod used in the experiments was an extremely exaggerated target with a reflection coefficient of 0.94 in water; thus, the strength of the echo was large enough such that weak high-frequency components of transmission were detectable. Due to the viscoelastic behavior of tissue, we would not expect to detect tissue harmonics in the high-frequency bandwidth at the transmission acoustic pressure levels (MI < 0.48 at 6.5 MHz) used in these experiments. Contrast agent detection in high fidelity is necessary for microbubble imaging strategies involved with both molecular imaging and vasa vasorum localization. Small aperture transducers with high CTR and high resolution capable of detecting contrast agents would promote the transition of advanced contrast imaging methods to intravascular ultrasound applications.

After preliminary characterization, the intravascular ultrasound method for visualizing microbubble contrast agents using higher-order super-harmonics was used to detect microvascular blood vessels in vivo. Dual-frequency images effectively suppressed tissue signal. Contrast images obtained using this method reject tissue well, enabling the production of 3-D renderings of vessels. The ability of this technique to detect contrast in 200 μm vessels smaller than 200 mm in diameter in vivo without using multiple pulses was described. Additionally, phantom studies revealed the feasibility of using a dual-frequency approach to detect *vasa vasorum*-sized vessels at depths up to 7 mm.

With the concept of the dual-frequency transducer design demonstrated by the preliminary results (6.5 MHz single crystal transmitter), further optimizations of the performance were elucidated. With similar peak negative pressure, the 5 MHz transducers generated higher CTR (23 dB) imaging results than that with the 6.5 MHz transmitter (15 dB). However, it is not preferable simply to reduce the transmission frequency by increasing the thickness of the PMN-PT single crystals because of vibration modes coupling. Replacing the single crystal material by the 1-3 composite, the low-frequency components disappeared, and a short pulse with a single negative pressure peak was generated. With 1-cycle excitation, the 5 MHz 1-3 composite transducers showed reasonable CTR (12 dB) and very short pulse length (70 μm). Such high resolution indicated the ability of detecting the second-order vasa vasorum (67.99 ± 2.72 μm) (Kwon et al. 1998).

In conclusion, the dual-frequency intravascular transducers developed in recent years demonstrated their capability of intravascular acoustic angiography, indicating a promising future for effective evaluation of the plaque vulnerability and diagnosis of atherosclerosis cardiovascular diseases.

References

Azuma T, Ogihara M, Kubota J, Sasaki A, S-i Umemura, Furuhata H (2010) Dual-frequency ultrasound imaging and therapeutic bilaminar array using frequency selective isolation layer. IEEE Trans Ultrason Ferroelectr Freq Control 57(5):1211–1224

Borden MA, Sarantos MR, Stieger SM, Simon SI, Ferrara KW, Dayton PA (2006) Ultrasound radiation force modulates ligand availability on targeted contrast agents. Mol Imaging 5(3):139–147

Bouakaz A, Krenning BJ, Vletter WB, Cate FJT, Jong ND (2003) Contrast superharmonic imaging: a feasibility study. Ultrasound Med Biol 29(4):547–553. https://doi.org/10.1016/s0301-5629(03)00012-7

Choudhury RP, Fuster V, Fayad ZA (2004) Molecular, cellular and functional imaging of atherothrombosis. Nat Rev Drug Discov 3(11):913–925. https://doi.org/10.1038/nrd1548
Constantinides P (1990) Cause of thrombosis in human atherosclerotic arteries. Am J Cardiol 66(16):G37–G40
Doinikov AA, Haac JF, Dayton PA (2009) Resonance frequencies of lipid-shelled microbubbles in the regime of nonlinear oscillations. Ultrasonics 49(2):263–268
Feinstein SB (2006) Contrast ultrasound Imaging of the carotid artery vasa vasorum and atherosclerotic plaque neovascularization. J Am Coll Cardiol 48(2):236–243. https://doi.org/10.1016/j.jacc.2006.02.068
Garcia-Garcia HM, Costa MA, Serruys PW (2010) Imaging of coronary atherosclerosis: intravascular ultrasound. Eur Heart J 31(20):2456–2469. https://doi.org/10.1093/eurheartj/ehq280
Gessner R, Dayton PA (2010) Advances in molecular imaging with ultrasound. Mol Imaging 9(3):117–127
Gessner R, Lukacs M, Lee M, Cherin E, Foster FS, Dayton PA (2010) High-resolution, high-contrast ultrasound imaging using a prototype dual-frequency transducer: in vitro and in vivo studies. IEEE Trans Ultrason Ferroelectr Freq Control 57(8):1772–1781. https://doi.org/10.1109/tuffc
Gessner RC, Aylward SR, Dayton PA (2012) Mapping microvasculature with acoustic angiography yields quantifiable differences between healthy and tumor-bearing tissue volumes in a rodent model. Radiology 264(3):733–740
Gessner RC, Frederick CB, Foster FS, Dayton PA (2013a) Acoustic angiography: a new imaging modality for assessing microvasculature architecture. Int J Biomed Imaging 2013
Gessner RC, Frederick CB, Foster FS, Dayton PA (2013b) Acoustic angiography: a new imaging modality for assessing microvasculature architecture. Int J Biomed Imaging 2013:14
Gessner RC, Frederick CB, Foster FS, Dayton PA (2013c) Acoustic angiography: a new imaging modality for assessing microvasculature architecture. Int J Biomed Imaging 2013:936593. https://doi.org/10.1155/2013/936593
Glaser R (2005) Clinical progression of Incidental, asymptomatic lesions discovered during culprit vessel coronary intervention. Circulation 111(2):143–149. https://doi.org/10.1161/01.cir.0000150335.01285.12
Go AS, Mozaffarian D, Roger VL, Benjamin EJ, Berry JD, Blaha MJ, Dai S, Ford ES, Fox CS, Franco S (2014) Heart disease and stroke statistics–2014 update: a report from the American Heart Association. Circulation 129(3):e28
Goertz DE, Frijlink ME, Tempel D, van Damme LC, Krams R, Schaar JA, Folkert J, Serruys PW, Jong ND, van der Steen AF (2006) Contrast harmonic intravascular ultrasound: a feasibility study for vasa vasorum imaging. Invest Radiol 41(8):631–638
Goertz DE, Frijlink ME, Tempel D, Bhagwandas V, Gisolf A, Krams R, de Jong N, van der Steen AF (2007) Subharmonic contrast intravascular ultrasound for vasa vasorum imaging. Ultrasound Med Biol 33(12):1859–1872. https://doi.org/10.1016/j.ultrasmedbio.2007.05.023
Jaffer FA, Libby P, Weissleder R (2006) Molecular and cellular imaging of atherosclerosis: emerging applications. J Am Coll Cardiol 47(7):1328–1338
Jang I-K, Bouma BE, Kang D-H, Park S-J, Park S-W, Seung K-B, Choi K-B, Shishkov M, Schlendorf K, Pomerantsev E (2002) Visualization of coronary atherosclerotic plaques in patients using optical coherence tomography: comparison with intravascular ultrasound. J Am Coll Cardiol 39(4):604–609
Jong N, Bouakaz A, Cate FJT (2002) Contrast harmonic imaging. Ultrasonics 40:567–573
Kasprzak JD, Paelinck B, Ten Cate FJ, Vletter WB, de Jong N, Poldermans D, Elhendy A, Bouakaz A, Roelandt JR (1999) Comparison of native and contrast-enhanced harmonic echocardiography for visualization of left ventricular endocardial border. Am J Cardiol 83(2):211–217
Kinsler LE, Frey AR, Coppens AB, Sanders JV (2000) Fundamentals of acoustics, 4th edn. Wiley, USA
Kolodgie FD, Gold HK, Burke AP, Fowler DR, Kruth HS, Weber DK, Farb A, Guerrero LJ, Hayase M, Kutys R, Narula J, Finn AV, Virmani R (2003) Intraplaque hemorrhage and progression of coronary atheroma. N Engl J Med 349(24):2316–2325

Kossoff G (1966) The effects of backing and matching on the performance of piezoelectric ceramic transducers. IEEE Trans Son Ultrason 13(1):20–30
Krimholtz R, Leedom DA, Mattaei GL (1970) New equivalent circuits for elementary piezoelectric transducers. Electron Lett 41(13):398–399
Kruse DE, Ferrara KW (2005) A new imaging strategy using wideband transient response of ultrasound contrast agents. IEEE Trans Ultrason Ferroelectr Freq Control 52(8):1320–1329
Kwon HM, Sangiorgi G, Ritman EL, McKenna C, Holmes DR Jr, Schwartz RS, Lerman A (1998) Enhanced coronary vasa vasorum neovascularization in experimental hypercholesterolemia. J Clin Invest 101(8):1551
Lerch R (1990) Simulation of piezoelectric devices by two-and three-dimensional finite elements. IEEE Trans Ultrason Ferroelectr Freq Control 37(3):233–247
Li Y, Ma J, Martin KH, Choi H, Dayton PA, Jiang X, Shung KK, Zhou Q (2014) A configurable dual-frequency transmit/receive system for acoustic angiography imaging. In: 2014 IEEE International ultrasonics symposium (IUS). IEEE, pp 731–733
Lindner JR (2004) Molecular imaging with contrast ultrasound and targeted microbubbles. J Nucl Cardiol 11(2):215–221. https://doi.org/10.1016/j.nuclcard.2004.01.003
Lockwood GR, Ryan LK, Hunt JW, Foster FS (1991) Measurement of the ultrasonic properties of vascular tissues and blood from 35–65 MHz. Ultrasound Med Biol 17:653–666
Ma J, Jiang X, Heath Martin K, Dayton PA (2013a) Small aperture, dual frequency ultrasound transducers for intravascular contrast imaging. In: 2013 IEEE Int Ultrason Symp (IUS), Chicago, IL. IEEE, pp 769–772
Ma J, Wang Z, Jiang X (2013b) Design, fabrication, and test of a small aperture, dual frequency ultrasound transducer. In: Kundu T (ed) Proceedings of SPIE, San Diego, p 86951H. https://doi.org/10.1117/12.2009716
Ma J, Jiang X, Martin KH, Dayton PA, Li Y, Zhou Q (2014a) Dual frequency transducers for intravascular ultrasound super-harmonic imaging and acoustic angiography. In: 2014 IEEE International Ultrasonics Symposium (IUS). IEEE, pp 675–678
Ma J, Martin KH, Dayton PA, Jiang X (2014b) A preliminary engineering design of intravascular dual-frequency transducers for contrast enhanced acoustic angiography and molecular imaging. IEEE Trans Ultrason Ferroelectr Freq Control 61(5):870–880
Ma J, Li S, Wang Z, Jiang X (2014c) Anti-matching design for wave isolation in dual frequency transducer for intravascular super-harmonic imaging. In: Proceedings of ASME 2014 IMECE, Montreal, Canada (in press)
Martin K, Lindsey B, Ma J, Lee M, Li S, Foster F, Jiang X, Dayton P (2014) Dual-frequency piezoelectric transducers for contrast enhanced ultrasound imaging. Sensors 14(11):20825–20842. https://doi.org/10.3390/s141120825
Mason WP (1930) The approaximate network of acoustic filters. Bell Syst Tech J 9(2):332–340
Milei J, Parodi JC, Alonso GF, Barone A, Grana D, Matturri L (1998) Carotid rupture and intraplaque hemorrhage: immunophenotype and role of cells involved. Am Heart J 136(6):1096–1105
Moreno PR, Fuster V (2004) New aspects in the pathogenesis of diabetic atherothrombosis. J Am Coll Cardiol 44(12):2293–2300
Moulton KS, Vakili K, Zurakowski D, Soliman M, Butterfield C, Sylvin E, Lo K, Gillies S, Javaherian K, Folkman J (2003) Inhibition of plaque neovascularization reduces macrophage accumulation and progression of advanced atherosclerosis. P Natl Acad Sci USA 100(8):4736–4741
Naghavi M (2003) From vulnerable plaque to vulnerable patient: a call for new definitions and risk assessment strategies: Part I. Circulation 108(14):1664–1672. https://doi.org/10.1161/01.cir.0000087480.94275.97
Naghavi M (2010) Asympotomatic atherosclerosis: pathophysiology, DETECTION and treatment. Humana Press
Nissen SE, Gurley JC, Grines CL, Booth DC, McClure R, Berk M, Fischer C, DeMaria AN (1991) Intravascular ultrasound assessment of lumen size and wall morphology in normal subjects and patients with coronary artery disease. Circulation 84(3):1087–1099

Schomann T, Qunneis F, Widera D, Kaltschmidt C, Kaltschmidt B (2013) Improved method for ex ovo-cultivation of developing chicken embryos for human stem cell xenografts. Stem Cells Int 2013:960958

Shung KK, Cannata JM, Zhou QF (2007) Piezoelectric materials for high frequency medical imaging applications: a review. J Electroceram 19(1):141–147. https://doi.org/10.1007/s10832-007-9044-3

Slager CJ, Wentzel JJ, Schuurbiers JC, Oomen JA, Kloet J, Krams R, Von Birgelen C, Van Der Giessen WJ, Serruys PW, De Feyter PJ (2000) True 3-dimensional reconstruction of coronary arteries in patients by fusion of angiography and IVUS (ANGUS) and its quantitative validation. Circulation 102(5):511–516

Sluimer JC, Daemen MJ (2009) Novel concepts in atherogenesis: angiogenesis and hypoxia in atherosclerosis. J Pathol 218(1):7–29. https://doi.org/10.1002/path.2518

Smith P (1995) Electronic applications of the smith chart. The Institution of Engineering and Technology

Staub D, Schinkel AFL, Coll B, Coli S, van der Steen AFW, Reed JD, Krueger C, Thomenius KE, Adam D, Sijbrands EJ, ten Cate FJ, Feinstein SB (2010) Contrast-enhanced ultrasound imaging of the vasa vasorum: from early atherosclerosis to the identification of unstable plaques. JACC: Cardiovasc Imaging 3(7):761–771. https://doi.org/10.1016/j.jcmg.2010.02.007

Steer MB (2013) Microwave and RF design: a systems approach, 2nd edn. SciTech, New Jersey

Tang MX, Eckersley RJ (2006) Nonlinear propagation of ultrasound through microbubble contrast agents and implications for imaging. IEEE Trans Ultrason Ferroelectr Freq Control 53:2406–2415

Ten Kate GL, Renaud GGJ, Akkus Z, van den Oord SCH, ten Cate FJ, Shamdasani V, Entrekin RR, Sijbrands EJG, de Jong N, Bosch JG, Schinkel AFL, van der Steen AFW (2012) Far-wall pseudoenhancement during contrast-enhanced ultrasound of the carotid arteries: clinical description and in vitro reproduction. Ultrasound Med Biol 38:593–600

ten Kate GL, Sijbrands EJG, Valkema R, ten Cate FJ, Feinstein SB, Steen AFW, Daemen MJAP, Schinkel AFL (2010) Molecular imaging of inflammation and intraplaque vasa vasorum: a step forward to identification of vulnerable plaques? J Nucl Cardiol 17(5):897–912. https://doi.org/10.1007/s12350-010-9263-x

Tobis JM, Mallery J, Mahon D, Lehmann K, Zalesky P, Griffith J, Gessert J, Moriuchi M, McRae M, Dwyer M-L (1991) Intravascular ultrasound imaging of human coronary arteries in vivo. Analysis of tissue characterizations with comparison to in vitro histological specimens. Circulation 83(3):913–926

Van Neer PLMJ, Matte G, Danilouchkine MG, Prins C, Van Den Adel F, Jong ND (2010) Super-harmonic imaging: development of an interleaved phase-array transducer. IEEE Trans Ultrason Ferroelectr Freq Control 57(2):455–468

Virmani R (2005) Atherosclerotic plaque progression and vulnerability to rupture: angiogenesis as a source of intraplaque hemorrhage. Arterioscler Thromb Vasc 25(10):2054–2061. https://doi.org/10.1161/01.atv.0000178991.71605.18

Wang H, Ritter T, Cao W, Shung KK (2001) High frequency properties of passive materials for ultrasonic transducers. IEEE Trans Ultrason Ferroelectr Freq Control 48(1):78–84

Chapter 6
Dual-Modality Fluorescence Lifetime and Intravascular Ultrasound for Label-Free Intravascular Coronary Imaging

Jennifer E. Phipps, Julien Bec and Laura Marcu

Introduction

Atherosclerosis in the coronary arteries accounts for nearly 50% of cardiovascular disease-related deaths. To characterize an atherosclerotic disease, imaging modalities that measure both the biochemical and structural properties of the lesions in the coronary arteries are necessary. To accomplish this, time-resolved fluorescence spectroscopy (TRFS) can be used to determine biochemical composition while adding diagnostic value to intravascular ultrasound (IVUS), which identifies structural features of the arterial wall. Healthy and diseased human arteries have distinct autofluorescent properties that allow specific biochemical features of atherosclerosis to be characterized with ultraviolet light excitation. Fluorophores native to the arterial wall include structural proteins (e.g., collagen and elastin) and lipid constituents (e.g., cholesterols and lipopigments). This chapter will present a brief history of autofluorescence studies of atherosclerosis, an evolution of TRFS instrumentation used for characterizing atherosclerosis, development of intravascular bimodal catheter systems combining TRFS technology with IVUS, data processing methods, and ex vivo and in vivo studies performed with this bimodal catheter.

J. E. Phipps · J. Bec · L. Marcu (✉)
Department of Biomedical Engineering, University of California, Davis, Davis, CA 95616, USA
e-mail: lmarcu@ucdavis.edu

J. Bec
e-mail: jbec@ucdavis.edu

Q. Zhou and Z. Chen (eds.), *Multimodality Imaging*,
https://doi.org/10.1007/978-981-10-6307-7_6

Autofluorescence of Atherosclerotic Arteries

Origin of Autofluorescence

Normal and diseased arterial vessels are known to exhibit autofluorescence properties upon ultraviolet (UV) light excitation. This is due to several types of intrinsic fluorophores present in the arterial wall. This includes structural proteins (e.g., collagen and elastin) and lipid constituents such as cholesterols (e.g., oxidized low-density lipoprotein, very low-density lipoproteins, cholesteryl linoleate, and cholesteryl oleate) and lipopigments (e.g., ceroid). A normal artery wall is dominated by elastin fluorescence, fibrous plaques and fibrous caps are dominated by collagen fluorescence, and lipid pools and foam cell-rich arterial walls are dominated by fluorescence from the lipid constituents.

History of Fluorescence Spectroscopy Studies of Atherosclerosis

The first studies of human atherosclerotic plaques to take advantage of the intrinsic fluorophores in the artery wall were fluorescence spectroscopy techniques that used a single excitation wavelength and measured fluorescence emission at a single wavelength range (Edholm and Jacobson 1965; Kittrell et al. 1985). The next major advance for fluorescence studies in atherosclerosis came with techniques developed to assist with laser ablation procedures. The goal of this work was to use arterial fluorescence to determine diseased locations for laser ablation. Fluorescence spectra from the normal arterial wall, noncalcified plaque, and calcified plaque differed in peak intensity, peak intensity wavelength, and shape in measurements from 268 locations in vivo in 48 patients undergoing open heart surgery or percutaneous catheterization with a 325 nm low power helium–cadmium laser excitation through a flexible 200 μm fiber (Bartorelli et al. 1991). Calcified and noncalcified plaques showed a decrease in intensity (especially calcified plaques), higher shape index, and a shift of peak wavelength location to longer wavelengths for noncalcified plaques all in comparison with the normal artery. These variables were combined in a classification algorithm to identify normal artery with 100% specificity and atherosclerotic arteries with 73% sensitivity (Bartorelli et al. 1991). In a separate study with a similar goal, a single XeCl excimer laser (308 nm excitation through a 600 μm silica fiber) was used to induce fluorescence and ablate tissue. Arterial media, lipid-rich plaques, and calcified plaques were identified and distinguished with this technique (Morguet et al. 1994). The strategy to use fluorescence spectroscopy to better identify laser ablation location has not been pursued since laser ablation for coronary artery disease fell out of favor in the mid-1990s.

Following these studies that showed diseased arteries could be identified from the normal artery wall, more careful studies of plaque composition were performed.

Fluorescence measurements from *en face* frozen sections of human coronary arteries and aorta were acquired with an Olympus BH-2 metallurgic bright-field microscope and an argon ion laser (476 nm excitation) (Fitzmaurice et al. 1989). The fluorescence properties of collagen, elastin, lipid, and calcium allowed for normal arteries and atherosclerotic arteries to be differentiated. Building on this work, Verbunt et al. more carefully studied the fluorescence of lipid and calcium-rich plaques by analyzing ceroid deposits in these specimens (Verbunt et al. 1992). Ceroid, a type of lipopigment, was found to have an autofluorescence spectrum that was wider and red-shifted in comparison with both collagen and elastin. In 1994, a study was done that showed autofluorescence spectra could distinguish fibrous, aneurysmal, and normal samples but not lipid and calcium-rich plaques. This work was performed on samples of the abdominal aorta from cadavers and femoral artery from bypass surgeries with an He-Cd laser (442 nm excitation) (Papazoglou et al. 1994). In 2000, a fluorescence spectroscopy study was performed in vivo to observe plaque disruption in a rabbit model (Christov et al. 2000).

During the 1990s, the characteristics of plaque that indicate higher levels of risk for causing heart attack were meticulously studied (Virmani et al. 2006). Subsequently, studies that focused on using fluorescence to assess features of plaque vulnerability were initiated. The first study to use fluorescence to look for a thin-capped fibroatheroma was performed by Arakawa et al. in human femoral and coronary arteries obtained from cadavers (Arakawa et al. 2002). Autofluorescence was induced with a He-Cd laser (excitation 325 nm); autofluorescence properties of collagen, elastin, and lipid were compared to the measurements from the human tissue to determine what fluorophores were fluorescing in the tissue samples. Plaques with lipid cores had fluorescence spectra that were red-shifted and broader compared to normal tissue and collagenous plaques. Lipid core fluorescence properties were similar to those of oxidized low-density lipoprotein (oxLDL), indicating the presence of this molecule in lipid cores. The fluorescence properties of fibrous plaques were similar to collagen and those of normal artery were similar to elastin. A classification algorithm was able to use the fluorescence spectra from these molecules (collagen, elastin, oxLDL) to distinguish normal artery, lipid cores, atheroma, or pre-atheroma with 86% accuracy. This work also showed that fibrous cap thickness correlated well with spectral collagen content ($R = 0.65$, $P < 0.0001$) (Arakawa et al. 2002).

Time-Resolved Fluorescence Spectroscopy and Imaging Studies of Atherosclerotic Lesions

Point Measurements of Time-Resolved Fluorescence Spectroscopy

Single wavelength measurements Due to the increased ability of time-resolved fluorescence spectroscopy (TRFS) to resolve fluorophores with overlapping fluorescence intensity spectra, these studies were used to determine if they could more accurately identify plaque composition than fluorescence intensity spectroscopy alone. The TRFS point measurement technique involved for this work consisted of exciting at a single wavelength and recording the time-resolved fluorescence decay of the emitted light at a single wavelength band. Fluorescence emission intensity from normal and atherosclerotic aorta was found to differ at 340 and 380 nm with 308 nm ultraviolet laser excitation (Baraga et al. 1989). A binary classification scheme was used to distinguish normal from the atherosclerotic aorta in 56 of 60 total cases by using these wavelength bands that characterized the tryptophan content (340 nm) and the elastin-to-collagen ratio (380 nm) (Baraga et al. 1990). Part of this study also used a time-correlated single photon counting technique (TCSPC). With this technique, fluorescence lifetime was shorter in the tryptophan region (340 nm) and longer at the collagen/elastin region (380 nm) (Baraga et al. 1989). A frequency-doubled mode-locked and cavity-dumped continuous wave dye laser for picosecond pulse generation at 320 nm was used by Andersson-Engels et al. with TCSPC to acquire time-resolved signal from the human aorta samples in vitro. Both collagenous and calcified plaques exhibited longer fluorescence lifetimes than normal artery at 400 and 480 nm (Andersson-Engels et al. 1990). Several studies by Maarek et al. reported the spectrally resolved time-resolved fluorescence spectroscopy characterization of arterial wall constituents, including collagen, elastin, and lipid as well as characterizing specific lesion types (Maarek et al. 1997, 1998; Marcu et al. 1998).

Multiple wavelength measurements Following the TRFS studies that interrogated single wavelength bands of emitted fluorescence light per measurement, newer studies employed a technique that allowed for multiple bands of emitted fluorescence to be temporally resolved. These newer studies could be faster and provide more information than earlier TRFS studies. Human aortic samples were first studied with this technique (94 samples, postmortem) (Maarek et al. 2000; Marcu et al. 2003). Excitation was performed with a 337 nm nitrogen laser, and emission was collected through silica fibers combined in a single probe. Wavelength selection occurred with a scanning monochromator, a gated micro-channel plate photomultiplier was used to measure the emission waveform, and sampling was performed with a digitizing oscilloscope. The intrinsic fluorescence decays were computed with a Laguerre deconvolution method, from which the time-integrated fluorescence emission spectrum and fluorescence lifetime were derived. TRFS spectra from the aortic samples varied with the progression of atherosclerosis–fluorescence intensity and average lifetime. The substantial differences in time-resolved properties between clinically relevant plaque compositions suggested that TRFS parameters could translate to arterial disease diagnostic markers. TRFS was also found sensitive to macrophage infiltration in plaques (Marcu et al. 2005) and the presence or absence of matrix metalloproteinases (Phipps et al. 2011). A TRFS study was also conducted in vivo in an atherosclerotic rabbit model, demonstrating that TRFS could discriminate between macrophage and collagen content in atherosclerotic plaques (Marcu et al. 2005).

A similar system as described above was also used to study time-resolved fluorescence emission of normal and atherosclerotic human coronary arteries (58 coronary segments from 11 subjects, postmortem) (Marcu et al. 2001, 2003). Similar to the study in the human aorta, this work showed parameters derived from time-resolved fluorescence spectra could be used to enhance discrimination between normal and different grades (I, II, III, IV, Va, and Vb) of atherosclerotic lesions as defined by the American Heart Association (AHA). Lipid-rich lesions were distinguishable from other lesion types, particularly fibrous lesions and normal artery wall. The fluorescence emission of lipid-rich lesions matched that of lipid components and the fluorescence emission of fibrous lesions matched that of collagen type I. This study demonstrated that time-resolved spectra varied significantly between lipid-rich (more unstable) and fibrous lesions (more stable), further validating the potential impact TRFS could have if implemented intravascularly.

Fluorescence Lifetime Imaging Microscopy (FLIM) Studies

Fluorescence lifetime imaging microscopy (FLIM) allows for time-resolved spectra to be acquired for an image versus a single point. An early fluorescence lifetime imaging microscopy (FLIM) system with 10 ps temporal resolution was introduced for biomedical applications in 1998. The system acquired fluorescence lifetime images to be acquired through excitation with a commercial Ti: Sapphire laser (415 nm), a time-gated image intensifier, and an intensified CCD camera (Dowling et al. 1998). Lifetime differences were found between elastin, collagen, and aorta samples. More recent advances in FLIM technology revolutionized the potential of this technique positive impact in clinically relevant biomedical applications (De Beule et al. 2007; Munro et al. 2005).

In 2011, atherosclerotic aortas were studied with a FLIM system implemented in a wide-field fiber bundle (Phipps et al. 2011). $N = 11$ human aorta samples ($N = 48$ locations) were imaged. A pulsed nitrogen laser was used for excitation (337 nm, 700 ps width), an imaging bundle with 10,000 optical fibers in a 0.6 mm outer diameter probe was used to collect fluorescence, and a fast-gated (up to 400 ps) intensified CCD camera generated images. Two emission wavelength bands were detected, F377: 377/50 nm and F460: 460/60 nm (center wavelength/bandwidth). This combination of wavelengths provided discrimination between intrinsic fluorophores related to plaque vulnerability. Different tissue types were identified with average lifetime and Laguerre deconvolution parameters (Jo et al. 2005). $N = 4$ distinct regions of interest (ROIs): Lipid-rich (LR), collagen-rich (CR), elastin-rich (ER), and elastin + macrophage-rich (E + M) were identified based on histopathology from $N = 81$ regions of interest (ROIs) within the FLIM images. FLIM-derived parameters were distinct in each group.

Another FLIM investigation with a wide-field fiber bundle was used to study ex vivo human coronary arteries and found discrimination between collagen-rich plaques and normal (elastin-rich) artery (Thomas et al. 2010). However, this study did not include lipid-rich plaques, and thus, the extensive comparison with TRFS results could not be made.

Wavelength-Multiplexed Fluorescence Lifetime Imaging (FLIm) of Planar Samples

FLIM enables the acquisition of whole images of the field of view, but relies on bulky and stiff fiber bundles. Multispectral TRFS measurements can be performed using a single fiber for excitation and collection, but the slow acquisition time is prohibitive for the acquisition of images. Here, we present a fluorescence lifetime imaging (FLIm) method that relies on a single fiber for excitation and collection and is able to acquired data at high speed (Sun et al. 2011). This approach was initially reported in 2008 (Sun et al. 2008) and allowed simultaneous acquisition and storage of fluorescence data from three wavelength bands in a few microseconds. When coupled with a compact *x*-*y*-*z* 3D positioning stage, this system also allowed the collection of point measurements from these wavelength bands in a fast-scanning mode that could then be reconstructed to create fluorescence lifetime imaging (FLIm) images (Sun et al. 2011). This scanning system was validated in ex vivo atherosclerotic aortas (Sun et al. 2011). Using the fast-scanning system, FLIm images were created that demonstrated the distinction between plaque types. A similar system has also been incorporated within optical coherence tomography (OCT) for arterial imaging (Jo et al. 2015). We note here the difference between FLIM and FLIm. The FLIm technique described here acquires multiple wavelengths of emitted fluorescence simultaneously through a single fiber in contrast to the FLIM technique that acquires high-resolution images of the emitted fluorescence (microscopy) at a single wavelength band per measurement. Moving forward, we will show how this single fiber, high-speed technique has been further developed to allow for the technology to be translated to an intravascular catheter.

Challenges for Implementing FLIm in an Intravascular Catheter

This chapter will discuss the necessary qualities of a catheter-based FLIm system and the challenges to achieving these specifications that include:

- Fast-scanning speed to account for the fact that blood will need to be flushed from the field of view, which can only be done for short amounts of time for patient safety
- Pairing with a morphologic imaging modality that will guide FLIm measurements from within the coronary artery tree
- A flexible fiber that is able to reach the coronary arteries and acquire images from tortuous vessels.

FLIm excitation and signal cover the near-UV to visible range. Over this wavelength range, blood presents strong absorption, it is therefore necessary to displace blood from the field of view during imaging. In a similar way to OCT, this can be achieved using a balloon located proximally to the imaging catheter field of view in combination of injection of clear fluid, or bolus injection of clear fluid without occlusion. In both cases, this restricts the imaging duration as blood flow interruption or fluid injection duration need to be limited to a few seconds. A system able to acquire FLIM images in vivo must therefore be able to measure locations with speeds in the kHz range.

Coronary arteries, in particular when diseased, present a small cross section and notable tortuosity. Successful navigation through lesions places a constraint on device cross section and flexibility. Additionally, FLIm is not depth-resolved and only provides information down to 200 μm from the lumen surface. To provide navigational information (lumen dimension, localization in the arterial tree) and structural and anatomical information (plaque burden, presence of necrotic or calcified cores, etc.), it needs to be combined with a morphological imaging technique such as IVUS. While the biochemical information provided by FLIm is ideally complementing the morphological information provided by IVUS, integration of FLIm with IVUS creates constraints for the design of the instrument. Namely, both FLIm and IVUS elements need to be integrated into a low profile catheter, and the imaging section needs to be both optically transparent in the near-UV to visible and present low sound impedance in addition to being flexible.

FLIm-IVUS Catheter System Instrumentation

The FLIm instrumentation dedicated to intracoronary imaging is composed of a multispectral high-speed time-domain fluorescence lifetime imaging system, integrated with a clinical 40 MHz IVUS system. The design of the FLIm instrumentation as well as its integration into a bimodal catheter system is described in the following sections.

Principle of the High-Speed Fluorescence Lifetime Imaging Instrumentation

The FLIm instrumentation (see Fig. 6.1) consists of a pulsed UV fiber laser (20 kHz, 80 ps pulses, Fianium, UK), a wavelength selection module (WSM) that directs the laser light from the laser to the sample via the excitation/collection fiber optic. Autofluorescence generated by the sample is then collected by the same fiber and transmitted to the WSM where it is split into four different bands matched with the emission wavelengths of fluorophores expected in diseased and healthy arteries.

Optical temporal multiplexing is performed by routing the signal from each band through different optical delay lines (1, 10, 19, and 28 m). The 45 ns delay between channels introduced by the fibers enables fluorescence contribution from each of

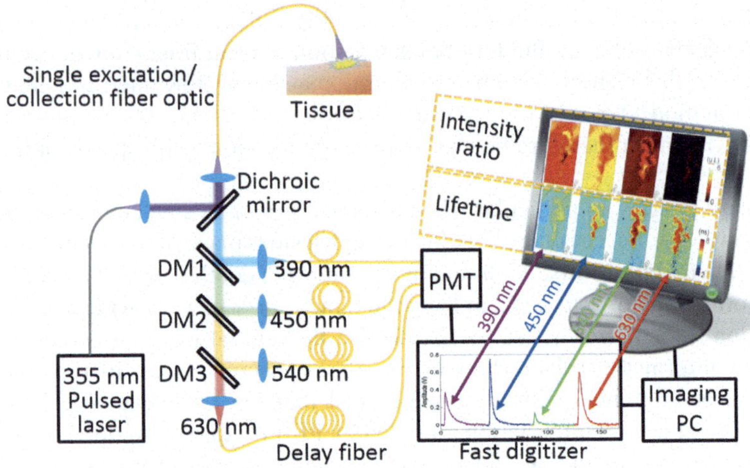

Fig. 6.1 Fast FLIm principle of operation: A pulsed UV source delivers excitation light to the vessel wall via a fiber optic channel (100–200 μm core silica fiber) integrated inside of the FLIm-IVUS motor drive and catheter. Autofluorescent light generated by the arterial wall is collected and transmitted via the same fiber and separated in four different wavelength bands. Temporal multiplexing achieved by transmitting the contribution of each channel to the photodetector using different length of fiber optic enables sampling of all four bands using a single channel 12.5 GS/s digitizer. For each excitation pulse, full fluorescence decays over four different wavelengths are acquired

the four spectral bands to be sent to a single multi-channel plate photomultiplier (R3809U-50, Hamamatsu, Japan), without overlap of the signals. The electrical signal is amplified (C5594 Hamamatsu, Japan) and sampled using a high-speed digitizer (12.5 GS/s, 3 GHz, 8-bit, 512 Mbytes, PXIe-5185, National Instruments, Austin, TX).

This detection method enables acquisition of fluorescence decay from multiple wavelength bands from a single excitation pulse, therefore enabling much higher acquisition speeds than alternative fluorescence lifetime methods (Stary et al. 1994). Successful measurements of fluorescence decay using a pulse sampling technique rely on a high enough number of photons reaching the photodetector following each excitation pulse. Therefore, excitation pulse energy, throughput of the excitation and collection optical paths, and width of each spectral band of the instrument need to be optimized.

FLIm-IVUS Intravascular Catheter System

The FLIm system described above is ideally suited to the implementation of FLIm in the intravascular setting since it is able to work at high speed and requires only a single multimode fiber for fluorescence excitation and collection. Nevertheless, challenges to implement intravascular FLIm highlighted previously, coupled with the high collection efficiency required for successful implementation of the pulse sampling method lead to several engineering challenges. The design of a FLIm-IVUS catheter system suitable for in vivo interrogation of coronary arteries required several design iterations until a configuration able to fulfill all requirements was identified.

The first design, characterized by IVUS and FLIm channels arranged in a parallel configuration, allowed for independent rotation of the ultrasonic and optical channels (using independent motor drives) within an 8 Fr outer diameter catheter sheath (Bec et al. 2012). No significant modification of the commercial IVUS system/catheter was required. The ultrasound imaging core was still introduced in the original IVUS sheath, while the FLIm fiber was housed in an acrylic tube. With this catheter, the twin tube imaging section was enclosed in a compliant siloxane balloon that was inflated with saline solution during imaging and displaced the blood from the field of view. This catheter enabled the initial technical performance evaluations of the rotational FLIm system and the first in vivo measurements in arterial vessels autofluorescence in pulsatile blood flow in swine peripheral arteries (Bec et al. 2014). In this configuration, the scanning pullback length was limited to 6 mm.

This design was further refined in a subsequent design using a parallel/sequential configuration (Design 2). It consisted of a catheter with a small cross-sectional profile (3.2 Fr) relying on sequential scanning of the field of view by parallel independent FLIm and IVUS imaging cores, advancing sequentially into a single sheath and imaging section (Ma et al. 2015) by a dedicated hardware and software interface. The use of polymethylpentene (TPX med-18) for the imaging section provided both high transparency and low autofluorescence in the near-UV-visible range, required for FLIm operation, and low sound impedance, required for IVUS operation. Similarly to Design 1, no significant modification of the commercial IVUS system/imaging core was required. This catheter enabled us to demonstrate for the first time the ability of our FLIm system to rapidly acquire spectrally resolved fluorescence lifetime data in helical scanning from tubular and tortuous structures (including ex vivo porcine coronary arteries in intact hearts) (Ma et al. 2014). Moreover, this design enabled the generation of experimental data from diseased human coronary arteries supporting the development of methods for IVUS image segmentation, FLIm data acquisition under IVUS guidance, and methods/algorithms for plaque diagnosis (Fatakdawala et al. 2015; Gorpas et al. 2015). However, in this configuration co-registration between FLIm and IVUS is only achieved in the absence of motion, and thus, this was not suitable for in vivo use (Ma et al. 2015).

These limitations were addressed in the third design of the instrument, reported in (Bec et al. 2017). It was built on the knowledge gained from the two earlier designs and consisted of a catheter that fully integrated FLIm imaging capability into an existing clinical IVUS platform (iLab™ Boston Scientific). A custom 3.7 Fr monorail bimodal imaging catheter includes both a 40 MHz ultrasound transducer and optical channel in tandem configuration (see Fig. 6.2). Additionally, the IVUS motor drive was modified to accommodate an optical channel. Thus, both the 40 MHz transducer and the fiberoptic are rotated by the same motor drive. This design enabled

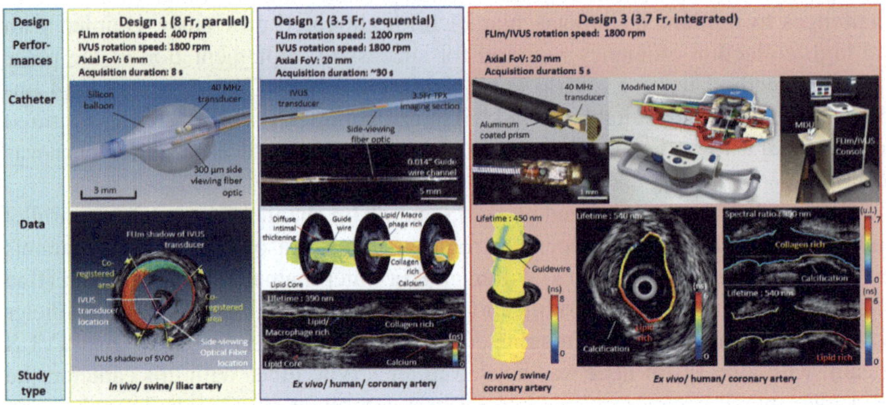

Fig. 6.2 **Design 1 (Parallel)**: enabled co-registered FLIm-IVUS data acquisition in vivo in swine peripheral arteries. **Design 2 (Sequential)**: allowed for co-registered FLIm-IVUS data acquisition in ex vivo excised human coronaries but co-registration is susceptible to motion artifacts. **Design 3 (Integrated)**: FLIm and IVUS fully integrated into a single 3.7 Fr imaging core. With Boston Scientific support, their IVUS motor drive (MDU5 PLUS) was modified to incorporate an optical channel. This system was used for co-registered FLIm-IVUS data acquisition in vivo in swine coronary arteries as well as ex vivo diseased human coronary arteries

the acquisition of directly co-registered FLIm-IVUS data in a single pullback of the catheter. With this system, 25,000 independent point spectroscopy measurements are acquired in 5 s, using an 1800 rpm rotation speed, and 4 mm/s pullback speed. The 20 mm axial field of view can be increased by using higher pullback speeds, thus increasing axial sampling. This novel catheter was found as being fully compatible with in vivo coronary interventions (Bec et al. 2016).

Methods for Data Processing

FLIm data were processed using a Laguerre expansion technique (Upston et al. 2002) due to the speed of this nonparametric technique and its ability to perform deconvolution without a priori information about the fluorescence emission. The acquired waveform of each channel was fitted with a set of Laguerre functions convolved with the instrument response function. The weights obtained from the fitting operation were then applied to the original Laguerre basis functions (without IRF convolution) to estimate the original fluorescence decay. The average fluorescence lifetime was then derived from this decay. For each measurement, the average lifetime and fluorescence intensity were computed and maps were generated where the location of each measurement corresponds to a distinct angular pullback location of the imaging core of the catheter. IVUS data were also processed to identify the vessel lumen (Spite and Serhan 2011) and to estimate the distance between catheter and vessel wall.

The distance map was used to correct the FLIm intensity images from variations of light excitation/collection due to changes in probe-to-wall distance (Angheloiu et al. 2011). This step enables quantification of the amount of fluorescence emitted in each location. For each channel, intensity ratio maps were created by normalizing each pixel of the intensity images of each channel by the summed intensities of the same pixel across all four channels. Intensity-weighted lifetime maps were then created where colors represent lifetime values, while brightness represents the amount of fluorescence. Spectral ratio weighted lifetime maps, where color represents lifetime while brightness represents spectral contribution, were created. This representation contains both time-resolved and spectral information and is independent of changes in excitation/collection efficiency as they are inherently ratiometric. Finally, fluorescence lifetime information was projected on the vessel surface recovered from IVUS to generate 3D renderings of the fluorescence lifetime maps.

FLIm-IVUS Imaging of In Vivo Swine Coronary Arteries

The ability of the catheter system to access tortuous anatomy, flush blood from the imaged location, image various sized vessels, recover the signal with sufficient signal-to-noise ratio, and discriminate targets based on lifetime and spectral information was tested in 60–70 kg swine, which present morphology similar to humans.

Flushing was performed through the catheter guide using a power injector for precise control of the injection parameters. 10% LMD solution (Dextran 40) with 5% glucose was used as a flushing medium. This solution, evaluated for human use as an alternative to iodinated contrast for OCT flushing (Ozaki et al. 2012; Frick et al. 2014), is well suited to FLIm imaging due to its high transparency in the near-UV, low toxicity, and high viscosity. Flushing flows between 2 and 10 cc/s were evaluated. It was observed that flow above 4 cc/s was sufficient to provide adequate clearing.

The ability of the system to recover lifetime information from the vessel wall in vivo was demonstrated by imaging a healthy section of coronary, expected to have a homogeneous composition, and computing the average lifetime (Bec et al. 2017). In spite of large variations of the amount of signal collected for different regions of the field of view, a low standard deviation was obtained (5.19 ± 0.12 ns). Subsequently, a stent with fluorescent markers was placed and imaged. Both spectral and temporal information enabled easy identification of the marker locations, as demonstrated in Fig. 6.3. Finally, the ability to recover lifetime measurements was evaluated in larger vessels (5 mm diameter). In that case, an increase in standard deviation of the measured lifetime was observed compared to smaller vessels, but the correct lifetime was still retrieved (5.14 ± 0.33 ns).

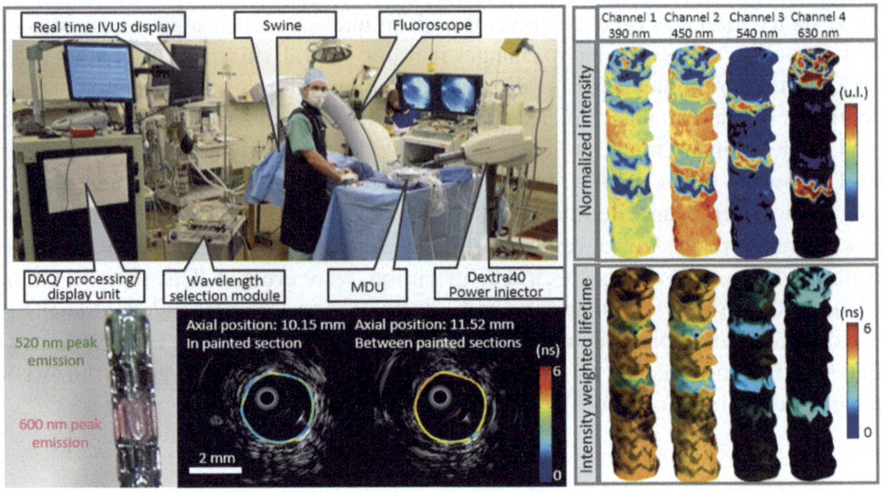

Fig. 6.3 View of large animal surgical facility at UC Davis (top left): Boston Scientific IVUS interface provides real-time IVUS display and enables identification of areas of interest for bimodal imaging. A commercial Medrad® Avanta™ fluid management system (power injector) is used to deliver 10% LMD (dextran 40) in 5% dextrose solution required to displace blood during bimodal acquisition (4–6 cc/s, 8 s). **Close-up of stent deployed in coronary vessels**: Fluorescent painted sections on stent struts were used to provide contrast in healthy pig coronary arteries. **Bimodal B-scans (from one** in vivo **image cross section)** showing easily differentiated fluorescence signal from painted section of stent and artery wall fluorescence between painted section. **Full spectral and average lifetime data**: 20 mm section imaged in 5 s. Both spectral and lifetime data sets enable straightforward identification of two marker sets from arterial wall (Bec et al. 2014)

FLIm-IVUS Imaging of Ex Vivo Human Coronary Arteries

To image ex vivo human samples, coronary arteries are imaged in circulating heated (37 °C) PBS using the FLIm-IVUS catheter system. IVUS and FLIm data sets are co-registered by shared time stamps during acquisition. IVUS data are processed twofold: (1) RF data are used to extract spectral parameters such as integrated backscatter and energy norm, and (2) gray-scale images are segmented to extract lumen and media boundaries. FLIm data are deconvolved using Laguerre-based techniques (Liu et al. 2012; Su et al. 2016) and for each spectral channel, integrated intensity, average lifetime, and Laguerre coefficients are computed. FLIm and IVUS data are then jointly displayed (3D rendering, cross and longitudinal sections) and registered with histology sections. Axial registration is achieved using pullback position information and angular registration is achieved using IVUS morphology (lumen shape, presence of calcifications). Classification groups are chosen based on histologic analysis: normal intima, fibrous cap, macrophage infiltration, and lipid core/extracellular lipid, as determined by H&E, Movat's pentachrome and CD68 citeRN192, RN32.

Table 6.1 Arterial wall morphology distinguishable by FLIm-IVUS

Arterial wall morphology	Composition	Fluorophores in luminal 200 μm depth	IVUS feature
Normal artery wall	Thin intima, normal media	Elastin fibers of media	Thin intima
Diffuse intimal thickening	Thickened intima	Collagen fibers of thickened intima, elastin fibers of media depending on depth of intima	Thicker intima
Pathologic intimal thickening	Thickened intima with macrophages and extracellular lipid deposits	Collagen fibers of thickened intima, lipid, and ceroid from foam cell macrophages and lipid pools	Thicker intima
Fibrocalcific plaque	Fibrotic plaque with calcified fibrous cap and/or calcified necrotic core	Collagen fibers of the fibrous plaque	Thickened intima with calcium
Thin-fibrous cap atheroma	Thin-fibrous cap (< 65 μm) infiltrated with macrophages over a large lipidic or necrotic core	Collagen fibers from thin-fibrous cap, ceroid, and lipid from foam cells and macrophages	Large plaque burden
Thick-fibrous cap atheroma	Thick-fibrous cap (> 65 μm) with or without macrophages over a large lipidic or necrotic core	Collagen fibers from thick-fibrous cap, ceroid, and lipid from foam cells if they are present	Large plaque burden

Table 6.1 shows how FLIm and IVUS complement each other to identify various plaque types. Figure 6.4 shows how IVUS and FLIm can be used together to determine plaque composition. When combined with FLIm, IVUS is predominantly useful for determining plaque burden and the presence of calcium. Table 6.1 also depicts how IVUS and FLIm are used together to identify specific plaque compositions.

The first studies in ex vivo human samples using a FLIM-IVUS catheter were performed with the sequential scanning catheter discussed above (Design 2) (Fatakdawala et al. 2015). $N = 16$ left anterior descending coronary artery segments ($N = 16$ cadavers) were imaged with this system in custom-built artery holders and correlated to 8 distinct pathological features: diffuse intimal thickening (DIT), pathologic intimal thickening (PIT), thick-capped fibroatheroma (ThCFA), thick-capped fibroatheroma with macrophages (ThCFAM), thin-capped fibroatheroma (TCFA), thin-capped fibroatheroma with macrophages (TCFAM), fibrocalcific plaque (FC),

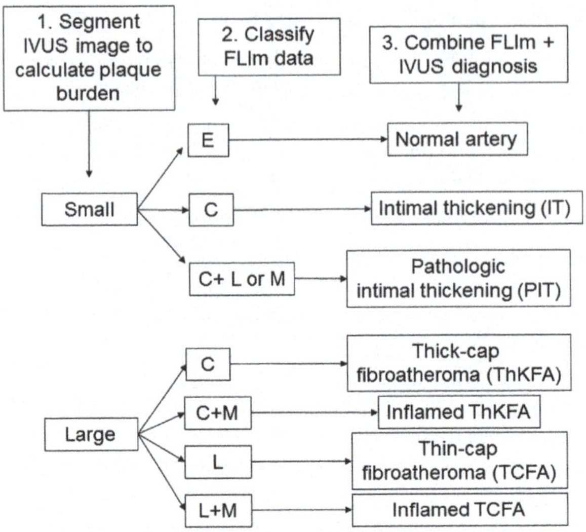

Fig. 6.4 Demonstration of classification scheme to combine FLIm and IVUS parameters for plaque characterization

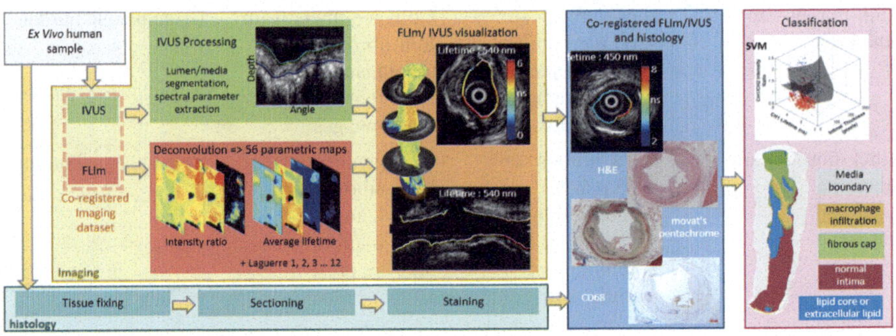

Fig. 6.5 Process for bimodal FLIm-IVUS imaging of ex vivo human coronary artery samples, validation with conventional histopathology, and tissue classification

and fibrotic tissue (FT). Histologic validation included trichrome staining and immunohistochemistry analysis (CD68 and CD45). Figure 6.5 depicts the work flow for artery imaging, histology co-registration, and data analysis.

Support vector machine (SVM) classification was employed in this study and allowed FLIm-IVUS to detect macrophages in fibrous caps with 86% sensitivity and distinguish between stable ThCFA and rupture-prone TCFA with 80% sensitivity (Fatakdawala et al. 2015). This study showed also that when combined, IVUS and FLIm perform better than when used independently for plaque characterization, verifying that this is a viable technique moving forward. Thus, the next steps in this work were to combine the modalities into a single catheter to allow for improved ease

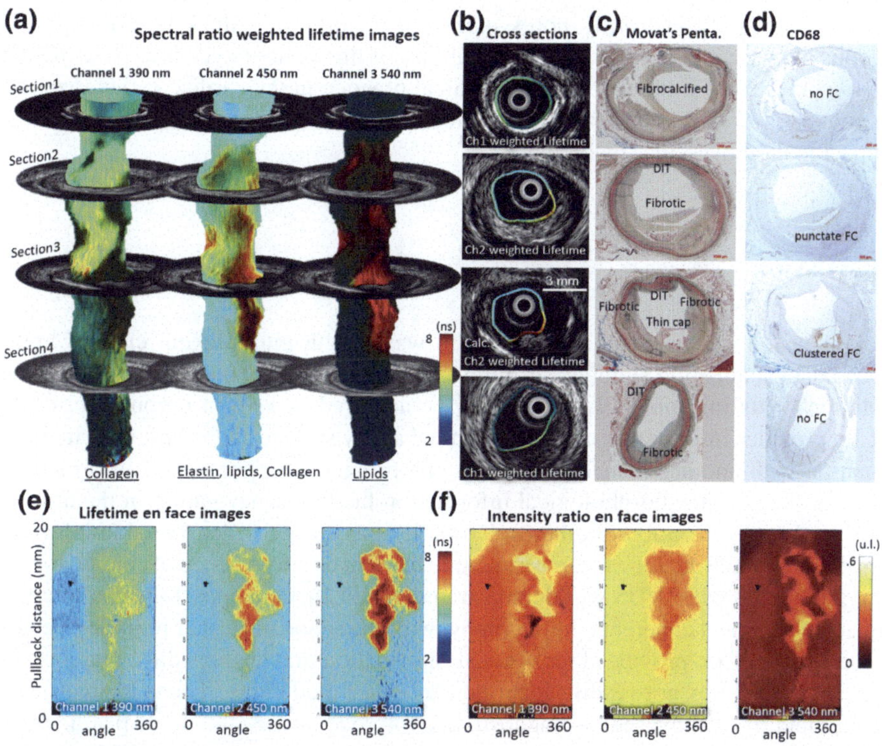

Fig. 6.6 Fluorescence lifetime information supplements IVUS in assessing atherosclerotic lesion pathophysiology. **a** Spectral ratio weighted lifetime images. **b** FLIm-IVUS cross sections. **c** Corresponding Movat's pentachrome. **d** Corresponding CD68. (3) *En face* lifetime images. **f** *En face* intensity ratio images. Figure 6.6 reprints with permissions (Bec et al. 2017)

of use and improved accuracy of co-registration between FLIm and IVUS data since they will be acquired simultaneously. This led to Design 3 that combines FLIm and IVUS in a single imaging core. This system was used to acquire the data presented in Fig. 6.6.

These results demonstrate how FLIm and IVUS complement each other to determine plaque type. Figure 6.6e displays FLIm maps of an ex vivo human coronary artery, angle (0–360 °C) on the x-axis and pullback distance (0–20 mm) on the y-axis for 3 channels. Figure 6.6f shows spectral intensity ratios for the same 3 channels. Figure 6.6a depicts 3D renderings of the arterial wall using the IVUS lumen segmentation to identify the luminal shape, and the spectral ratio weighted lifetime maps as the color map. Individual FLIm-IVUS frames shown in Fig. 6.6a are shown in Fig. 6.6b. Movat's pentachrome (Fig. 6.6c) and CD68 (Fig. 6.6d), enable plaque type identification and highlight the presence of macrophages in this vessel. Additional ex vivo arteries have been imaged with this generation of the catheter. Results from these studies found that this generation of the FLIm-IVUS catheter is able to discriminate lipid and necrotic cores, macrophage infiltration, thick-fibrous caps and fibrous plaques, and normal artery. The ability of this system to characterize plaque in a manner compatible with current clinical practice in the cardiac catheterization laboratory makes this an exciting new technique with great potential for clinical impact in cardiovascular medicine.

Discussion

In conclusion, pulse sampling FLIm with wavelength multiplexing enables high-speed data acquisition suitable for intravascular use. In combination with a dextran solution bolus injection, FLIm data can be acquired in vivo in coronary arteries. FLIm provides information about the luminal artery surface and is best used in combination with another imaging modality that provides morphological information, such as IVUS. This morphological information facilitates navigation in the arterial tree, enables localization of atherosclerotic lesions and provides valuable information including the degree of stenosis, plaque burden, or the presence of calcifications. A combination with IVUS was demonstrated, but FLIm could possibly also be combined with optical coherence tomography, with the scope of further reducing device dimensions. FLIm provides biochemical information that enables differentiation of plaque phenotypes not readily identified by other modalities. There is great need in the field of cardiovascular diagnostics to improve understanding of plaque morphology and features that predispose a plaque to cause future cardiovascular events. FLIm-IVUS can recognize features of thin-capped fibroatheroma and may also be able to detect other vulnerable features, such as erosion. Thus, FLIm-IVUS is a promising research tool for the study of atherosclerosis. The first study in patients will be needed to demonstrate the benefits of technology in clinical practice.

References

Andersson-Engels S, Johansson J, Stenram U, Svanberg K, Svanberg S (1990) Time-resolved laser-induced fluorescence spectroscopy for enhanced demarcation of human atherosclerotic plaques. J Photochem Photobiol B Biology 4(4):363–369

Angheloiu GO, Haka AS, Georgakoudi I, Arendt J, Muller MG, Scepanovic OR, Evanko SP, Wight TN, Mukherjee P, Waldeck DH, Dasari RR, Fitzmaurice M, Kramer JR, Feld MS (2011) Detection of coronary atherosclerotic plaques with superficial proteoglycans and foam cells using real-time intrinsic fluorescence spectroscopy. Atherosclerosis 215(1):96–102

Arakawa K, Isoda K, Ito T, Nakajima K, Shibuya T, Ohsuzu F (2002) Fluorescence analysis of biochemical constituents identifies atherosclerotic plaque with a thin fibrous cap. Arterioscler Thromb Vasc Biol 22(6):1002–1007

Baraga JJ, Taroni P, Park YD, An K, Maestri A, Tong LL, Rava RP, Kittrell C, Dasari RR, Feld MS (1989) Ultraviolet-laser induced fluorescence of human aorta. Spectrochimica Acta Part A Mol Biomol Spectrosc 45(1):95–99

Baraga JJ, Rava RP, Taroni P, Kittrell C, Fitzmaurice M, Feld MS (1990) Laserinduced fluorescence spectroscopy of normal and atherosclerotic human aorta using 306-310 nm excitation. Lasers Surg Med 10(3):245–261

Bartorelli AL, Leon MB, Almagor Y, Prevosti LG, Swain JA, McIntosh CL, Neville RF, House MD, Bonner BF (1991) In vivo human atherosclerotic plaque recognition by laser-excited fluorescence spectroscopy. J Am Coll Cardiol 17(6, Suppl B):160–168

Bec J, Xie H, Yankelevich DR, Zhou F, Sun Y, Ghata N, Aldredge R, Marcu L (2012) Design, construction, and validation of a rotary multifunctional intravascular diagnostic catheter combining multispectral fluorescence lifetime imaging and intravascular ultrasound. J Biomed Opt 17(10):106012

Bec J, Ma DM, Yankelevich DR, Liu J, Ferrier WT, Southard J, Marcu L (2014) Multispectral fluorescence lifetime imaging system for intravascular diagnostics with ultrasound guidance: in vivo validation in swine arteries. J Biophotonics 7(5):281–285

Bec J, Gorpas D, Ma D, Fatakdawala H, Phipps JE, Margulies KB, Southard JA, Marcu L (2016) Integrated intravascular ultrasound and multispectral fluorescence lifetime catheter system for label-free simultaneous structural and biochemical imaging of arteries: a study in vivo swine heart coronaries and in explanted diseased human coronaries. Circ Cardiovasc Interv. (vol. in review)

Bec J, Phipps J, Gorpas D, Ma D, Fatakdawala H, Margulies K, Southard J, Marcu L (2017) In vivo label-free structural and biochemical imaging of coronary arteries using an integrated ultrasound and multispectral fluorescence lifetime catheter system. Scientific Reports 7(1):8960

Christov A, Dai E, Drangova M, Liu L, Abela GS, Nash P, McFadden G, Lucas A (2000) Optical detection of triggered atherosclerotic plaque disruption by fluorescence emission analysis. Photochem Photobiol 72(2):242–252

De Beule P, Owen DM, Manning HB, Talbot CB, Requejo-Isidro J, Dunsby C, McGinty J, Benninger RK, Elson DS, Munro I, John Lever M, Anand P, Neil MA, French PM (2007) Rapid hyperspectral fluorescence lifetime imaging. Microsc Res Tech 70(5):481–484

Dowling K, Dayel MJ, Lever MJ, French PM, Hares JD, Dymoke-Bradshaw AK (1998) Fluorescence lifetime imaging with picosecond resolution for biomedical applications. Opt Lett 23(10):810–812

Edholm P, Jacobson B (1965) Detection of aortic atheromatosis in vivo by reflection spectrophotometry. J Atherosclerosis Research 5(6):592–595

Fatakdawala H, Gorpas D, Bishop JW, Bec J, Ma D, Southard JA, Margulies KB, Marcu L (2015) Fluorescence lifetime imaging combined with conventional intravascular ultrasound for enhanced assessment of atherosclerotic plaques: an ex vivo study in human coronary arteries. J Cardiovasc Transl Res 8(4):253–263

Fitzmaurice M, Bordagaray JO, Engelmann GL, Richards-Kortum R, Kolubayev T, Feld MS, Ratliff NB, Kramer JR (1989) Argon ion laser-excited autofluorescence in normal and atherosclerotic aorta and coronary arteries: morphologic studies. Am Heart J 118(5 Pt 1):1028–1038

Frick K, Michael TT, Alomar M, Mohammed A, Rangan BV, Abdullah S, Grodin J, Hastings JL, Banerjee S, Brilakis ES (2014) Low molecular weight dextran provides similar optical coherence tomography coronary imaging compared to radiographic contrast media. Catheter Cardiovasc Interv 84(5):727–731

Gorpas D, Fatakdawala H, Bec J, Ma D, Yankelevich DR, Qi J, Marcu L (2015) Fluorescence lifetime imaging and intravascular ultrasound: co-registration study using ex vivo human coronaries. IEEE Trans Med Imaging 34(1):156–166

Jo JA, Fang Q, Marcu L (2005) Ultrafast method for the analysis of fluorescence lifetime imaging microscopy data based on the laguerre expansion technique. IEEE J Quant Electron 11(4):835–845

Jo JA, Park J, Pande P, Shrestha S, Serafino MJ, Rico Jimenez Jde J, Clubb F, Walton B, Buja LM, Phipps JE, Feldman MD, Adame J, Applegate BE (2015) Simultaneous morphological and biochemical endogenous optical imaging of atherosclerosis. Eur Heart J Cardiovasc Imaging 16(8):910–918

Kittrell C, Willett RL, de los Santos-Pacheo C, Ratliff NB, Kramer JR, Malk EG, Feld MS (1985) Diagnosis of fibrous arterial atherosclerosis using fluorescence. Applied Optics 24(15):2280–2281

Liu J, Sun Y, Qi J, Marcu L (2012) A novel method for fast and robust estimation of fluorescence decay dynamics using constrained least-squares deconvolution with laguerre expansion. Phys Med Biol 57(4):843–865

Ma D, Bec J, Yankelevich DR, Gorpas D, Fatakdawala H, Marcu L (2014) Rotational multispectral fluorescence lifetime imaging and intravascular ultrasound: bimodal system for intravascular applications. J Biomed Opt 19(6):066004

Ma D, Bec J, Gorpas D, Yankelevich D, Marcu L (2015) Technique for real-time tissue characterization based on scanning multispectral fluorescence lifetime spectroscopy (ms-trfs). Biomed Opt Express 6(3):987–1002

Maarek JMI, Snyder WJ, Grundfest WS (1997) Time-resolved laser-induced fluorescence of arterial wall constituents: deconvolution algorithm and spetrotemporal characteristics. Proc SPIE 2980:278–285

Maarek JMI, Marcu L, Grundfest WS (1998) Characterization of atherosclerotic lesions with laser-induced time-resolved fluorescence spectroscopy. Proc SPIE 3250:181–187

Maarek JMI, Marcu L, Fishbein MC, Grundfest WS (2000) Time-resolved fluorescence of human aortic wall: use for improved identification of atherosclerotic lesions. Lasers Surg Med 27(3):241–254

Marcu L, Maarek JMI, Grundfest WS (1998) Time-resolved laser-induced fluorescence of lipids involved in development of atherosclerotic lesion lipid-rich core. Proc SPIE 3250:158–167

Marcu L, Fishbein MC, Maarek JMI, Grundfest WS (2001) Discrimination of human coronary artery atherosclerotic lipid-rich lesions by time-resolved laser-induced fluorescence spectroscopy. Arterioscler Thromb Vasc Biol 21(7):1244–1250

Marcu L, Grundfest WS, Fishbein M (2003) Time-resolved laser-induced fluorescence spectroscopy for staging atherosclerotic lesions. In: Book section 12. Marcel Dekker, Inc., New York, pp 397–430

Marcu L, Fang QY, Jo JA, Papaioannou T, Dorafshar A, Reil T, Qiao JH, Baker JD, Freischlag JA, Fishbein MC (2005) In vivo detection of macrophages in a rabbit atherosclerotic model by time-resolved laser-induced fluorescence spectroscopy. Atherosclerosis 181(2):295–303

Morguet AJ, Korber B, Abel B, Hippler H, Wiegand V, Kreuzer H (1994) Autofluorescence spectroscopy using a xecl excimer-laser system for simultaneous plaque ablation and fluorescence excitation. Lasers Surg Med 14(3):238–248

Munro I, McGinty J, Galletly N, Requejo-Isidro J, Lanigan PM, Elson DS, Dunsby C, Neil MA, Lever MJ, Stamp GW, French PM (2005) Toward the clinical application of time-domain fluorescence lifetime imaging. J Biomed Opt 10(5):051403

Ozaki Y, Kitabata H, Tsujioka H, Hosokawa S, Kashiwagi M, Ishibashi K, Komukai K, Tanimoto T, Ino Y, Takarada S, Kubo T, Kimura K, Tanaka A, Hirata K, Mizukoshi M, Imanishi T, Akasaka T (2012) Comparison of contrast media and low-molecular-weight dextran for frequency-domain optical coherence tomography. Circ J 76(4):922–927

Papazoglou TG, Liu WQ, Katsamouris A, Fotakis C (1994) Laser-induced fluorescence detection of cardiovascular atherosclerotic deposits via their natural emission and hypocrellin (ha) probing. J Photochem Photobiol B Biology 22(2):139–144

Phipps JE, Hatami N, Galis ZS, Baker JD, Fishbein MC, Marcu L (2011a) A fluorescence lifetime spectroscopy study of matrix metalloproteinases-2 and-9 in human atherosclerotic plaque. J Biophotonics 4(9):650–658

Phipps J, Sun YH, Saroufeem R, Hatami N, Fishbein MC, Marcu L (2011) Fluorescence lifetime imaging for the characterization of the biochemical composition of atherosclerotic plaques. J Biomed Opt 16(9)

Spite M, Serhan CN (2011) Lipid signatures of unstable atheromas: fossils or a step toward personalized lipidomics-metabolomics? Circ Cardiovasc Genet 4(3):215–217

Stary HC, Chandler AB, Glagov S, Guyton JR, Insull W, Rosenfeld ME, Schaffer SA, Schwartz CJ, Wagner WD, Wissler RW (1994) A definition of initial, fatty streak, and intermediate lesions of atherosclerosis. a report from the committee on vascular lesions of the council on arteriosclerosis, american heart association. Circulation 89(5):2462–2478

Su J, Greiner C, Grainger S, Saybolt M, Pickering W, Wilensky R, Raichlen J, He V, Sum S, Muller J, Madden S (2016) Tct-575 combined near-infrared spectroscopy and intravascular ultrasound (nirs-ivus) coronary imaging as a means to improve prediction of events by ivus plaque burden alone. J Am Coll Cardiol 68(18S):B232–B233

Sun Y, Liu R, Elson DS, Hollars CW, Jo JA, Park J, Marcu L (2008) Simultaneous time- and wavelength-resolved fluorescence spectroscopy for near real-time tissue diagnosis. Opt Lett 33(6):630–632

Sun Y, Chaudhari AJ, Lam M, Xie HT, Yankelevich DR, Phipps J, Liu J, Fishbein MC, Cannata JM, Shung KK, Marcu L (2011a) Multimodal characterization of compositional, structural and functional features of human atherosclerotic plaques. Biomed Opt Express 2(8):2288–2298

Sun Y, Stephens D, Xie H, Phipps J, Saroufeem R, Southard J, Elson DS, Marcu L (2011b) Dynamic tissue analysis using time- and wavelength-resolved fluorescence spectroscopy for atherosclerosis diagnosis. Opt Express 19(5):3890–3901

Thomas P, Pande P, Clubb F, Adame J, Jo JA (2010) Biochemical imaging of human atherosclerotic plaques with fluorescence lifetime angioscopy. Photochem Photobiol 86(3):727–731

Upston JM, Niu X, Brown AJ, Mashima R, Wang H, Senthilmohan R, Kettle AJ, Dean RT, Stocker R (2002) Disease stage-dependent accumulation of lipid and protein oxidation products in human atherosclerosis. Am J Pathol 160(2):701–710

Verbunt RJAM, Fitzmaurice MA, Kramer JR, Ratliff NB, Kittrell C, Taroni P, Cothren RM, Baraga J, Feld M (1992) Characterization of ultraviolet laser-induced autofluorescence of ceroid deposits and other structures in atherosclerotic plaques as a potential diagnostic for laser angiosurgery. Am Heart J 123(1):208–216

Virmani R, Burke AP, Farb A, Kolodgie FD (2006) Pathology of the vulnerable plaque. J Am Coll Cardiol 47(8 Suppl):13–18

Chapter 7
Intravascular Dual-Modality Imaging (NIRF/IVUS, NIRS/IVUS, IVOCT/NIRF, and IVOCT/NIRS)

Yan Li and Zhongping Chen

Introduction

Coronary artery disease (CAD) is the leading cause of global mortality (Kolodgie et al. 2001; White and Chew 2008; Bentzon et al. 2014). Atherosclerosis, a chronic disease typically asymptomatic at early stages, is characterized by the thickening of the arterial vessel wall due to the buildup of atherosclerotic plaque in the inner lining of arteries. Vulnerable atherosclerotic plaque, which is composed of a large lipid-rich necrotic core (NC) infiltrated with abundant macrophages and a thin fibrous cap, is widely recognized to be the main cause of underlying acute coronary artery disease (Muller et al. 1989; Virmani et al. 2000). Computed tomography (CT) angiography has been the gold standard technology for evaluating coronary arterial disease for the past 50 years. However, due to the inability to supply information in regard to the coronary wall, intravascular imaging, such as intravascular ultrasound (IVUS), intravascular optical coherence tomography (IVOCT), near-infrared fluorescence (NIRF) imaging, and near-infrared reflectance spectroscopy (NIRS), has been developed to provide supplementary information for plaque characterization (Brezinski et al. 1996; Yang et al. 2010; Puri et al. 2011; Benni et al. 1995; Fard et al. 2013; Yamada et al. 1995; Brezinski et al. 1997).

IVOCT based on low-coherence interferometry provides three-dimensional microscopic images of blood vessels with high resolution which are used to capture arterial microstructural detail, such as thin fibrous cap and microvasculature (Brezinski et al. 1996; Yaqoob et al. 2006). IVOCT has been demonstrated by several groups for imaging and evaluation of vulnerable plaques (Li et al. 2017a; Brezinski et al. 1996, 1997; Yaqoob et al. 2006; Fard et al. 2013). However, due to its limited

Y. Li · Z. Chen (✉)
Beckman Laser Institute, University of California, Irvine, Irvine, CA 92697, USA
e-mail: z2chen@uci.edu

Y. Li
e-mail: yanl30@uci.edu

Q. Zhou and Z. Chen (eds.), *Multimodality Imaging*,
https://doi.org/10.1007/978-981-10-6307-7_7

penetration depth, it cannot resolve the full depth of a large lipid pool (the key characteristic of vulnerable plaque) in plaque.

IVUS based on echo delay of high-frequency sound waves from different depths of the biological tissue is able to provide large penetration depth, cross-sectional images of the coronary vessel in vivo. In daily clinical practice, IVUS is increasingly used for visualization of the coronary lumen, vessel wall, and atherosclerotic plaque formation (Yamada et al. 1995; Nissen and Yock 2001). However, current IVUS has limited resolution to evaluate the thickness of the thin fibrous cap (the key characteristic of vulnerable plaque) for plaque classifications.

Both IVUS and IVOCT provide structural information of the arterial wall but lack molecular specificity for identification of lipid core-containing coronary plaques (LCP) (Li et al. 2010; Yang et al. 2010; Yin et al. 2010; Li et al. 2013, 2014, 2015). NIRF imaging utilizes molecular probes or autofluorescence to provide complementary information with regard to plaque activity and inflammation (Giovanni et al. 2016; Lee et al. 2014; Abran et al. 2015). In addition, a NIRS method has been developed which has the capability of providing chemical components assessment related to the presence of cholesterol esters in lipid cores and generating spectra that distinguish cholesterol from collagen in coronary plaques through their unique spectroscopic fingerprints (Benni et al. 1995; Waxman et al. 2009; Brilakis and Banerjee 2015). Each intravascular imaging technology has its unique advantages and is able to provide partial features of vulnerable plaque, but it is still insufficient to obtain accurate diagnosis if only one imaging technology is applied. In order to have a better characterization of atherosclerotic plaque, a dual-modality intravascular imaging system (such as integrated NIRS/IVUS, NIRF/IVUS, IVOCT/NIRS, and IVOCT/NIRF imaging systems) (Roleder et al. 2014; Fard et al. 2013; Lee et al. 2014; Abran et al. 2015) have been developed with the aim of identifying multiple features of the arterial wall.

This chapter outlines several representative dual-modality intravascular imaging systems which combine IVOCT or IVUS with NIRS or NIRF imaging technologies. In addition, the in vivo and ex vivo experimental results obtained by these dual-modality imaging systems are presented and discussed.

Principle

OCT is based on low-coherence interferometry (Huang et al. 1991; Brezinski et al. 1996). Light from a low-coherence source is split into two light beams by a fiber optic coupler. The light beam with low energy will go through a circulator and then a reference arm including a collimator, a lens, and a mirror. Another light beam will go through a circulator, optics rotary joint and imaging probe to illuminate the biological tissue as a sample arm. The backscattered light from the sample arm and backreflected light from reference arm generate an interference signal through a 50:50 fiber optic coupler. Then, the interference signal is detected by a balanced photodetector. For intravascular imaging (Brezinski et al. 1996; Jang et al. 2002), the light is scanned sideways to perform cross-sectional imaging. Therefore, an optical rotary joint is often used to allow uninterrupted transmission of an optical signal during the scanning. A rotary motor and translation stage are also incorporated to drive the imaging probe to perform three-dimensional (3D) imaging.

Ultrasound imaging is based on the oscillatory movement (expansion and contraction) of an acoustic transducer which is able to generate acoustic waves when electrically excited and receive acoustic waves from the biological tissue (Zacharatos et al. 2010; Stahli et al. 2017). When the acoustic waves penetrate biological tissue with different impedances, some acoustic waves are reflected back to the transducer (echo signal) and some continue to penetrate deeper. The returned echo signals are detected by the same acoustic transducer, and an ultrasound image can be reconstructed based on the time delay of echo signals from different layers. For IVUS imaging, a single-element piezoelectric transducer is often applied to generate and detect ultrasound signals. During imaging, the acoustic transducer is rotated in order to obtain cross-sectional images.

NIRF based on exogenous or endogenous biomarkers is used to provide molecular contrast of biological tissue. Several works have reported the autofluorescence signal in cadaver coronary arteries excited by a 633-nm wavelength (Wang et al. 2015; Giovanni et al. 2016). According to the intensity of detected NIRF signals from endogenous biomarkers, different plaque types (normal vessel wall, fibrotic tissue, fibrocalcific plaque, thick-cap fibroatheroma, thin-cap fibroatheroma, and ruptured plaque) can be identified. In addition, several groups have proposed that Food and Drug Administration–approved indocyanine green (ICG) is able to bind to lipoproteins and also accumulates in inflamed tissues (Lee et al. 2014; Abran et al. 2015; Yoneya et al. 1998; Fischer et al. 2006; Vinegoni et al. 2011). Therefore, atherosclerotic plaque can be identified using the exogenous biomarker ICG.

NIRS (Benni et al. 1995; Chen et al. 2011) is a spectroscopic method based on molecular overtone and combination vibrations. Due to the unique combinations of carbon–hydrogen (C–H), nitrogen–hydrogen (N–H), and oxygen–hydrogen (O–H) bonds that are responsible for the major absorption of NIR light, different compositions have unique absorption patterns which are able to provide quantitative composition characterization (Moreno and Muller 2002; Kilic et al. 2015). For intravascular imaging, NIRS has been investigated for the identification of atherosclerotic plaque composition by analyzing absorption spectra. In addition, due to the low absorption of hemoglobin at the near-infrared range, NIRS is capable of identifying plaque composition.

Methods and Results

Integrated IVOCT/NIRS and IVOCT/NIRF Imaging Systems

OCT has superior spatial resolution and is a promising tool for imaging the microstructure of coronary artery walls. However, it is insufficient for accurate characterization of atherosclerotic plaque because of an inability to supply a molecular contrast that is a key factor for determining vulnerability. NIRF is able to provide molecular contrast for the identification of inflammation reaction and lipid components using an endogenous or exogenous biomarker. Therefore, a dual-modality IVOCT/NIRS (or IVOCT/NIRF) imaging system which can perform IVOCT and NIRS (NIRF) simultaneously is capable of providing both microstructure and compositional information of biological tissue for better plaque characterization.

Integrated IVOCT/NIRS System

A dual-modality IVOCT/NIRS imaging system has been reported by Tearney's group (Fard et al. 2013). The system applies a single-swept light source (center wavelength: 1282 nm; bandwidth: ~106 nm; repetition rate: 100 kHz) to perform both IVOCT and NIRS imaging. In addition, the utilization of a swept light source allows one to detect the spectrum in the time domain using a single-pixel photoreceiver. The schematic of the dual-modality IVOCT/NIRS is shown in Fig. 7.1. For IVOCT, the light from the swept light source IVOCT is divided into two beams by a 90/10 optical power splitter. The light with 10% power is delivered to the reference arm. Another one is delivered to the sample arm and then goes through the core of a double-clad fiber (DCF) coupler, rotary joint and dual-modality IVOCT/NIRS imaging probe to illuminate the tissue. The function of the DCF coupler is to propagate the IVOCT light by a single-mode core of DCF with low light loss and separate the IVOCT backscattered light and NIRF diffuse light from the sample. The backscattered light from IVOCT and backreflected light from the reference arm will generate interference through a 50:50 fiber coupler. Then, the generated signal will be detected by a balanced photodetector. For NIRS detection, the backscattered light is collected through the inner cladding of the DCF and then detected by a photoreceiver.

The IVOCT-NIRS imaging probe consists of a single-mode fiber (SMF), multimode fiber (MMF), and DCF. The light propagates through the single-mode core of the DCF, a power combiner, and then goes through a single-mode fiber to illuminate the tissue. For transmission and detection, OCT light always propagates in the single-mode core. Regarding the NIRS, the backscattered light is collected by a MMF. Then, the multimode light and single-mode light are combined into a DCF through the power combiner. At the distal tip, both SMF and MMF are terminated by two angle-cleaved ball lenses as shown in Fig. 7.2. In addition. This configuration allows the light to penetrate the biological tissue deeper and make the NIRS detection in the diffuse regime possible. In order to perform cross-sectional imaging, the

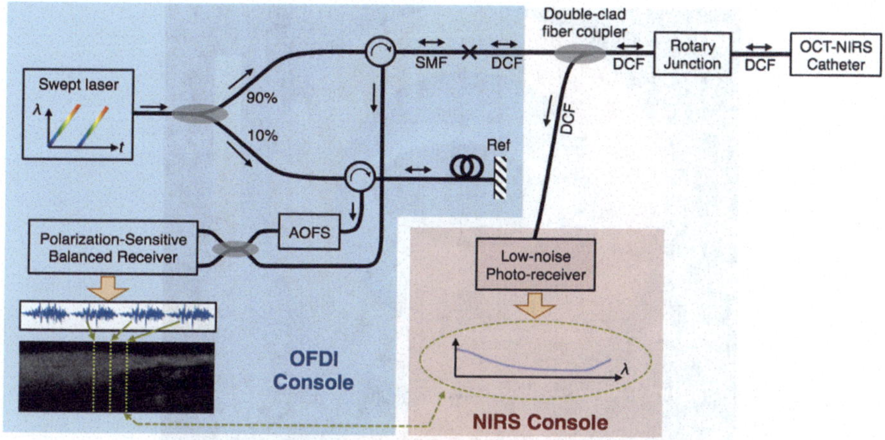

Fig. 7.1 Schematic of a dual-modality IVOCT and NIRS imaging system. OFDI: optical frequency domain imaging; SMF: single-mode fiber; DCF: double-clad fiber; AOFS: acousto-optic frequency shifter

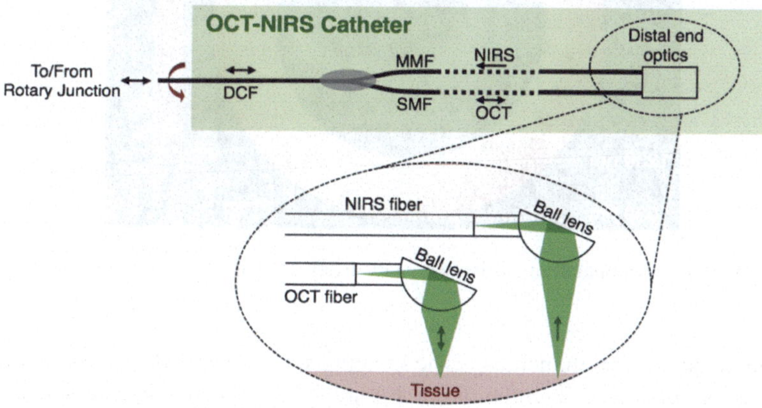

Fig. 7.2 Schematic for IVOCT/NIRS imaging probe. SMF delivers and collects OCT backscattered light to/from tissue and an MMF collects NIRS light. The two fibers are combined into the DCF using a commercial power combiner

fibers and power combiner are housed in a cardiovascular drive shaft and stainless steel tubing. A transparent plastic sheath is used to enclose the drive shaft and distal optics.

Ex Vivo Studies

In order to demonstrate the capability of this dual-modality IVOCT/NIRS imaging system, human coronary arteries were imaged ex vivo. IVOCT and NIRS data

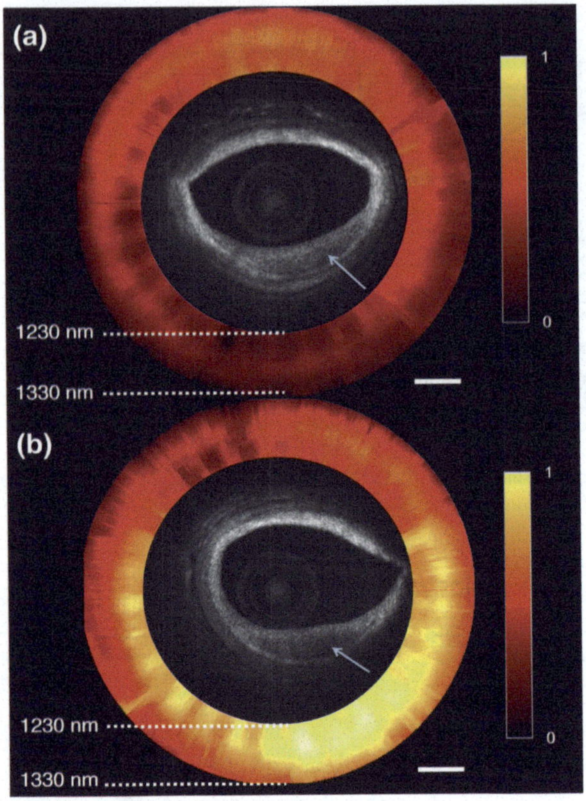

Fig. 7.3 IVOCT/NIRS images of cadaver coronary artery ex vivo

sets were acquired simultaneously. The imaging was performed in water in order to minimize the reflection from the tissue. Figure 7.3 shows two representative cross-sectional IVOCT/NIRS images of human coronary artery obtained from ex vivo imaging. The NIRS image (outer) shows the absorption spectrum versus wavelengths for each A-line. Figure 7.3a shows a lesion indicated by the white arrow that has a low OCT signal. The corresponding NIRS image also demonstrates a homogenous flat absorption spectrum which indicates the low possibility of a high lipid concentration. In Fig. 7.3b, a lesion with a low OCT signal can be visualized, denoted by the white arrow. According to reported studies (Fleming et al. 2013; Madder et al. 2013), a negative absorption slope in the near-infrared wavelengths (1200–1400 nm) corresponds to lipid content. In Fig. 7.3b, a high absorption at a shorter wavelength is shown in the NIRS images which suggests that the plaque in Fig. 7.3b contains abundant lipids. The ex vivo imaging result shows the dual-modality imaging system is capable of providing structural information and chemical contrast.

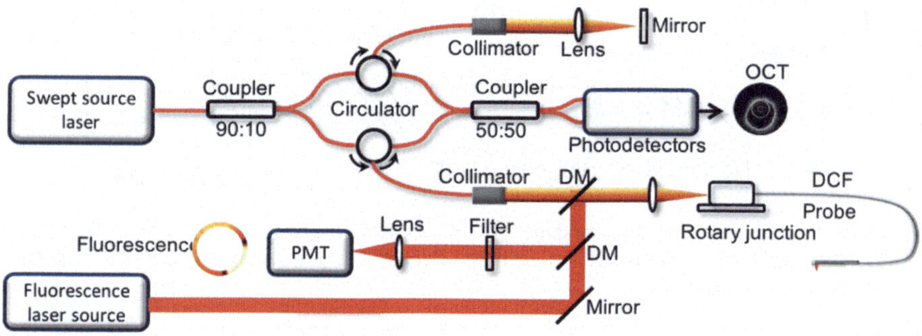

Fig. 7.4 Overall design of the dual-modality NIRF/OCT imaging system. DM: dichroic mirror. PMT: photomultiplier tubes

Integrated IVOCT/NIRF System

NIRF based on exogenous or endogenous biomarkers has demonstrated the capability of identifying vulnerable plaque. Integrated with IVOCT, dual-modality IVOCT/NIRF has the capability to provide microstructure and molecular contrast (Yoo et al. 2011; Li et al. 2017b; Wang et al. 2015; Giovanni et al. 2016). Figure 7.4 illustrates the overall setup of the dual-modality IVOCT/NIRF imaging system. Typically, a dichroic mirror is utilized to combine the OCT and fluorescence laser beams together. The combined beam is delivered to tissue through the imaging probe. For the IVOCT system, the output light is split by a 90:10 coupler. The reference arm (low energy) has a fixed delay. The sample arm (high energy) includes a rotary junction and imaging probe for transmitting the light to the tissue. The backscattered light of OCT comes back through the same path. Then, the collected light from the sample arm and reference arm generates interference through a 50/50 coupler. The interference signal is detected by a balanced photodetector which is able to remove the noise by subtracting the two optical input signals from each other. For NIRF imaging, a continuous laser is normally used to excite the biological tissue. The center wavelength of the laser source depends on the absorption spectrum of an exogenous or endogenous biomarker. For an endogenous biomarker, a 633-nm wavelength has been demonstrated to be able to identify atherosclerosis plaque. For NIRF imaging based on an exogenous biomarker, ICG is often used to target the lipid-rich inflamed plaque so a 785-nm laser source which corresponds to the absorption peak of the ICG is used as the excitation source. The emission light can be separated from the excitation light by a dichroic mirror, then filtered by a bandpass filter, and detected by photomultiplier tubes (PMT).

The IVOCT/NIRF imaging probe is similar to the typical IVOCT imaging probe, as shown in Fig. 7.5. The difference is that a DCF is applied instead of a single-mode fiber to propagate the illumination light. The combined light can be delivered to tissue through the single-mode core of the DCF, and OCT backscattered light will

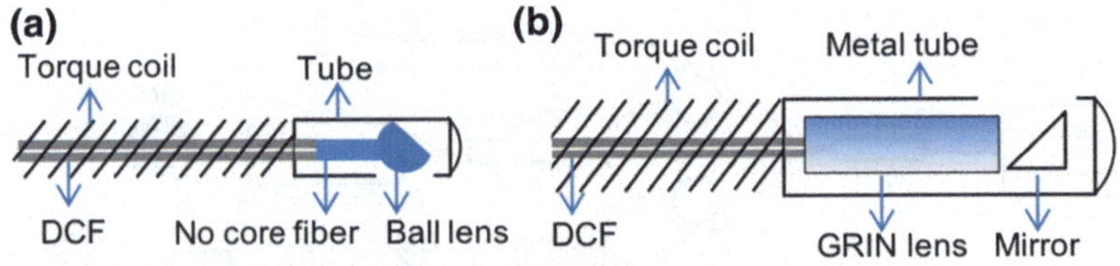

Fig. 7.5 Schematic of the dual-modality IVOCT/NIRF imaging probe. **a** Ball-lens-based IVOCT/NIRF imaging probe. **b** GRIN-lens-based IVOCT/NIRF imaging probe. GRIN: gradient index. DCF: double-clad fiber

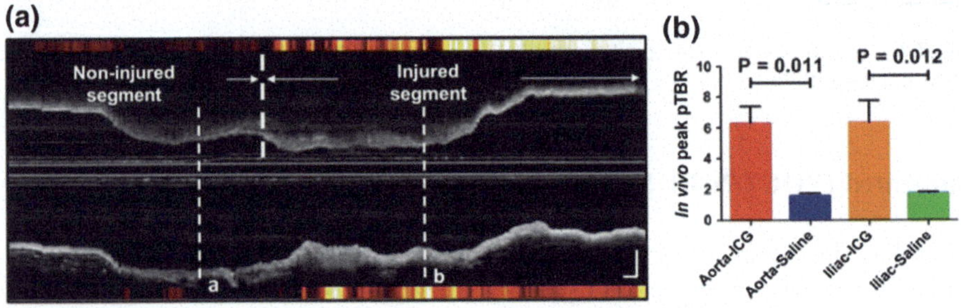

Fig. 7.6 Combined IVOCT/NIRF images of rabbit aorta. **a** IVOCT/NIRF cross-sectional image along longitudinal direction. **b** Comparison of peak plaque target-to-background ratio (TBR) signals in vivo between ICG-injected and saline-injected rabbits of aorta and iliac artery

return through the same path. Fluorescence emission light can be collected by the inner cladding of the DCF.

Animal and Clinical Validation Studies

For the integrated IVOCT/NIRF imaging system, in vivo rabbit studies have been reported by Lee et al. using an ICG-based integrated IVOCT/NIRF imaging system (Lee et al. 2014). Figure 7.6 shows representative co-registered IVOCT and NIRF images. In the IVOCT images, the microstructure of the coronary artery can be visualized clearly. From the NIRF images, the intensity corresponds to the compositional data of coronary artery plaque. The high intensity represents the high possibility of the existence of a lipid component. Figure 7.6b shows the plaque target-to-background ratio (TBR) of four groups (Aorta-ICG, Aorta-Saline, Iliac-ICG, and Iliac-Saline). From the results, it can be seen that the ICG-injected group has higher TBR compared with the saline-injected group at the injured section, which demonstrates the capability of this NIRF to identify the lesions from normal biological tissue.

Based on an endogenous biomarker, Ughi et al. have reported the first clinical imaging of human coronary arteries in vivo using a dual-modality IVOCT and NIRF imaging system in which IVOCT and NIRF data were obtained from 12 patients with coronary artery disease (Ughi et al. 2016). Figure 7.7 shows representative

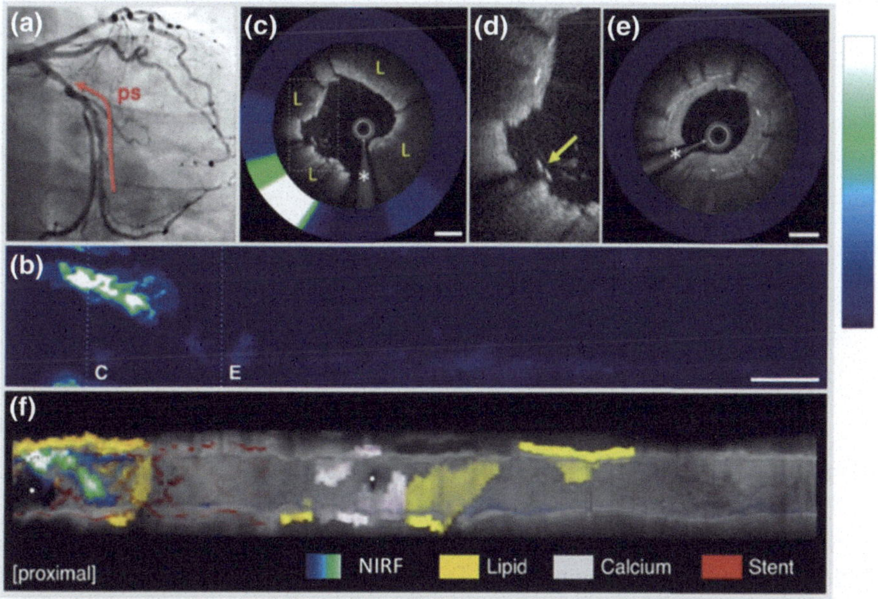

Fig. 7.7 IVOCT/NIRF imaging of human coronary arteries. **a** Angiography. **b** NIRF image. **c** Combined IVOCT/NIRF image of a diseased artery. **d** Enlarged IVOCT image. **e** Combined IVOCT/NIRF image of a healthy artery. **f** 3D cutaway rendering

IVOCT and fluorescence images. In the IVOCT image, microstructure of the arterial wall can be visualized. With respect to fluorescence imaging, it is excited by the 633-nm laser and can be used to further identify plaque types based on maximum NIRF intensities. From Fig. 7.7c, the lipid lesions are marked, and the NIRF signal is significantly high at one of these lesions. For a normal blood vessel (as is shown in Fig. 7.7e), NIRF intensity is low and homogenous, which agrees with IVOCT image results well. According to the maximum intensity change of the fluorescence signal, IVOCT/NIRF imaging has the potential to differentiate plaque types. Figure 7.8 shows the maximum fluorescence intensity of different plaque types.

The ex vivo and in vivo imaging results verify that the dual-modality IVOCT and NIRF system is able to provide microstructure and molecular contrast simultaneously which shows the potential for enhanced visualization, identification, and quantification of tissue contents. However, for NIRS, basic spectra for chemical composition need to be further studied in order to extract the different compositions of biological tissue. In addition, the mechanistic details with regard to ICG accumulation, such as factors determining its uptake by macrophages and binding to lipids, still require further study.

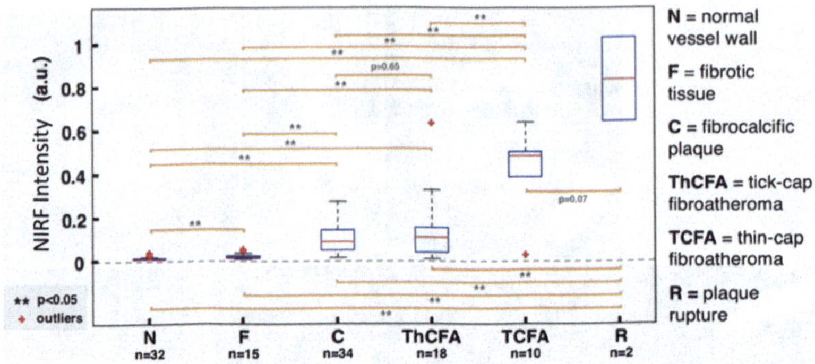

Fig. 7.8 Analysis of maximum NIRF plaque intensities for different plaque types

Integrated IVUS/NIRS and IVUS/NIRF Imaging Systems

NIRS/NIRF is a promising method for the identification of chemical components of atherosclerotic plaque composition. IVUS imaging offers direct visualization of the morphology of the arterial structures. Considering that both composition and structure are highly associated with vulnerability of atherosclerotic plaque, a dual-modality imaging system combining NIRS or NIRF with IVUS is essential for accurate diagnosis in the clinic. Currently, a hybrid technology combining NIRS and IVUS has been developed by InfraRedx Inc., TVC Imaging System^{TM}, which is able to obtain structural and compositional data of coronary artery plaques simultaneously. In addition, a dual-modality imaging system which integrates IVUS and NIRF together has been reported. This section will introduce combined NIRF/IVUS and NIRS/IVUS imaging systems and discuss the corresponding imaging results.

Integrated IVUS and NIRF System

The imaging system consisting of an NIRF system, IVUS system, and an imaging probe is shown in Fig. 7.9. For the NIRF part, a DCF coupler is often applied to propagate excitation light and collect the emission light. Alternatively, a free-space optical path can also be used to separate the excitation light and emission light. The center wavelength of the fluorescence laser is determined by the biomarker. For transmission, the fluorescence light goes through the single-mode core of the DCF from port A to port S, and the small diameter of the single-mode core contributes to high fluence on the surface tissue which enables a high-efficiency excitation. The emission light comes back from the first clad of the DCF and core (port S) to a multimode fiber (port B) whose larger diameter and higher numeric number (NA) enhance the capability of collecting emission light, filtered by a bandpass filter, and then detected by a photomultiplier tube. For IVUS imaging, a Pulser/Receiver is used

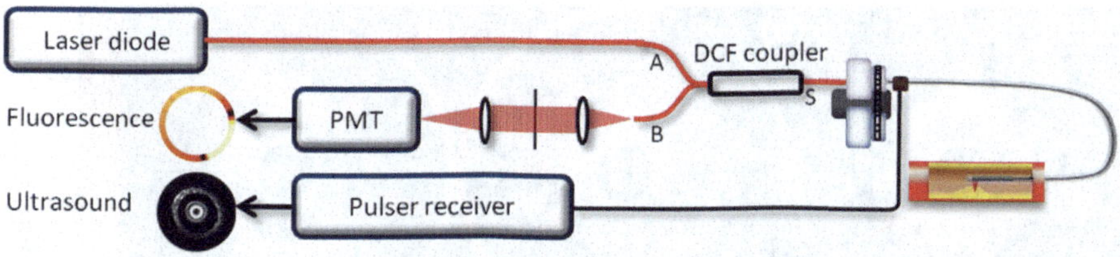

Fig. 7.9 Overall design of the dual-modality NIRF/IVUS imaging system. DCF: double-clad fiber. PMT: photomultiplier tube

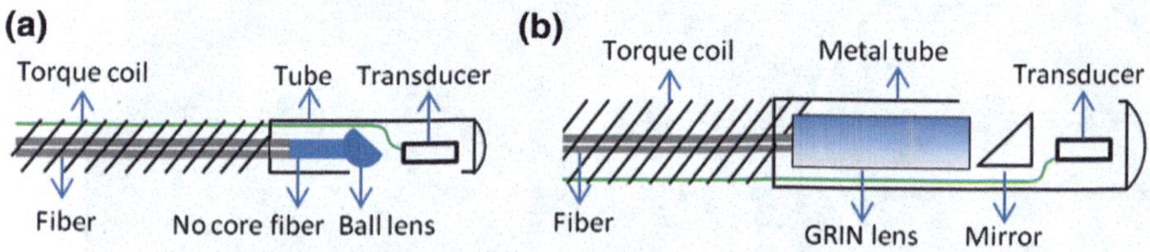

Fig. 7.10 Schematic of dual-modality NIRF/IVUS imaging probe. **a** Ball-lens-based NIRF/IVUS imaging probe. **b** GRIN-lens-based NIRF/IVUS imaging probe. GRIN: gradient index

to generate and detect ultrasound signals. In order to obtain co-registered IVUS and NIRF images, a trigger signal from the Pulser/Receiver is used to generate ultrasound signals and synchronize the data acquisition of IVUS and NIRF signals.

Figure 7.10 is the schematic of a typical NIRF/IVUS imaging probe. Optical components are similar to the IVOCT/NIRF imaging probe. In order to perform ultrasound imaging, an ultrasound transducer is sequentially aligned with the optical components.

In Vivo Animal Study

Abran et al. have imaged healthy and atherosclerosis rabbit aortas in vivo after ICG injection using a combined IVUS/NIRF imaging system (Abran et al. 2015). Figure 7.11 shows representative co-registered IVUS/NIRF images. Figures 7.11a–e show co-registered images from an atherosclerosis rabbit in which the echogenicity and intimal thickening can be visualized. In addition, an enhanced intensity of the NIRF signal can be found. Figure 7.11f shows co-registered images from a healthy rabbit in which a typical layered structure and relatively low and homogenous NIRF signal are demonstrated.

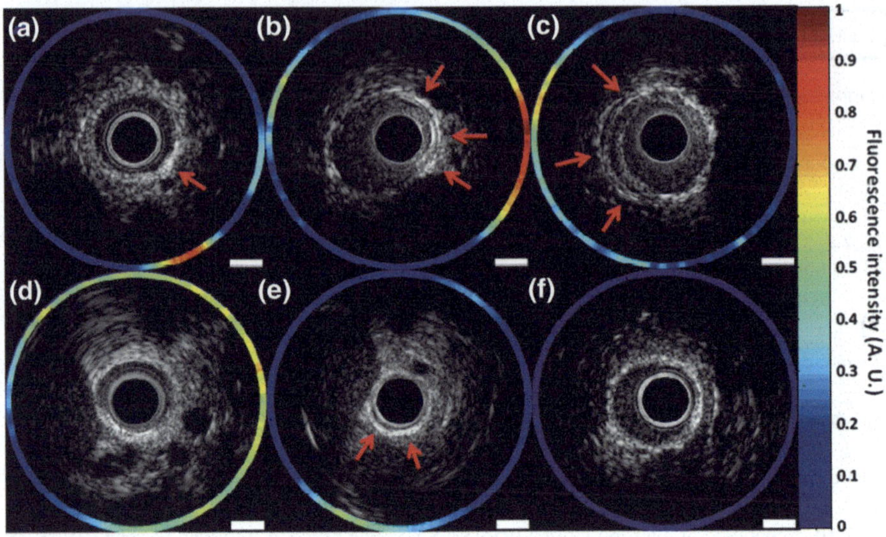

Fig. 7.11 Integrated IVUS and NIRF images from rabbit aorta. **a–e** Cross-sectional IVUS/NIRF images from cholesterol-enriched diet rabbit; **f** Cross-sectional IVUS/NIRF images from a regular diet rabbit. The contour in the IVUS image represents the fluorescence signal intensity. Red arrows show increased echogenicity in (**a**, **e**) and intimal thickening in (**b**, **f**)

Integrated IVUS and NIRS System

Figure 7.12 shows the commercial NIRS/IVUS imaging system. It consists of a 3.2-French rapid exchange catheter which is compatible with 6F-guiding catheters, a pullback and rotation device, and a console that houses the scanning NIR laser, computer and two monitors. For the NIRS/IVUS catheter, an optical fiber is applied to deliver near-infrared light onto biological tissue and collect the backscattered light in the form of an imaging spectrum. A single-element ultrasound transducer (center frequency: 40 MHz; axial resolution: 100 μm) was applied for IVUS imaging. The commercial NIRS/IVUS imaging system is able to perform imaging with a rotation speed of 16 rps and pullback speed of 0.5 mm/s. Currently, an upgraded version of the TVC Catheter Imaging System™ utilizes a new ultrasound transducer with an extended bandwidth to generate IVUS images at frequencies between 30 and 70 MHz, which contributes to improved resolution and imaging depth.

Clinical Validation Studies

Several clinical studies using NIRS/IVUS have been reported. Figure 7.13 is a combined NIRS/IVUS in vivo image from a human subject. The outer ring image is a

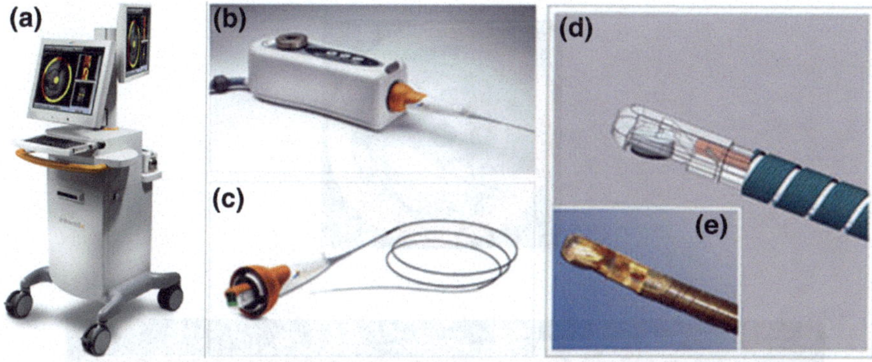

Fig. 7.12 Commercial NIRS/IVUS imaging system. **a** The console, which houses the scanning NIR laser, computer and two monitors. **b** Pullback and rotation device. **c** NIRS/IVUS imaging catheter. **d** Schematic of imaging catheter. **e** Photo of imaging catheter

near-infrared spectroscopy chemogram that indicates the location and intensity of the lipid core with the probability of lipid content presence. The inner gray image is a simultaneously acquired ultrasound image in which the structure of the entire arterial wall can be visualized.

The combined NIRS/IVUS imaging system is able to obtain IVUS and NIRS images simultaneously which provide both plaque composition and morphology of the arterial wall. In addition, combined NIRS/IVUS has shown the potential to provide improved LCP detection and fibroatheroma detection compared to either IVUS or NIRS used alone (Brilakis and Banerjee 2015; Puri et al. 2015). In the future, multiple ongoing clinical trials will hopefully validate this tool for vulnerable plaque detection as well as treatment management.

Summary

Dual-modality imaging technologies, including IVOCT/NIRF, IVOCT/NIRS, NIRF/IVUS, and NIRS/IVUS, are valuable tools that are able to provide both structure and molecular contrast for the characterization and quantification of cardiovascular tissue. Currently, IVOCT/NIRF and IVUS/NIRS have been applied in a clinical trial and demonstrate improved capability of diagnosis of cardiovascular disease. Such dual-modality technologies that combine structural and functional imaging together will further improve the capability of diagnosis and managment of cardiovascular diseases.

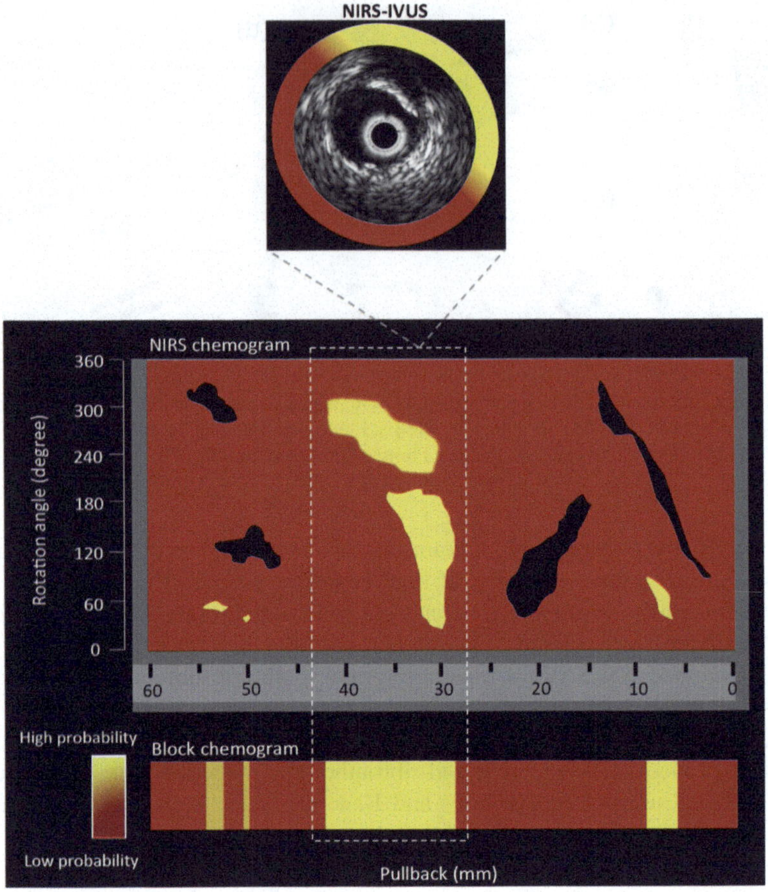

Fig. 7.13 Example of NIRS chemogram and IVUS

References

Abran M, Stahli BE, Merlet N, Mihalache-Avram T, Mecteau M, Rheaume E, Busseuil D, Tardif JC, Lesage F (2015) Validating a bimodal intravascular ultrasound (IVUS) and near-infrared fluorescence (NIRF) catheter for atherosclerotic plaque detection in rabbits. Biomed Opt Express 6(10):3989–3999. https://doi.org/10.1364/BOE.6.003989

Benni PB, Chen B, Amory D, Li JKJ (1995) A novel near-infrared spectroscopy (NIRS) system for measuring regional oxygen saturation. In: Proceedings of the IEEE 21st annual Northeast bioengineering conference, 105–107. https://doi.org/10.1109/nebc.1995.513751

Bentzon JF, Otsuka F, Virmani R, Falk E (2014) Mechanisms of plaque formation and rupture. Circ Res 114(12):1852–1866. https://doi.org/10.1161/CIRCRESAHA.114.302721

Brezinski ME, Weissman N, Tearney GJ, Boppart SA, Bouma B, Swanson E, Fujimoto JG (1996) High resolution intraarterial imaging with optical coherence tomography. In: Alfano RR, Fujimoto JG (eds), pp 200–202

Brezinski ME, Tearney GJ, Weissman NJ, Boppart SA, Bouma BE, Hee MR, Weyman AE, Swanson EA, Southern JF, Fujimoto JG (1997) Assessing atherosclerotic plaque morphology: comparison of optical coherence tomography and high frequency intravascular ultrasound. Heart 77(5):397–403

Brilakis ES, Banerjee S (2015) How to detect and treat coronary fibroatheromas the synergy between IVUS and NIRS. Jacc-Cardiovasc Imag 8(2):195–197. https://doi.org/10.1016/j.jcmg.2014.11.009

Chen QQ, Liu XD, Liu WQ, Jiang S (2011) Near infrared reflectance spectroscopy (NIRS): a novel approach to reconstructing historical changes of primary productivity in Antarctic Lake. Spectrosc Spect Anal 31(10):2688–2691. https://doi.org/10.3964/j.issn.1000-0593(2011)10-2688-04

Fard AM, Vacas-Jacques P, Hamidi E, Wang H, Carruth RW, Gardecki JA, Tearney GJ (2013) Optical coherence tomography–near infrared spectroscopy system and catheter for intravascular imaging. Opt Express 21(25):30849–30858. https://doi.org/10.1364/OE.21.030849

Fischer T, Gemeinhardt I, Wagner S, Stieglitz DV, Schnorr J, Hermann KG, Ebert B, Petzelt D, Macdonald R, Licha K, Schirner M, Krenn V, Kamradt T, Taupitz M (2006) Assessment of unspecific near-infrared dyes in laser-induced fluorescence imaging of experimental arthritis. Acad Radiol 13(1):4–13. https://doi.org/10.1016/j.acra.2005.07.010

Fleming CP, Eckert J, Halpern EF, Gardecki JA, Tearney GJ (2013) Depth resolved detection of lipid using spectroscopic optical coherence tomography. Biomed Opt Express 4(8):1269–1284. https://doi.org/10.1364/Boe.4.001269

Huang D, Swanson EA, Lin CP, Schuman JS, Stinson WG, Chang W, Hee MR, Flotte T, Gregory K, Puliafito CA, Fujimoto JG (1991) Optical Coherence Tomography. Science 254(5035):1178–1181

Jang IK, Bouma BE, Kang DH, Park SJ, Park SW, Seung KB, Choi KB, Shishkov M, Schlendorf K, Pomerantsev E, Houser SL, Aretz HT, Tearney GJ (2002) Visualization of coronary atherosclerotic plaques in patients using optical coherence tomography: comparison with intravascular ultrasound. J Am Coll Cardiol 39:604–609

Kilic ID, Caiazzo G, Fabris E, Serdoz R, Abou-Sherif S, Madden S, Moreno PR, Goldstein J, Di Mario C (2015) Near-infrared spectroscopy-intravascular ultrasound: scientific basis and clinical applications. Eur Heart J-Card Img 16(12):1299–1306. https://doi.org/10.1093/ehjci/jev208

Kolodgie FD, Burke AP, Farb A, Gold HK, Yuan J, Narula J, Finn AV, Virmani R (2001) The thin-cap fibroatheroma: a type of vulnerable plaque: the major precursor lesion to acute coronary syndromes. Curr Opin Cardiol 16(5):285–292

Lee S, Lee MW, Cho HS, Song JW, Nam HS, Oh DJ, Park K, Oh WY, Yoo H, Kim JW (2014) Fully integrated high-speed intravascular optical coherence tomography/near-infrared fluorescence structural/molecular imaging in vivo using a clinically available near-infrared fluorescence-emitting indocyanine green to detect inflamed lipid-rich atheromata in coronary-sized vessels. Circ Cardiovasc Interv 7(4):560–569. https://doi.org/10.1161/CIRCINTERVENTIONS.114.001498

Li X, Yin J, Hu C, Zhou Q, Shung KK, Chen Z (2010) High-resolution coregistered intravascular imaging with integrated ultrasound and optical coherence tomography probe. Appl Phys Lett 97(13):133702. https://doi.org/10.1063/1.3493659

Li J, Ma T, Jing J, Zhang J, Patel PM, Kirk Shung K, Zhou Q, Chen Z (2013) Miniature optical coherence tomography-ultrasound probe for automatically coregistered three-dimensional intracoronary imaging with real-time display. J Biomed Opt 18(10):100502. https://doi.org/10.1117/1.JBO.18.10.100502

Li X, Li JW, Jing J, Ma T, Liang SS, Zhang J, Mohar D, Raney A, Mahon S, Brenner M, Patel P, Shung KK, Zhou QF, Chen ZP (2014) Integrated IVUS-OCT Imaging for Atherosclerotic Plaque Characterization. IEEE J Sel Top Quant Electron 20(2). Artn 7100108. https://doi.org/10.1109/jstqe.2013.2274724

Li J, Ma T, Mohar D, Steward E, Yu M, Piao Z, He Y, Shung KK, Zhou Q, Patel PM, Chen Z (2015) Ultrafast optical-ultrasonic system and miniaturized catheter for imaging and characterizing atherosclerotic plaques in vivo. Sci Rep 5:18406. https://doi.org/10.1038/srep18406

Li Y, Jing J, Heidari E, Zhu J, Qu Y, Chen Z (2017a) Intravascular optical coherence tomography for characterization of atherosclerosis with a 1.7 Micron Swept-Source Laser. Sci Rep 7 (1):14525. https://doi.org/10.1038/s41598-017-15326-4

Li Y, Jing J, Qu Y, Miao Y, Zhang B, Ma T, Yu M, Zhou Q, Chen Z (2017b) Fully integrated optical coherence tomography, ultrasound, and indocyanine green-based fluorescence tri-modality system for intravascular imaging. Biomed Opt Express 8(2):1036–1044. https://doi.org/10.1364/BOE.8.001036

Madder RD, Steinberg DH, Anderson RD (2013) Multimodality direct coronary imaging with combined near-infrared spectroscopy and intravascular ultrasound: Initial US experience. Catheter Cardio Inte 81(3):551–557. https://doi.org/10.1002/ccd.23358

Moreno PR, Muller JE (2002) Identification of high-risk atherosclerotic plaques: a survey of spectroscopic methods. Curr Opin Cardiol 17(6):638–647

Muller JE, Tofler GH, Stone PH (1989) Circadian variation and triggers of onset of acute cardiovascular disease. Circulation 79(4):733–743

Nissen SE, Yock P (2001) Intravascular ultrasound: novel pathophysiological insights and current clinical applications. Circulation 103:604–616

Puri R, Worthley MI, Nicholls SJ (2011) Intravascular imaging of vulnerable coronary plaque: current and future concepts. Nat Rev Cardiol 8(3):131–139. https://doi.org/10.1038/nrcardio.2010.210 nrcardio.2010.210 [pii]

Puri R, Madder RD, Madden SP, Sum ST, Wolski K, Muller JE, Andrews J, King KL, Kataoka Y, Uno K, Kapadia SR, Tuzcu EM, Nissen SE, Virmani R, Maehara A, Mintz GS, Nicholls SJ (2015) Near-infrared spectroscopy enhances intravascular ultrasound assessment of vulnerable coronary plaque a combined pathological and in vivo study. Arterioscl Throm Vas 35(11):2423–2431. https://doi.org/10.1161/Atvbaha.115.306118

Roleder T, Kovacic JC, Ali Z, Sharma R, Cristea E, Moreno P, Sharma SK, Narula J, Kini AS (2014) Combined NIRS and IVUS imaging detects vulnerable plaque using a single catheter system: a head-to-head comparison with OCT. Eurointervention 10(3):303–311. https://doi.org/10.4244/Eijv10i3a53

Stahli BE, Riedel M, Lauten A, Leistner DM (2017) Intravascular ultrasound imaging for the therapy of coronary artery disease. Technical and clinical basic principles. Kardiologe 11(6):501–512. https://doi.org/10.1007/s12181-017-0205-6

Ughi GJ, Wang H, Gerbaud E, Gardecki JA, Fard AM, Hamidi E, Vacas-Jacques P, Rosenberg M, Jaffer FA, Tearney GJ (2016) First-in-human dual-modality oct and near-infrared autofluorescence imaging of coronary artery disease. JACC Cardiovascular Imaging 9(11):1304–1314

Vinegoni C, Botnaru I, Aikawa E, Calfon MA, Iwamoto Y, Folco EJ, Ntziachristos V, Weissleder R, Libby P, Jaffer FA (2011) Indocyanine green enables near-infrared fluorescence imaging of lipid-rich, inflamed atherosclerotic plaques. Sci Transl Med 3 (84):84ra45. https://doi.org/10.1126/scitranslmed.3001577

Virmani R, Kolodgie FD, Burke AP, Farb A, Schwartz SM (2000) Lessons from sudden coronary death: a comprehensive morphological classification scheme for atherosclerotic lesions. Arterioscler Thromb Vasc Biol 20(5):1262–1275

Wang H, Gardecki JA, Ughi GJ, Jacques PV, Hamidi E, Tearney GJ (2015) Ex vivo catheter-based imaging of coronary atherosclerosis using multimodality OCT and NIRAF excited at 633 nm. Biomed Opt Express 6(4):1363–1375. https://doi.org/10.1364/Boe.6.001363

Waxman S, Dixon SR, L'Allier P, Moses JW, Petersen JL, Cutlip D, Tardif JC, Nesto RW, Muller JE, Hendricks MJ, Sum ST, Gardner CM, Goldstein JA, Stone GW, Krucoff MW (2009) In vivo validation of a catheter-based near-infrared spectroscopy system for detection of lipid core coronary plaques: initial results of the SPECTACL study. JACC Cardiovascular imaging 2(7):858–868. https://doi.org/10.1016/j.jcmg.2009.05.001 S1936-878X(09)00345-3 [pii]

White HD, Chew DP (2008) Acute myocardial infarction. Lancet 372(9638):570–584. https://doi.org/10.1016/S0140-6736(08)61237-4

Yamada E, Matsumura M, Kyo S, Omoto R (1995) Usefulness of a prototype intravascular ultrasound imaging in evaluation of aortic dissection and comparison with angiographic study, trans-

esophageal echocardiography, computed tomography, and magnetic resonance imaging. Am J Cardiol 75:161–165
Yang HC, Yin J, Hu C, Cannata J, Zhou Q, Zhang J, Chen Z, Shung KK (2010) A dual-modality probe utilizing intravascular ultrasound and optical coherence tomography for intravascular imaging applications. IEEE Trans Ultrason Ferroelectr Freq Control 57(12):2839–2843. https://doi.org/10.1109/TUFFC.2010.1758
Yaqoob Z, Wu J, McDowell EJ, Heng X, Yang C (2006) Methods and application areas of endoscopic optical coherence tomography. J Biomed Opt 11(6):063001. https://doi.org/10.1117/1.2400214
Yin J, Yang HC, Li X, Zhang J, Zhou Q, Hu C, Shung KK, Chen Z (2010) Integrated intravascular optical coherence tomography ultrasound imaging system. J Biomed Opt 15(1):010512. https://doi.org/10.1117/1.3308642
Yoneya S, Saito T, Komatsu Y, Koyama I, Takahashi K, Duvoll-Young J (1998) Binding properties of indocyanine green in human blood. Invest Ophthalmol Vis Sci 39(7):1286–1290
Yoo H, Kim JW, Shishkov M, Namati E, Morse T, Shubochkin R, McCarthy JR, Ntziachristos V, Bouma BE, Jaffer FA, Tearney GJ (2011) Intra-arterial catheter for simultaneous microstructural and molecular imaging in vivo. Nat Med 17(12):1680–1684. https://doi.org/10.1038/nm.2555 nm.2555 [pii]
Zacharatos H, Hassan AE, Qureshi A (2010) Intravascular Ultrasound: Principles and Cerebrovascular Applications. Am J Neuroradiol 31(4):586–597. https://doi.org/10.3174/ajnr.A1810

Chapter 8
Tri-Modality Intravascular Imaging System

Yan Li and Zhongping Chen

Introduction

Atherosclerosis is a progressive disease that is characterized by the accumulation of lipids, cholesterol, fibrous constituents, monocytes, and various other inflammatory cells in the arterial wall. Atherosclerosis is one of the major causes of morbidity and mortality in developed countries. The major cause of deaths from heart attacks (86%) and brain aneurysm (45%) is due to "vulnerable plaques" that rupture suddenly and trigger a blood clot or thrombus that blocks blood flow (Narula and Strauss 2005).

Early detection of plaque lesions is the first and most necessary step in preventing the lethal consequences of atherosclerosis (Narula and Strauss 2005). Unfortunately, atherosclerosis exhibits an asymptomatic nature as vulnerable plaques grow without causing any detrimental side effects until rupture. Due to this complication, the information provided by current clinical arterial imaging techniques is often insufficient to diagnose vulnerable plaque formation at an early stage. Diagnosis of the latent vulnerability of a plaque lesion relies on both tissue structural and chemical compositions. Many clinical studies have shown that there are three critical characteristics of vulnerable plaques. They are (a) large lipid pool, (b) thin fibrous cap, and (c) inflammatory reaction (Naghavi et al. 2003; Grech 2003).

Intravascular ultrasound (IVUS) and optical coherence tomography (OCT) are currently the two most commonly used modalities in the clinic for diagnosing cardiovascular diseases which allow direct tomographic visualization of cross-sectional images from inside the vessel lumen (Potkin et al. 1990; Landini and Verrazzani 1990; Huang et al. 1991; Tearney et al. 2006; Puri et al. 2011; Li et al. 2013, 2015a).

Y. Li · Z. Chen (✉)
Department of Biomedical Engineering, Beckman Laser Institute, University of California, Irvine, Irvine, CA 92697, USA
e-mail: z2chen@uci.edu

Y. Li
e-mail: yanl30@uci.edu

Q. Zhou and Z. Chen (eds.), *Multimodality Imaging*,
https://doi.org/10.1007/978-981-10-6307-7_8

IVUS is a catheter-based technique that provides large imaging depth cross-sectional images of the coronary vessel in vivo. In daily clinical practice, IVUS is increasingly used for the visualization of coronary lumen, vessel wall, and atherosclerotic plaque formation (Nissen and Yock 2001). Recent work in IVUS backscattering analysis demonstrates the feasibility of using IVUS to characterize specific lesions and identify plaques that lead to various clinical syndromes (Mintz and Weissman 2006; Bermejo et al. 1998; Hanekamp et al. 1999). However, current IVUS has limited resolution for measuring the thickness of the thin fibrous cap and limited sensitivity for plaque classifications (Sawada et al. 2008; Puri et al. 2011).

OCT is a recently developed optical-based technology that can provide real-time, high-resolution, and three-dimensional (3D) images similar to that of IVUS with micron-scale resolution (Huang et al. 1991; Fujimoto et al. 1995). OCT can achieve resolution on the order of several microns, which is 10–100 times more detailed than current commercially available imaging modalities, such as IVUS, magnetic resonance imaging, and computed tomography (Fujimoto 2003). Intravascular OCT has been demonstrated by several groups for imaging and evaluation of vulnerable plaques (Fujimoto 2003; Yun et al. 2006; Brezinski et al. 1996; Jang et al. 2002, 2005; Fujimoto et al. 1995; Brezinski 2006, 2007; Raffel et al. 2008). Although OCT has been used for vulnerable plaque evaluation and is capable of measuring microscopic features with high resolution, it has limited penetration depth and cannot image the full depth of a large lipid pool in plaques (Puri et al. 2011; Sawada et al. 2008).

In addition, both IVUS and OCT lack molecular specificity for the identification of inflammatory reaction and lipid composition. Recently, near-infrared fluorescence (NIRF) imaging has emerged as an ionizing radiation-free high-resolution imaging approach to provide molecular specificity. For intravascular imaging, NIRF imaging can be used to identify inflammatory reactions and lipid composition by utilizing a contrast agent. For example, indocyanine green (ICG), an NIRF contrast agent that is approved by the Food and Drug Administration (FDA), has been shown that it has the capability to bind to lipoproteins and accumulate at inflamed tissues, so it is often used to detect inflammatory reaction and lipid composition. Currently, most intravascular imaging systems focus on single- or dual-modality imaging (Abran et al. 2015; Lee et al. 2014; Li et al. 2015a, b; Sethuraman et al. 2007; Ughi et al. 2015; Piao et al. 2015) which are not enough to provide an accurate evaluation of the key characteristics of vunlerable plaque (large lipid pool, thin fibrous cap, and inflammatory reaction) because only one characteristic can be identified by one imaging technology. For example, it is difficult to identify the existence of a large lipid pool by using a combined NIRF and OCT system due to limited penetration depth. Therefore, a tri-modality system which has the capability of imaging all three characteristics is essential for clinical application. Recently, several groups reported on tri-modality imaging systems, such as integrated IVUS/OCT/photoacoustic, integrated IVUS/NIRF/photoacoustic, and integrated OCT/US/NIRF (Yang et al. 2011; Liang et al. 2014; Abran et al. 2014; Li et al. 2017). Among them, the IVUS/OCT/NIRF tri-modality imaging system (Li et al. 2017; Liang et al. 2014) is able to provide more comprehensive information on vulnerable plaque.

This chapter outlines a representative tri-modality intravascular imaging system which combine IVUS, OCT, and NIRF imaging together. Then, it introduces the design of a tri-modality system and probe. Finally, experimental results using an IVUS/OCT/NIRF tri-modality imaging system are presented and discussed.

Method

The tri-modality intravascular imaging system is able to perform IVUS, OCT, and NIRF imaging simultaneously for the identification of a large lipid pool, thin fibrous cap, and inflammatory reaction. First, OCT and IVUS provide structural imaging with different scales. IVUS can image more than 5 mm deep inside tissues to cover the full thickness of plaques (Low et al. 2005). On the other hand, OCT images provide details with a high resolution on the order of 5–20 microns, which can be used to accurately assess the thickness of a thin fibrous cap. The addition of NIRF to the IVUS/OCT imaging system would allow for improved differentiation of chemical composition of the vessel wall which will provide the physician a powerful tool for imaging and diagnosing vulnerable plaques and monitoring therapeutic efficacy at an early stage. Although one can perform IVUS, OCT, NIRF separately, the cost of separate disposable guide wires and catheters for IVUS, OCT, and NIRF is unnecessarily high. In addition, it is time-consuming to conduct three separate catheter procedures. Finally, real-time co-registration of three separate images is essential for clinical application. It would be impossible to obtain real-time co-registered images if separate catheter procedures were performed. Therefore, it is significantly advantageous to integrate these three technologies into a single system to exploit the various features of these high-resolution technologies.

The development of an IVUS/OCT/NIRF tri-modality imaging system includes systems integration, signals synchronization, a miniature integrated imaging probe, and data acquisition and processing.

Tri-Modality Imaging System Design

In this section, a tri-modality intravascular imaging system based on fiber optics will be demonstrated. The schematic of the integrated system is shown in Fig. 8.1, which is an integration of a swept-source OCT (SS-OCT) system, a NIRF imaging system, and an IVUS imaging system. For the SS-OCT system, a swept light source is applied to perform OCT imaging. For the NIRF system, a semi-conductive continuous wavelength (CW) laser is used as the NIRF excitation source. The filter set can be chosen based on the absorption and emission spectrum of the NIRF contrast agent. A wavelength division multiplexer (WDM) is used to combine two optical beams. The combined beam will go through the double-clad fiber (DCF) coupler, a rotary joint, and imaging probe and then illuminate on the tissue surface. For OCT, the

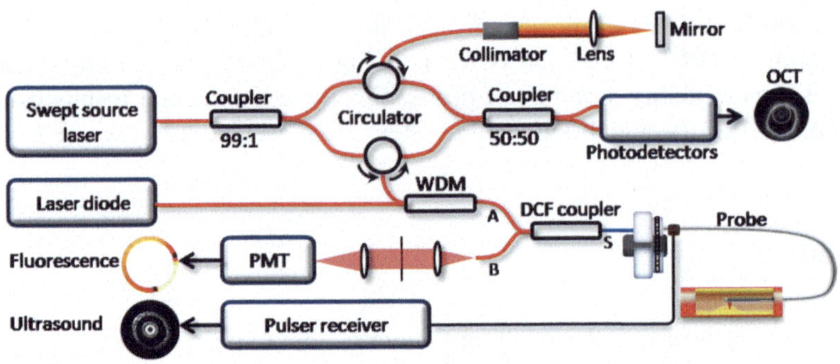

Fig. 8.1 Schematic of the tri-modality imaging system. PMT: photomultiplier tube. OCT: optical coherence tomography

backscattered light goes back through the single-mode core of the DCF coupler, WDM, and is then detected by the photodetector. For NIRF imaging, the emission light is collected by the single-mode core and inner clad of the DCF coupler (Port S) and detected by the photomultiplier tube (PMT). For IVUS imaging, a US pulser/receiver is used for US signal generation and detection.

Synchronization of Tri-Modality Imaging System

System synchronization is crucial for co-registration of tri-modality images. Typically, a trigger signal from the swept light source is used as a main trigger to synchronize the entire imaging system. For each trigger, an acoustic pulse is generated and transmitted by an acoustic transducer. Immediately after the transmission, the pulser/receiver switches to receive mode to receive the IVUS signal. For NIRF imaging, a continuous near-infrared laser is often used. Therefore, the light is always on, and the acquisition system will start to acquire an NIRF emission signal once it is triggered by the main trigger. The schematic of synchronization of the tri-modality system is shown in Fig. 8.2.

Integration of Optical Beams

The integration of OCT and NIRF involves a combination of incident optical beams and separation of NIRF emission light and OCT backscattered light. A commercial WDM is often used to combine OCT and NIRF light into one single-mode fiber with low insertion loss. A 2 × 2 DCF coupler (as shown in Fig. 8.3) is used for transmission and separation of two optical beams. The DCF coupler consists of a DCF and multimode fiber. The DCF is necessary since OCT requires single-mode propagation to maintain coherence, and NIRF requires a large propagation area for

Fig. 8.2 Synchronization of tri-modality imaging system. DAQ: data acquisition. PC: personal computer

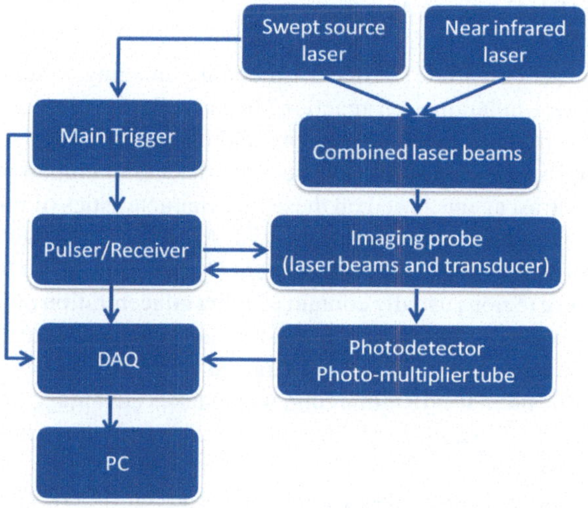

collecting emission light. For transmission, the combined beam from WDM goes through a single-mode core of DCF coupler (from Port A to Port S) and an imaging probe for imaging. For collection, the OCT backscattered light and NIRF emission light go back through the imaging probe and then are seperated by the DCF coupler. The OCT backscattered light goes back through the same way (from Port S to Port A) and is then detected by the photodetector. The NIRF emission light returns from the single-mode core and inner cladding (Port S) to the multimode fiber (Port B) of the DCF coupler and is then detected by the photomultiplier tube.

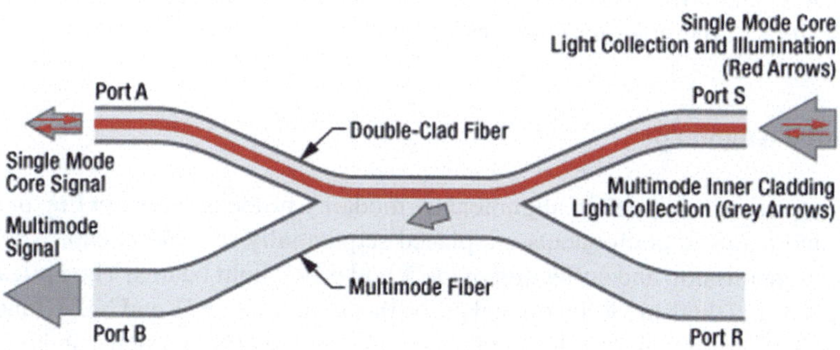

Fig. 8.3 Double-clad fiber coupler. The 2 × 2 double-clad fiber coupler combines a double-clad fifiber (DCF) with a standard step-index multimode fiber. First, the light in the single-mode core of the DCF is able to propagate bidirectionally with low insertion loss. In addition, for the light in the multimode fiber and inner clading of DCF, transmission efficiency is greater than 60% from Port S to Port B

NIRF Contrast Agents

In an IVUS/OCT/NIRF tri-modality imaging system, NIRF imaging is used to detect inflammation reaction. In the inflammation reaction area, large amounts of macrophage infiltrate the core and undergo apoptosis. Therefore, both macrophage and apoptosis are able to be used to detect inflammation reaction. In this section, two contrast agents based on these two characteristics of inflammation will be presented.

Annexin V-conjugated Cy5.5 Annexin V-conjugated Cy5.5 is a type of annexin V connected to the contrast agent Cy 5.5. During the formation of the necrotic core, these regions usually contain a higher concentration of phosphatidylserine, which can be used to target apoptosis. Moreover, Annexin V is usually used for the identification of apoptosis. Annexin V-conjugated Cy5.5 has been reported as the contrast agent for intravascular NIRF imaging, which takes advantage of the high excitation efficiency of Cy 5.5 (Liang et al. 2014) and is able to identify inflammatory reaction with high sensitivity.

Indocyanine green Indocyanine green (ICG), an amphiphilic NIRF agent that is approved by the Food and Drug Administration (FDA) for ophthalmologic NIRF imaging, has been demonstrated to be able to bind to lipoproteins and accumulate in inflamed tissues (Stanga et al. 2003; Yoneya et al. 1998; Fischer et al. 2006). Recently, Vinegoni et al. have reported that ICG could bind to lipid-rich macrophage-harboring atheroma (Vinegoni et al. 2011).

Tri-Modality Imaging Probe

One of the challenges of the tri-modality intravascular imaging system is to integrate IVUS, OCT, and NIRF into a single miniature probe. In this section, two representative designs of the imaging probe are presented.

Sequential Alignment

The schematic of a sequential alignment tri-modality probe is shown in Fig. 8.4. The optical and acoustic components are placed sequentially in a metal cap. A DCF is used for transmission and collection of OCT and NIRF light beams. The optical part of this probe is a dual-modality optical probe that combines OCT and NIRF functions together. The incident light is focused by a gradient index (GRIN) lens, and reflected onto the surface of the sample by a mirror. The acoustic sensor used is a single-element US transducer, which is sequentially aligned with the optical components and tilted at a slight angle in order to obtain optimum overlap between the optical beams and acoustic wave. A proximal-end scanning mechanism is used to obtain cross-sectional images of the sample by rotating the entire imaging probe. Torque from the motor is translated to the distal end of the probe by a double-wrapped torque coil. The rotational system is mounted on a linear translation stage which is used

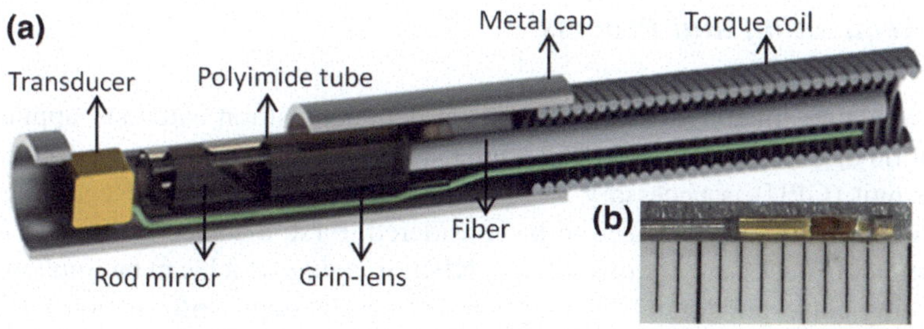

Fig. 8.4 Tip of the sequential alignment tri-modality probe. **a** Overall schematic. **b** Top view of the probe

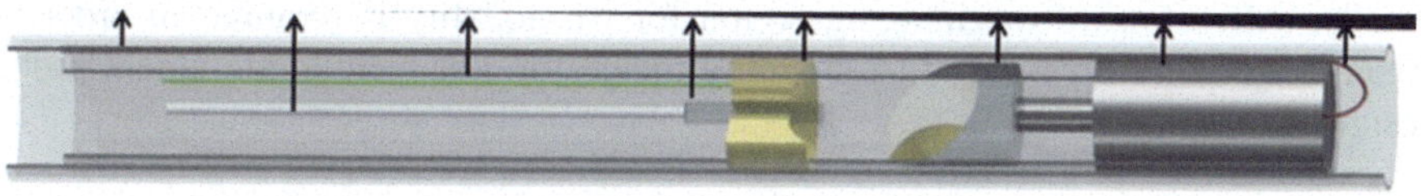

Fig. 8.5 Tip of the coaxial alignment tri-modality imaging probe

for pulling back. The rotational and linear scanning mechanisms allow the system to provide a 3D helical scan. The advantage of sequential alignment is that the probe can be manufactured with a small diameter and short rigid length. The downside is the limited scanning speed. In addition, non-uniform rotation distortion may be observed when the imaging probe travels through the branching points of the cardiovascular system. Finally, sequential alignment will cause longitudinal offset of tri-modality images, so extra data processing is required to obtain a perfect co-registered image.

Coaxial Alignment

Another alternative design is the coaxial alignment of optical and acoustic components (Wei et al. 2011; Wang et al. 2014), as shown in Fig. 8.5. This design applies a micro motor to drive the mirror instead of the entire imaging probe to scan optical beams and the acoustic wave. The imaging probe consists of a ring-shaped ultrasonic transducer, a mirror, a motor, a GRIN lens, and a DCF which is mounted in the central hole of the ultrasonic transducer. The innovation is that the ultrasonic wave and optical illumination beam share the same path, which provides enhanced sensitivity within the entire imaging range and a perfect co-registered image. In addition, it allows more steady rotation and a higher frame rate. However, the current reported minimum size of the ring-shaped ultrasonic transducer is ~2 mm, which makes it difficult for in vivo imaging.

Data Acquisition and Process

For tri-modality imaging, two synchronized data acquisition cards are applied for acquisition of OCT, NIRF, and IVUS signals. For real-time display, a graphics processing unit (GPU) is necessary for data processing and display. Regarding data processing, several aspects need to be considered. First, distance calibration of the NIRF signal is necessary for quantitative NIRF imaging which can be implemented by utilizing corresponding OCT images to extract the depth information for NIRF signal intensity compensation. In addition, the small size of the ultrasonic transducer causes a limited imaging range so an advanced signal processing algorithm, such as chirp coded excitation, also needs to be implemented for further improvement. Finally, calibration of the imaging region for tri-modality is required in order to obtain a perfect co-registered tri-modality image for a sequential alignment probe design.

Experiments

Intravascular imaging of atherosclerotic plaques from an animal model and human cadaver with the tri-modality system has been conducted. The corresponding results have been divided into two parts. The first part shows representative tri-modality images using Cy 5.5 (Liang et al. 2014). The second part demonstrates representative tri-modality images with ICG (Li et al. 2017).

Ex Vivo *Experiment of the Tri-Modality Imaging System with Cy 5.5*

Phantom Experiment

For evaluating the tri-modality imaging system's performance, ex vivo imaging of a normal New Zealand white rabbit aorta in which two model plaques had been planted inside the blood vessel wall was first performed. The two model plaques were injected side by side next to each other: one of them contained only the fatty mixture, and the other one was mixed with 0.1 μmol/L Cy5.5 dye.

Images of the model plaques from a normal New Zealand rabbit are shown in Fig. 8.6. Figure 8.6a is the structure diagram of the model plaques and shows the relative location of the two model plaques. Figures 8.6b–d are the combined OCT/NIRF, combined IVUS/NIRF, and tri-modality images, respectively. Figures 8.6e–g are the corresponding 3D images.

From the combined image, it is easy to identity two types of plaques. Meanwhile, the IVUS and OCT images show the whole and fine structure of the aorta.

Coronary Artery Experiment

An ex vivo experiment for imaging a coronary artery sample from a human cadaver has been conducted. Figure 8.7 illustrates the combined images from a human cadaver coronary artery sample stained with Annexin V-conjugated Cy5.5. The IVUS images show the whole structure including the entire shape of the blood vessel wall. In contrast to the IVUS image, the OCT image shows the fine structure of the vessel wall and provides a much clearer layered structure. The NIRF signal indicates the area where the Annexin V-conjugated Cy5.5 exists. These regions usually contain a higher concentration of phosphatidylserine that can be used to target apoptosis.

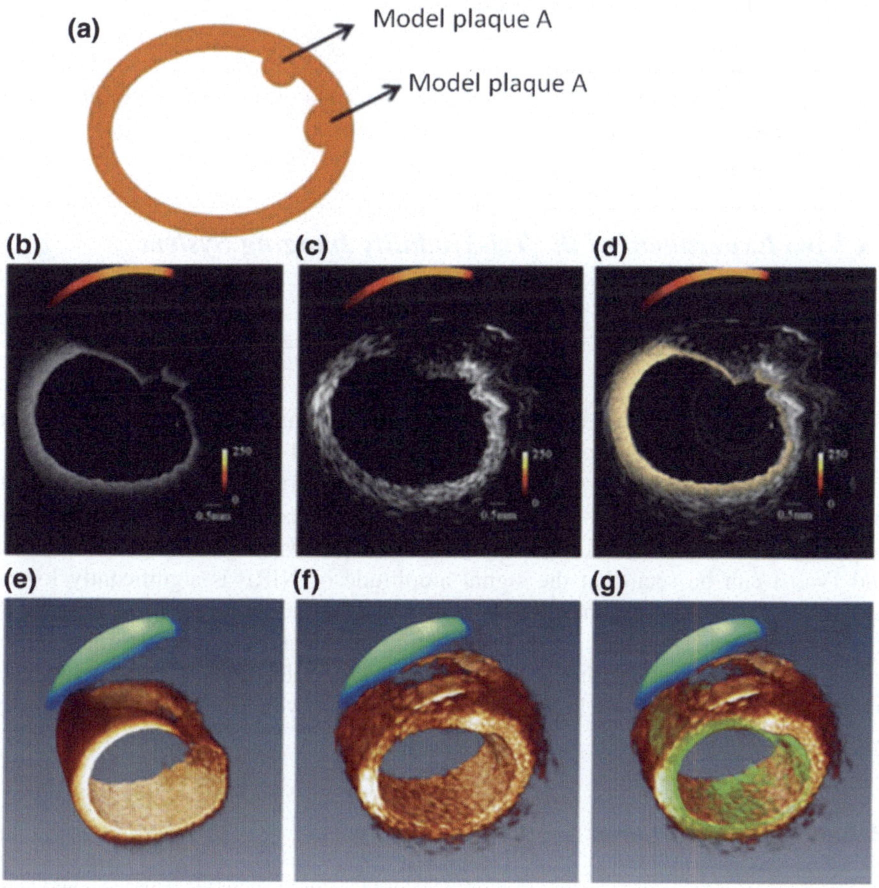

Fig. 8.6 Ex vivo images from two model plaques injected in rabbit aorta: **a** schematic of two model plaques, **b** OCT combined NIRF image, **c** US structure image combined NIRF, **d** combined OCT, IVUS, and NIRF image, **e** 3D reconstruction of OCT combined NIRF, **f** 3D US and NIRF image, and **g** 3D combined tri-modality image

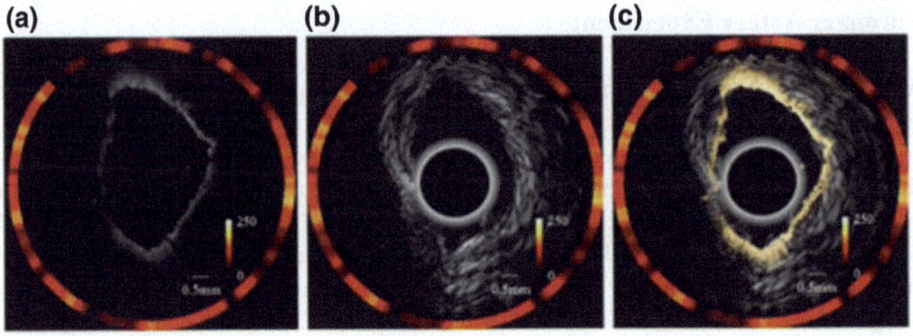

Fig. 8.7 Ex vivo images from human coronary artery: **a** combined OCT and NIRF image, **b** combined IVUS and NIRF image, and **c** tri-modality image

During the formation of a necrotic core, large amounts of macrophage infiltrate the core and undergo apoptosis.

Ex Vivo *Experiment of the Tri-Modality Imaging System with ICG*

Phantom Experiment

For evaluating the performance of the tri-modality imaging system with ICG, a lipid-mimicking phantom was fabricated by injecting 0.1 μmol/L ICG into a healthy pig artery. Figures 8.8I–IV show tri-modality images of the phantom at different sites. Figures 8.8Ia–IVa, Ib–IVb, and Ic–IVc are the combined OCT (inner) and NIRF (outer), IVUS, and tri-modality images, respectively. From Figs. 8.8Ia, IIIa, and IVa, it can be seen that the signal amplitude of NIRF is significantly lower and homogenous, corresponding to the sites without ICG. From OCT and IVUS images, the whole structural and microstructural information can be obtained. From Fig. 8.8IIa, it can be seen that the amplitude of the NIRF signal (indicated by the white arrow) is significantly higher than other regions, which corresponds to the site with injected ICG.

Rabbit Aorta Experiment

To demonstrate the capability of assessing vulnerable plaques, an aorta from an atherosclerotic rabbit was imaged. The experimental rabbit was anesthetized and ICG (2.25 mg/Kg) was injected. Twenty minutes after injection, the rabbit was sacrificed. The aorta was excised and conserved in 4% formaldehyde for ex vivo experiments.

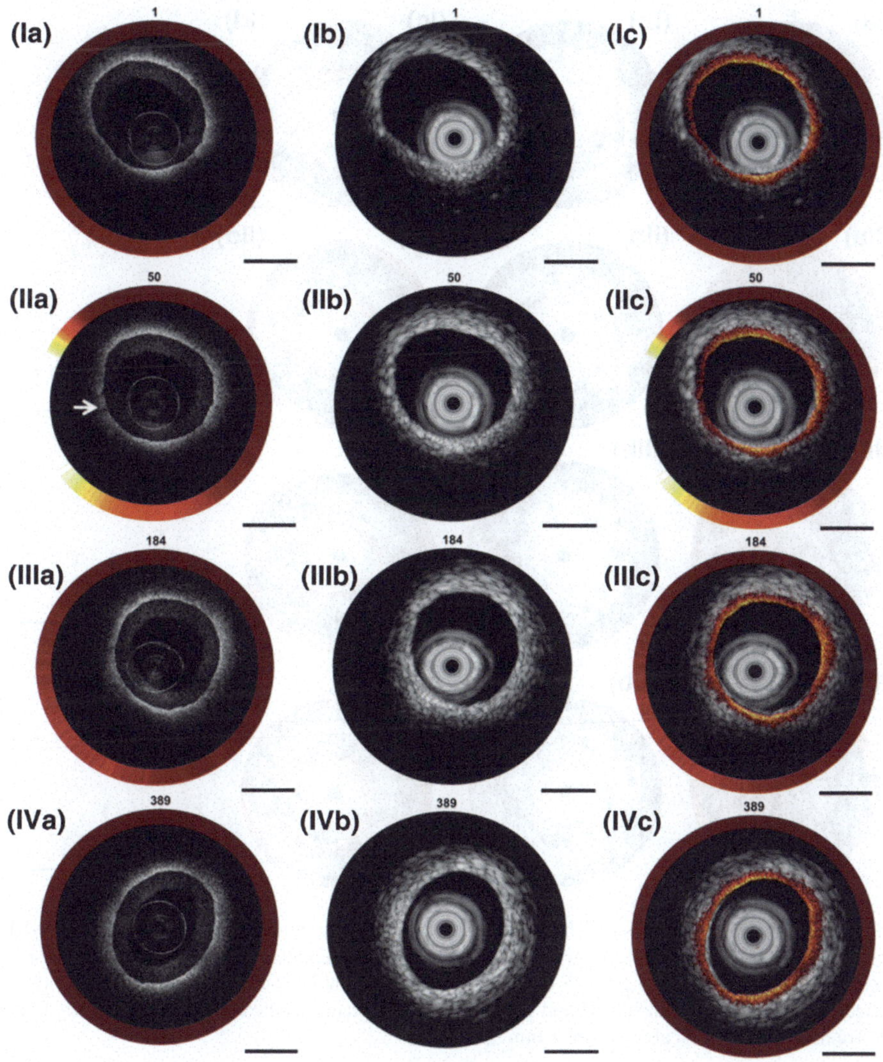

Fig. 8.8 Tri-modality images of lipid-mimicking phantom: **Ia–IVa** combined OCT (inner) and NIRF (outer), **Ib–IVb** US, and **Ic–IVc** tri-modality images

Representative OCT, IVUS, and NIRF image pairs and corresponding H&E staining of rabbit aorta segments with different pathological features are shown in Fig. 8.9.

From Figs. 8.9IIb and IIIb, intimal thickening and a low-density acoustic signal region (denoted by the white arrow) can be found, which demonstrates the existence of plaque. At the same site in the OCT image (Fig. 8.9IIIa), a homogenous high signal region also indicates intimal thickening. Moreover, the high signal region is also

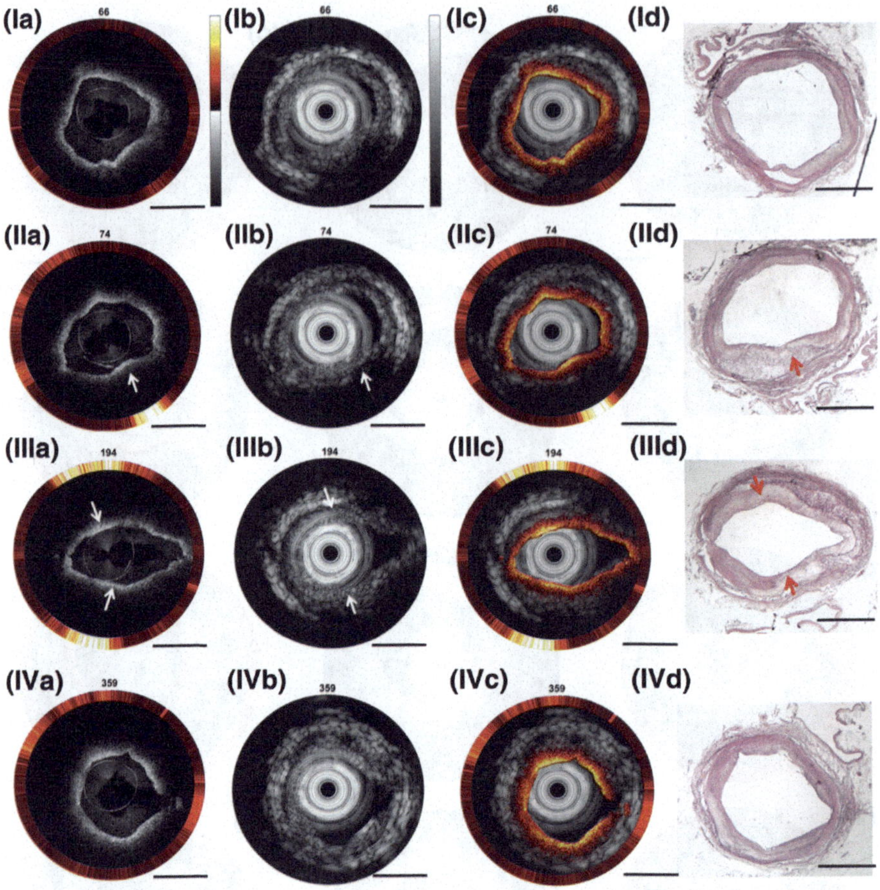

Fig. 8.9 Tri-modality images of atherosclerotic rabbit: **Ia–IVa** combined OCT (inner) and NIRF (outer), **Ib–IVb** US, **Ic–IVc** fused tri-modality, and **Id–IVd** hematoxylin and eosin (H&E) histology. The artifact circles in the IVUS images are caused by the US pulse ring-down effect and the reflection of the catheter sheath. **II** and **III** are aorta with plaque, indicated by white arrows. **I** and **IV** are healthy aortas. Scale bars are 1 mm

found at the same site in the NIRF images, which indicates inflammatory reaction. From the combined tri-modality images, it can be concluded that this aorta as shown in Fig. 8.9IIIc is in the early stage of plaque formation. The classification of plaque type is validated by the corresponding histology photographs. From Fig. 8.9IIa, the diffuse boundary and weak signal region under the high signal region indicates the existence of the lipid pool, and the thickness of the fiber cap is 150 μm. Furthermore, the high signal at the same site in the NIRF images indicates the inflammatory reaction. Moreover, a lipid pool can be found in the corresponding H&E histology photograph, which agrees with tri-modality images well. Therefore, we can conclude that the aorta shown in Fig. 8.9IIa is thick-cap (>65 μm) fibroatheroma (ThCFA) and in the stage of plaque progression. In Fig. 8.9Ib, some low echo signals can be found, which means that this region may have plaque. However, based on the combined OCT and NIRF results, it can be concluded that this aorta is normal. Histology further supports this conclusion. From Figs. 8.9IVc and IVd, the tri-modality images and the H&E histology all show that this aorta is healthy.

Summary

In this chapter, a tri-modality imaging system with an integrated imaging probe was presented. Both the phantom and the ex vivo experiments demonstrated that this tri-modality system has the capability of obtaining high-resolution OCT, deep-penetration-depth IVUS, and molecular-specific ICG-based NIRF images simultaneously while displaying images in real time. Furthermore, H&E staining validated the ex vivo experiment results. In the near future, the tri-modality system and fully integrated tri-modality imaging probes will provide a powerful tool for clinical management of cardiovascular disease.

References

Abran M, Cloutier G, Cardinal MH, Chayer B, Tardif JC, Lesage F (2014) Development of a photoacoustic, ultrasound and fluorescence imaging catheter for the study of atherosclerotic plaque. IEEE Trans Biomed Circuits Syst 8(5):696–703. https://doi.org/10.1109/TBCAS.2014.2360560

Abran M, Stahli BE, Merlet N, Mihalache-Avram T, Mecteau M, Rheaume E, Busseuil D, Tardif JC, Lesage F (2015) Validating a bimodal intravascular ultrasound (IVUS) and near-infrared fluorescence (NIRF) catheter for atherosclerotic plaque detection in rabbits. Biomed Opt Express 6(10):3989–3999. https://doi.org/10.1364/BOE.6.003989

Bermejo J, Botas J, Garcia E, Elizaga J, Osende J, Soriano J, Abeytua M, Delcan JL (1998) Mechanisms of residual lumen stenosis after high-pressure stent implantation: a quantitative coronary angiography and intravascular ultrasound study. Circulation 98:112–118

Brezinski ME (2006) Optical coherence tomography for identifying unstable coronary plaque. Int J Cardiol 107(2):154–165. https://doi.org/10.1016/j.ijcard.2005.07.066

Brezinski ME (2007) Applications of optical coherence tomography to cardiac and musculoskeletal diseases: bench to bedside? J Biomed Opt 12(5):051705. https://doi.org/10.1117/1.2795689

Brezinski ME, Tearney GJ, Bouma BE, Izatt JA, Hee MR, Swanson EA, Southern JF, Fujimoto JG (1996) Optical coherence tomography for optical biopsy. Properties and demonstration of vascular pathology. Circulation 93(6):1206–1213

Fischer T, Gemeinhardt I, Wagner S, Stieglitz DV, Schnorr J, Hermann KG, Ebert B, Petzelt D, Macdonald R, Licha K, Schirner M, Krenn V, Kamradt T, Taupitz M (2006) Assessment of unspecific near-infrared dyes in laser-induced fluorescence imaging of experimental arthritis. Acad Radiol 13(1):4–13. https://doi.org/10.1016/j.acra.2005.07.010

Fujimoto JG (2003) Optical coherence tomography for ultrahigh resolution in vivo imaging. Nat Biotechnol 21:1361–1367

Fujimoto JG, Brezinski ME, Tearney GJ, Boppart SA, Bouma B, Hee MR, Southern JF, Swanson EA (1995) Optical biopsy and imaging using optical coherence tomography. Nat Med 1:970–972

Grech ED (2003) Pathophysiology and investigation of coronary artery disease. BMJ 326(7397):1027–1030. https://doi.org/10.1136/bmj.326.7397.1027

Hanekamp C, Koolen J, Pijls J, Michels H, Bonnier H (1999) Comparison of quantitative coronary angiography, intravascular ultrasound, and coronary pressure measurement to assess optimum stent deployment. Circulation 99:1015–1021

Huang D, Swanson EA, Lin CP, Schuman JS, Stinson WG, Chang W, Hee MR, Flotte T, Gregory K, Puliafito CA, Fujimoto JG (1991) Optical coherence tomography. Science 254(5035):1178–1181

Jang IK, Bouma BE, Kang DH, Park SJ, Park SW, Seung KB, Choi KB, Shishkov M, Schlendorf K, Pomerantsev E, Houser SL, Aretz HT, Tearney GJ (2002) Visualization of coronary atherosclerotic plaques in patients using optical coherence tomography: comparison with intravascular ultrasound. J Am Coll Cardiol 39:604–609

Jang IK, Tearney GJ, MacNeill B, Takano M, Moselewski F, Iftima N, Shishkov M, Houser S, Aretz HT, Halpern EF, Bouma BE (2005) In vivo characterization of coronary atherosclerotic plaque by use of optical coherence tomography. Circulation 111:1551–1555

Landini L, Verrazzani L (1990) Spectral characterization of tissues microstructure by ultrasounds: a stochastic approach. IEEE Trans Ultrason Ferroelectr Freq Control 37(5):448–456. https://doi.org/10.1109/58.105251

Lee S, Lee MW, Cho HS, Song JW, Nam HS, Oh DJ, Park K, Oh WY, Yoo H, Kim JW (2014) Fully integrated high-speed intravascular optical coherence tomography/near-infrared fluorescence structural/molecular imaging in vivo using a clinically available near-infrared fluorescence-emitting indocyanine green to detect inflamed lipid-rich atheromata in coronary-sized vessels. Circ Cardiovasc Interv 7(4):560–569. https://doi.org/10.1161/CIRCINTERVENTIONS.114.001498

Li J, Ma T, Jing J, Zhang J, Patel PM, Kirk Shung K, Zhou Q, Chen Z (2013) Miniature optical coherence tomography-ultrasound probe for automatically coregistered three-dimensional intracoronary imaging with real-time display. J Biomed Opt 18(10):100502. https://doi.org/10.1117/1.JBO.18.10.100502

Li J, Ma T, Mohar D, Steward E, Yu M, Piao Z, He Y, Shung KK, Zhou Q, Patel PM, Chen Z (2015a) Ultrafast optical-ultrasonic system and miniaturized catheter for imaging and characterizing atherosclerotic plaques in vivo. Sci Rep 5:18406. https://doi.org/10.1038/srep18406

Li Y, Gong X, Liu C, Lin R, Hau W, Bai X, Song L (2015b) High-speed intravascular spectroscopic photoacoustic imaging at 1000 A-lines per second with a 0.9-mm diameter catheter. J Biomed Opt 20(6):065006. https://doi.org/10.1117/1.jbo.20.6.065006

Li Y, Jing J, Qu Y, Miao Y, Zhang B, Ma T, Yu M, Zhou Q, Chen Z (2017) Fully integrated optical coherence tomography, ultrasound, and indocyanine green-based fluorescence tri-modality system for intravascular imaging. Biomed Opt Express 8(2):1036–1044. https://doi.org/10.1364/BOE.8.001036

Liang S, Ma T, Jing J, Li X, Li J, Shung KK, Zhou Q, Zhang J, Chen Z (2014) Trimodality imaging system and intravascular endoscopic probe: combined optical coherence tomography, fluorescence imaging and ultrasound imaging. Opt Lett 39(23):6652–6655. https://doi.org/10.1364/OL.39.006652

Low AF, Tearney GJ, Bouma BE, Jang IK (2005) Technology Insight: optical coherence tomography—current status and future development. Nat Clin Pract Cardiovasc Med 3:154–162

Mintz GS, Weissman NJ (2006) Intravascular ultrasound in the drug-eluting stent era. J Am Coll Cardiol 48:421

Naghavi M, Libby P, Falk E, Casscells SW, Litovsky S, Rumberger J, Badimon JJ, Stefanadis C, Moreno P, Pasterkamp G, Fayad Z, Stone PH, Waxman S, Raggi P, Madjid M, Zarrabi A, Burke A, Yuan C, Fitzgerald PJ, Siscovick DS, de Korte CL, Aikawa M, Airaksinen KE, Assmann G, Becker CR, Chesebro JH, Farb A, Galis ZS, Jackson C, Jang IK, Koenig W, Lodder RA, March K, Demirovic J, Navab M, Priori SG, Rekhter MD, Bahr R, Grundy SM, Mehran R, Colombo A, Boerwinkle E, Ballantyne C, Insull W Jr, Schwartz RS, Vogel R, Serruys PW, Hansson GK, Faxon DP, Kaul S, Drexler H, Greenland P, Muller JE, Virmani R, Ridker PM, Zipes DP, Shah PK, Willerson JT (2003) From vulnerable plaque to vulnerable patient: a call for new definitions and risk assessment strategies: Part II. Circulation 108(15):1772–1778. https://doi.org/10.1161/01.CIR.0000087481.55887.C9

Narula J, Strauss HW (2005) Imaging of unstable atherosclerotic lesions. Eur J Nucl Med Mol Imaging 32(1):1–5. https://doi.org/10.1007/s00259-004-1580-3

Nissen SE, Yock P (2001) Intravascular ultrasound: novel pathophysiological insights and current clinical applications. Circulation 103:604–616

Piao Z, Ma T, Li J, Wiedmann MT, Huang S, Yu M, Kirk Shung K, Zhou Q, Kim CS, Chen Z (2015) High speed intravascular photoacoustic imaging with fast optical parametric oscillator laser at 1.7 m. Appl Phys Lett 107(8):083701. https://doi.org/10.1063/1.4929584

Potkin BN, Bartorelli AL, Gessert JM, Neville RF, Almagor Y, Roberts WC, Leon MB (1990) Coronary artery imaging with intravascular high-frequency ultrasound. Circulation 81(5):1575–1585

Puri R, Worthley MI, Nicholls SJ (2011) Intravascular imaging of vulnerable coronary plaque: current and future concepts. Nat Rev Cardiol 8(3):131–139. https://doi.org/10.1038/nrcardio.2010.210

Raffel OC, Merchant FM, Tearney GJ, Chia S, Gauthier DD, Pomerantsev E, Mizuno K, Bouma BE, Jang IK (2008) In vivo association between positive coronary artery remodelling and coronary plaque characteristics assessed by intravascular optical coherence tomography. Eur Heart J 29(14):1721–1728. https://doi.org/10.1093/eurheartj/ehn286

Sawada T, Shite J, Garcia-Garcia HM, Shinke T, Watanabe S, Otake H, Matsumoto D, Tanino Y, Ogasawara D, Kawamori H, Kato H, Miyoshi N, Yokoyama M, Serruys PW, Hirata KI (2008) Feasibility of combined use of intravascular ultrasound radiofrequency data analysis and optical coherence tomography for detecting thin-cap fibroatheroma. Eur Heart J 29:1136–1146

Sethuraman S, Aglyamov SR, Amirian JH, Smalling RW, Emelianov SY (2007) Intravascular photoacoustic imaging using an IVUS imaging catheter. IEEE Trans Ultrason Ferroelectr Freq Control 54(5):978–986

Stanga PE, Lim JI, Hamilton P (2003) Indocyanine green angiography in chorioretinal diseases: indications and interpretation: an evidence-based update. Ophthalmology 110(1):15–21; quiz 22-13

Tearney GJ, Jang IK, Bouma BE (2006) Optical coherence tomography for imaging the vulnerable plaque. J Biomed Opt 11(2):021002. https://doi.org/10.1117/1.2192697

Ughi GJ, Verjans J, Fard AM, Wang H, Osborn E, Hara T, Mauskapf A, Jaffer FA, Tearney GJ (2015) Dual modality intravascular optical coherence tomography (OCT) and near-infrared fluorescence (NIRF) imaging: a fully automated algorithm for the distance-calibration of NIRF signal intensity for quantitative molecular imaging. Int J Cardiovasc Imaging 31(2):259–268. https://doi.org/10.1007/s10554-014-0556-z

Vinegoni C, Botnaru I, Aikawa E, Calfon MA, Iwamoto Y, Folco EJ, Ntziachristos V, Weissleder R, Libby P, Jaffer FA (2011) Indocyanine green enables near-infrared fluorescence imaging of lipid-rich, inflamed atherosclerotic plaques. Sci Transl Med 3(84):84ra45. https://doi.org/10.1126/scitranslmed.3001577

Wang P, Ma T, Slipchenko MN, Liang S, Hui J, Shung KK, Roy S, Sturek M, Zhou Q, Chen Z, Cheng JX (2014) High-speed intravascular photoacoustic imaging of lipid-laden atherosclerotic plaque enabled by a 2-kHz barium nitrite raman laser. Sci Rep 4:6889. https://doi.org/10.1038/srep06889

Wei W, Li X, Zhou Q, Shung KK, Chen Z (2011) Integrated ultrasound and photoacoustic probe for co-registered intravascular imaging. J Biomed Opt 16(10):106001. https://doi.org/10.1117/1.3631798

Yang Y, Li X, Wang T, Kumavor PD, Aguirre A, Shung KK, Zhou Q, Sanders M, Brewer M, Zhu Q (2011) Integrated optical coherence tomography, ultrasound and photoacoustic imaging for ovarian tissue characterization. Biomed Opt Exp 2(9):2551–2561. https://doi.org/10.1364/BOE.2.002551

Yoneya S, Saito T, Komatsu Y, Koyama I, Takahashi K, Duvoll-Young J (1998) Binding properties of indocyanine green in human blood. Invest Ophthalmol Vis Sci 39(7):1286–1290

Yun SH, Tearney GJ, Vakoc BJ, Shishkov M, Oh WY, Desjardins AE, Suter MJ, Chan RC, Evans JA, Jang IK, Nishioka NS, de Boer JF, Bouma BE (2006) Comprehensive volumetric optical microscopy in vivo. Nat Med 12:1429–1433

Chapter 9
Acoustic Radiation Force Optical Coherence Elastography

Yueqiao Qu, Youmin He, Teng Ma, Qifa Zhou and Zhongping Chen

Introduction

Mechanical properties, such as the elasticity and viscosity, are often major indicators of diseases. The stiffness of tissue changes in unison with the onset of pathology in the cases of cardiovascular diseases, ocular diseases, and tumor formations. The cellular composition of the tissues is altered over time, in tune with disease progression. However, the reported stiffness of a specific type of cell or tissue differs greatly depending on the type of imaging modality used and the experimental conditions. In order to accurately distinguish the diseased tissues from healthy ones, it is necessary to validate the results through both theoretical and experimental methods.

Y. Qu · Y. He · Z. Chen (✉)
Department of Biomedical Engineering, Beckman Laser Institute, University of California, Irvine, Irvine, CA 92697, USA
e-mail: z2chen@uci.edu

Y. Qu
e-mail: rachelyqu@gmail.com

Y. He
e-mail: youminh1@uci.edu

T. Ma
Paul C. Lauterbur Research Center for Biomedical Imaging, Institute of Biomedical and Health Engineering, Shenzhen Institutes of Advanced Technology, Chinese Academy of Sciences, Shenzhen 518055, China
e-mail: teng.ma@siat.ac.cn

Q. Zhou
Roski Eye Institute, University of Southern California, Los Angeles, CA 90033, USA
e-mail: qifazhou@usc.edu

Department of Biomedical Engineering, University of Southern California, Los Angeles, CA 90089, USA

Q. Zhou and Z. Chen (eds.), *Multimodality Imaging*,
https://doi.org/10.1007/978-981-10-6307-7_9

Cardiovascular diseases have the highest rate of fatalities and account for 30.8% of all deaths in the USA (Mozaffarian et al. 2016; Dariush et al. 2016). Atherosclerosis, accounting for 41 deaths per day, is a cardiovascular condition that is associated with changes in the composition of the blood vessel walls. During the early onset of disease, the walls of the artery thicken due to fatty deposits, inflammation, cells, and scar tissue build up (Ross 1999; Hansson 2005). Eventually, the lesions that form, called plaques, are composed of distinctive necrotic cores and a fibrous cap. If the plaque is stable with a relatively thick cap and small lipid core, there may be varying degrees of obstruction to blood flow. However, in the case of a vulnerable plaque, the cap, containing collagen and smooth muscle cells, becomes less than 65 μm in thickness and can rupture easily. When there is a plaque rupture, the inflammatory elements of the necrotic core burst into the artery and can cause blocked arterial flow, angina, or even myocardial infarction (Virmani et al. 2003; Cheruvu et al. 2007).

Early detection of vulnerable plaques is essential to the health and safety of cardiovascular patients. The structure and composition of the plaque are largely used currently to determine its vulnerability. Current clinical imaging techniques include angiography, angioscopy, ultrasound, and magnetic resonance imaging (MRI) (Amirbekian 2007; Waxman et al. 2006). Angiography allows the physician to visualize the region of blockage, by inserting a dye into the bloodstream and observing the mechanisms of flow (Little et al. 1988). Angioscopy helps to examine the surface of the interior blood vessel to identify areas of damage and abnormality (Sherman et al. 1986; Takano et al. 2001). Ultrasound and MRI allow for visualization through the depth of the blood vessel walls, at the expense of resolution and cost, respectively (LaMuraglia et al. 1996). Due to these limitations, current imaging modalities cannot effectively identify vulnerable plaques with high sensitivity and specificity (Amirbekian 2007; Waxman et al. 2006). Since the change in the composition of the blood vessel wall is indicative of the early onset of atherosclerosis, it is possible to classify vulnerability according to the composition. Plaques can be differentiated into three different types based on their composition: lipid, fibrous, and calcified. The mechanical stiffness of these three components differs by nearly one order of magnitude (Ebenstein et al. 2009; Inagaki et al. 2006; Baldewsing et al. 2005). Therefore, if the stiffness of the tissue can be measured, the composition can be determined, and vulnerable plaques can be isolated.

Mechanical testing methods have been used to observe the differences in the stiffness of lipid, fibrous, and calcified plaque components (Loree et al. 1994; Chai et al. 2014; Walsh et al. 2014). However, these tests require extraction and manipulation of the tissue, which is not possible for in vivo imaging. It is necessary to understand the change in tissue elasticity in vivo during the early onset and formation of plaques in order to accurately assess the mechanical properties under the influence of natural environmental factors (Takano et al. 2001; Ebenstein et al. 2009; Inagaki et al. 2006; Baldewsing et al. 2006). The feasibility of such measurements is limited by the resolution and accuracy of the measurement device, size of the device, and the accessibility of the plaque in question.

Tissue elastography is a method that has been developed to map out the mechanical properties of tissues (Schaar et al. 2003; Ophir et al. 1991). There are typically

Table 9.1 Elastography in three steps

Excitation		Detection	Parameter estimation
Internal	External	Mechanical test	Quantitative
Static	Dynamic	Magnetic resonance Ultrasound Optical	Qualitative

three steps involved as depicted in Table 9.1: excitation, detection, and parameter estimation (Sun et al. 2011; Manduca et al. 2001; Greenleaf et al. 2003). The tissue is first excited using an internal or external mechanism, where the tissue itself or an outside force causes deformation (Greenleaf et al. 2003). The force can be either static or dynamic in nature, depending on the variable to be measured. For example, a few external methods include piezoelectric elements, air puff devices, and acoustic radiation force (ARF) using ultrasound. All of these devices operate by giving a static, continuing force to analyze a stable deformation state, or by providing a single or modulated dynamic signal of pulses to analyze the change in deformation over time. Once the tissue is deformed, a technique is used to visualize and measure the amount of deformation. Traditionally, mechanical testing using pressure sensors was implemented to obtain data in ex vivo samples. Magnetic resonance and ultrasound methods have also been used to detect tissue deformation at the expense of high cost and low resolution, respectively. In recent years, optical imaging methods, such as optical coherence elastography, have been developed to detect tissue response (Khalil et al. 2005; Wang et al. 2006, 2007; Qi et al. 2012, 2013, 2014; Zhu et al. 2015; Qu et al. 2016, 2018; He et al. 2019; Kennedy et al. 2015; Liang et al. 2010; Manapuram et al. 2012; Rogowska et al. 2004; van Soest et al. 2007; Wang and Larin 2015). In particular, phase resolved Doppler optical coherence tomography (OCT) has been widely used for detection (Chen et al. 1997a, b; Zhao et al. 2000a, b), with its main advantages being its high resolution and high displacement sensitivity.

Most parameter estimation methods target the extraction of elasticity by the means of elastograms or elasticity maps (Khalil et al. 2005; Wang et al. 2006, 2007; Qi et al. 2012, 2014; Qu et al. 2016, 2018; Kennedy et al. 2015; Liang et al. 2010; Manapuram et al. 2012; Rogowska et al. 2004; van Soest et al. 2007; Wang and Larin 2015; He et al. 2019). Research has also been done to observe other mechanical properties, such as the viscosity (Sinkus et al. 2005a, b; Catheline et al. 2004). The parameter estimation can be either quantitative or qualitative. Qualitative methods allow users to obtain relative values for mechanical properties and can be beneficial for the comparison between healthy and diseased tissues. However, there are often problems with the calibration accuracy of the system as well as environmental and systematic changes between measurements that limit the functions of qualitative data. Due to these factors, quantitative measurements with strong theoretical evidence are always preferred. A few examples of quantification include shear wave velocity calculations, strain imaging, and tissue frequency response (Nightingale et al. 2003; Evans et al. 2010; O'Donnell et al. 1994; Qi et al. 2012, 2013, 2014; Ahmad et al. 2015; Liang et al. 2008; Qu et al. 2018; He et al. 2019). Select methods will be discussed in detail in the next section.

Intravascular elastography using ultrasound has been widely studied in the past 20 years (Takano et al. 2001; de Korte et al. 1998, 2000; Baldewsing et al. 2004a). In general, a pressure is applied to the artery, and ultrasound imaging is used for the detection of tissue displacement, which is then converted to strain measurements and an elastogram can be generated. Examples of ultrasound elastography techniques include compression strain imaging and phase-sensitive speckle tracking methods based on cross-correlation analyses (de Korte et al. 2000). In vivo intravascular ultrasound elastography studies have also taken place in the past years, along with modeling methods such as finite element analysis (de Korte et al. 2002; Baldewsing et al. 2004b, c). However, these methods are often limited by the low ultrasound resolution of typically 150–300 μm, which allows for the detection of homogenous plaque types, but are limited in the observation of heterogeneity within small regions, which is the case for most human plaques (Prati et al. 2001). In addition, most vulnerable plaques are characterized by thin fibrous caps, as little as 65 μm in thickness, which cannot be accurately measured using ultrasound (Virmani et al. 2003). Using optical methods, with micron-level resolution, it is possible to detect minute changes in tissue elasticity within a small region. Finally, due to the nanometer sensitivity of phase-resolved OCT, only small forces are necessary to induce vibrations, which is critical in in vivo clinical applications.

Compressional and Shear Wave Methods Using OCE

Optical coherence elastography (OCE) is a technology that uses the principles of optical coherence tomography (OCT) to detect the tissue response to excitation (Huang et al. 1991; Fujimoto 2001). OCT is based on the interference of backscattered light signals of the sample and a reference mirror. In regard to OCE, an excitation force, most often external, is applied to the tissue, while the optical interference information is extracted (Sun et al. 2011; Kennedy et al. 2015; Liang et al. 2010; Manapuram et al. 2012; Rogowska et al. 2004; van Soest et al. 2007; Wang and Larin 2015; Qi et al. 2012, 2014; Qu et al. 2016). In summary, the A-line interference signal, $I(z)$, can be summarized using Eq. 9.1, where its magnitude and phase, denoted by $\phi(z)$ at a certain depth z, can be separated. The information provided can be used to measure the tissue response by using certain parameters such as done in the phase-resolved method and Doppler variance methods (Zhao et al. 2000a, b).

$$I_{\text{Interference}}(z) = |I_{\text{Interference}}(z)|e^{i\phi(z)} \tag{9.1}$$

There are primarily two types of tissue responses, which rely on elastic wave properties, that are studied using OCE: the p-wave and the s-wave (Catheline et al. 1999). When a force is exerted on a sample, the first response consists of the p-wave, also known as the compressional wave, traveling across the sample parallel to the direction of the force. The p-wave travels at a high speed and essentially compresses the sample as it passes. The s-wave, also referred to as the secondary or shear wave, travels perpendicularly to the direction of the initial force and is approximately three orders of magnitude slower than the p-wave (Gennisson et al. 2005). The s-wave is directly related to the shear modulus. We will now introduce two different methods of parameter estimation: (1) Doppler OCE using the p-wave measurements to obtain the elastic modulus and (2) velocity extraction using s-waves to obtain the shear modulus.

Doppler OCE

With the extracted phase information shown in Eq. 9.1, the phase shift between 2 A-lines can be calculated. The Doppler frequency shift, f_D, is by definition directly proportional to the axial velocity denoted by $v_r \cos\theta$ and the measured phase shift, $\Delta\phi(z)$, as shown in Eq. 9.2 (Chen et al. 1997a, b; Zhao et al. 2000a, b):

$$f_D = \frac{2v_r n \cos\theta}{\lambda_0} = \frac{\Delta\phi(x, z, t)}{2\pi\,\Delta t} \tag{9.2}$$

The variable n refers to the refractive index of the sample, λ_0 represents the central wavelength of the light source, and Δt is the period between the A-lines. By rearranging Eq. 9.2, the Doppler velocity can be defined as a function of the phase shift between A-lines. The displacement of the sample can be obtained by integrating the velocity over time as done in Eq. 9.3:

$$\Delta d = \int_{t_1}^{t_2} v_r \mathrm{d}t = \int_{t_1}^{t_2} \frac{\Delta\phi(x, z, t)\lambda_0}{4\pi n \Delta t \cos\theta} \mathrm{d}t \tag{9.3}$$

In order to calculate the mechanical elasticity, it is necessary to associate the displacement from the Doppler relationship to the elastic modulus. By definition, the strain, ε, is linearly proportional to the displacement and inversely proportional to the change in sample thickness or the compression of the sample in the axial direction, denoted by Δz, as shown in Eq. 9.4:

$$\varepsilon = \frac{\Delta d}{\Delta z} \tag{9.4}$$

Young's modulus, Y, is linearly proportional to the stress, σ, and inversely proportional to the strain, ε. In Eq. 9.5 below, the stress can be written as the force per area, while the strain is defined as in Eq. 9.4.

$$Y = \frac{F/A}{\Delta d/\Delta z} = \frac{\sigma}{\varepsilon} \tag{9.5}$$

In compressional OCE, an elastogram is generated based on the inverse relationship between the displacement and Young's modulus. For excitation methods such as ARF or air puff, it is difficult to quantify the force applied per area. Although we can calculate ARF in a well-defined geometry, it is difficult to get the precise value of ARF for in vivo applications where the distance between the transducer and tissue changes. In other words, the stress is difficult to be quantified, so a qualitative map is produced. For qualitative imaging purposes, the elasticity of healthy tissue and plaque areas can be differentiated by the differences in displacement values, with a high displacement corresponding to softer tissue. Since the difference in elasticity is often at least one order of magnitude, qualitative information is helpful in disease diagnosis.

An example of an experiment done using compressional OCE is shown in Fig. 9.1 (Qi et al. 2012). The excitation mechanism was ARF. Since the resonance frequency of phantom is within 300 Hz, a pulsed excitation was used with a modulation of 500 Hz to avoid the effects of resonance. The force was applied to a side-by-side agarose phantom with Young's moduli of 83.6 kPa on the right-hand side and 265.7 kPa on the left side. Figure 9.1a shows the structural OCT image, where the boundary between the two different phantoms cannot be identified. Using Eq. 9.1 to isolate the phase shift, an elastogram is generated in Fig. 9.1b. It is evident that the 500 Hz modulation can be clearly observed, and the right side had a much higher phase response than the left, as shown in the amplitude plot in 1c. The measured response ratio between the left to the right side is 1:3.05. Figure 9.1d–f shows the 3D reconstruction of the same data. Since phase and displacement are proportional, it can be concluded that the right-hand side is approximately three times softer than its counterpart.

Although the compressional OCE method allows users to approximate the ratio between sample compositions, it is unable to directly offer quantitative elasticity. This is problematic when comparisons and diagnoses must be made between two different images or between different time points. Due to changes in experimental conditions and noise within the system, the displacement map cannot be effectively used to make conclusions between different samples and at different acquisition times. Because the ARF on the sample cannot be accurately measured for in vivo application, the absolute Young's modulus cannot be extracted. This leads to methods of quantification, which will be discussed in the next section.

Shear Velocity Estimation

The shear wave is the second parameter that has been widely studied in OCE techniques (Fujimoto 2001; Catheline et al. 1999; Gennisson et al. 2005). Since it travels

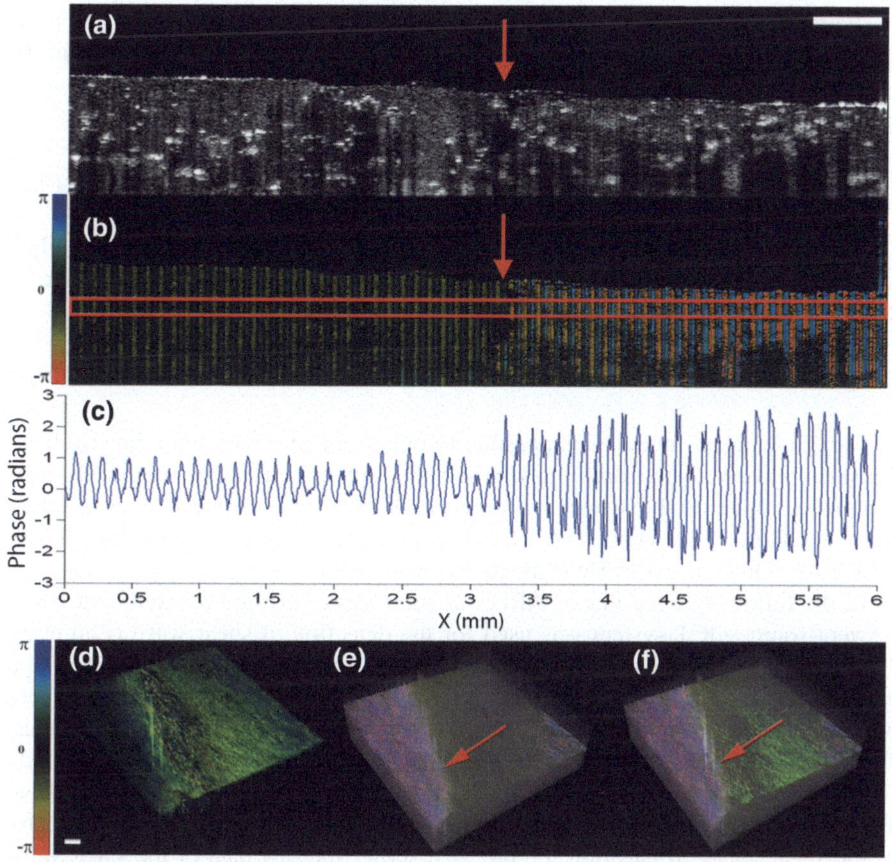

Fig. 9.1 Side-by-side agarose phantom results using compressional OCE method. **a** OCT intensity image. **b** OCE phase image. **c** OCE amplitude plot of data in the red box. **d** 3D OCT reconstruction. **e** 3D OCE reconstruction. **f** Fused 3D OCT and OCE images. Scale bar: 500 μm (Qi et al. 2012)

much slower than compressional waves, it is possible to analyze its speed of motion in different mediums. We can approximate the shear wave speed, $v_s(\omega)$, using a Voigt model for a homogeneous medium consisting of a single spring and damper. In Eq. 9.6 below, μ represents the shear modulus, ω is the shear wave angular frequency, η is the shear viscosity, and ρ is the tissue density. ω can also be defined to be twice the shear wave frequency (Razani et al. 2012).

$$v_s(\omega) = \sqrt{\frac{2\left(\mu^2 + \omega^2\eta^2\right)}{\rho\left(\mu + \sqrt{\mu^2 + \omega^2\eta^2}\right)}} \tag{9.6}$$

During OCE experiments, it is possible to use the phase maps to calculate the displacement of the shear wave at different locations. In order to calculate the mechanical elasticity, we can first relate the shear modulus to the shear wave velocity information. Intuitively, the stiffer the material, the faster the shear wave propagation. In addition, the tissue density must also be considered. This relationship is shown in Eq. 9.7.

$$\mu = \rho C_{\mathrm{s}}^{2} \tag{9.7}$$

Assuming that the tissue in question is incompressible, the Young's modulus, which is the direct measure of elasticity, is approximately three times the value of the shear modulus.

$$E \approx 3\mu \tag{9.8}$$

Using the above model, it is possible to obtain the elasticity map directly from the shear wave velocity. Both phase resolved Doppler and Doppler variance measurements can be used to detect the propagation of the shear wave (Zhu et al. 2015; Xu et al. 2016; Zhang et al. 2009; Razani et al. 2012; Chen et al. 2004; Yamakoshi et al. 1990; Qu et al. 2018; He et al. 2019).

In the following experiment, ARF was used as the method of excitation while a swept-source OCT system was used for the detection of shear wave (Zhu et al. 2015). The sample was an ex vivo rabbit cornea. Excitation ARF was applied in a diagonal direction to the cornea while the detection occurred from the top. There was a shear wave that propagated from the middle of the cornea to the two sides. The results are shown in Fig. 9.2. Figure 9.2a shows the B-mode OCT image, while Fig. 9.2b represents the propagation of the shear wave over time at each X location. The slope of the curve in Fig. 9.2b represents the distance over time or the velocity of the shear wave propagation. Figure 9.2c shows the raw data of the shear wave location at different sampling times, where the shear wave moves from the center of the cornea to the outer boundaries continuously. In order to gather all the necessary parameters, M-mode imaging was performed at each x location.

Two limitations of the above setup include the speed of the imaging device and the effects of tissue boundary conditions. The sampling rate must be faster than the shear wave propagation speed according to the Nyquist theory. For stiffer tissues, such as calcified atherosclerotic plaques, the wave propagation is much faster, and tracing the wave with a faster light source is necessary. The location of the excitation makes a difference in the boundary conditions, which is also determined by the geometry of the structure. There may be interference between different waves that are induced and also with the interfaces of the tissue layers. The alignment of the excitation and detection is also crucial in generating shear waves in the intended direction with minimal boundary influences. Several studies have co-aligned the excitation and detection so that it can be adapted to in vivo tissue imaging (Nguyen et al. 2014; Wang and Larin 2014). In addition, modeling and simulations may be necessary to study the wave dynamics in different tissues. In the case of intravascular and cardiovascular imaging with live pathological tissue, heartbeat and breathing motions may have a

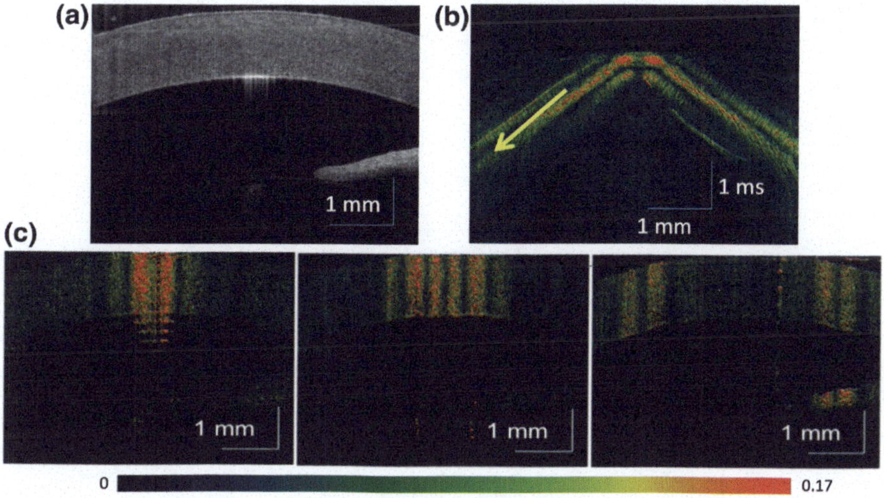

Fig. 9.2 Ex vivo rabbit cornea imaging. **a** OCT B-mode scan. **b** Spatial-temporal map of shear wave propagation. **c** Raw spatial data of shear wave propagation over time

large effect on data acquisition, and the mechanical structure may prove to be much more complex.

However, since only a single-excitation pulse is necessary to scan the entire field of view, shear wave OCE is suitable for cardiovascular applications. The pulse power is much lower than that of compressional OCE and can be kept within the federal safety limit for the Mechanical Index (MI). Also, the imaging time required is much shorter when detecting a single pulse, so large area intravascular acquisition can be performed to identify the pathological vessels within a long stretch. The successful translation of shear wave OCE imaging to in vivo studies and clinical trials would have the potential to make a great leap in the diagnosis of cardiovascular diseases.

Quantitative ARF-OCE Using Compressional Wave

The methodology of the compressional wave ARF-OCE has been outlined in the previous section. The feasibility of this method for vascular imaging will be examined. The schematic diagram of an ARF-OCE system is shown in Fig. 9.3a (Qi et al. 2013). A 4 MHz ultrasound transducer was used for excitation, driven by a function generator, and a radiofrequency amplifier. Individual pulses were given to the sample with a 50% duty cycle, 60 V excitation voltage, and 500 Hz. The OCT detection of the tissue response occurs on the opposite side of the sample. The OCT system consists of an 890 nm light source, with a high axial resolution of 3.5 μm. The light is split into the reference arm, where it is reflected back with a mirror, and the sample arm, where it interacts with the tissue sample. Galvanometer mirrors are used for

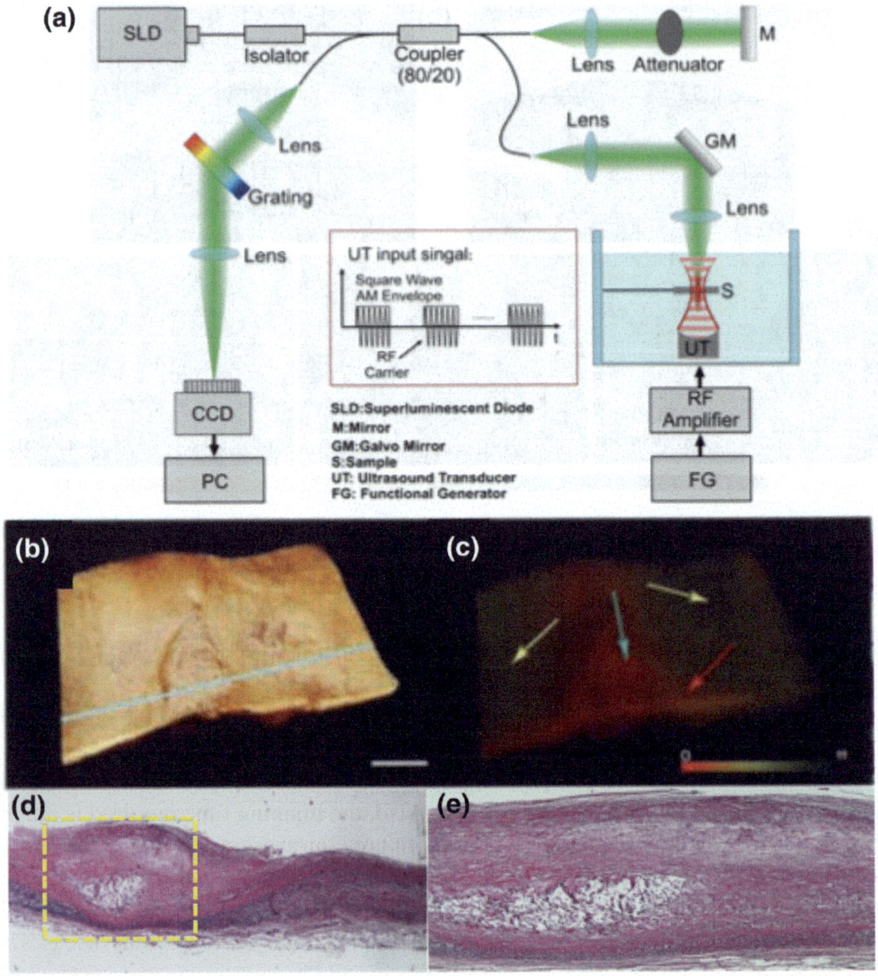

Fig. 9.3 Vascular imaging using compressional wave ARF-OCE system. **a** System schematic diagram. **b** OCT structural image of cadaver coronary artery. **c** OCE phase image under compressional wave excitation. **d** H&E histology of corresponding segment. **e** Close-up view of lesion in yellow box. Scale bar: 1 mm (Qi et al. 2012, 2013)

scanning the sample. The backscattered light from both the sample and reference arms travel back through the same path, into the detector arm, where their spectrums are detected using a line scan CCD camera. The interference signal is analyzed, and each A-line is obtained accordingly. The intensity and phase information for each A-line can be extracted by means of Eq. 9.1, so the OCT and Doppler OCE images can be obtained.

Using the system outlined in Fig. 9.3a, human cadaver coronary arteries were studied (Qi et al. 2012). The sample was placed between the ultrasound transducer and the OCT scanner, with water as the propagation medium for ARF. Figure 9.3b illustrates the 3D OCT image of the coronary artery, from which it is difficult to differentiate the abnormalities. Figure 9.3c shows the OCE phase response of the same artery. The blue arrow shows the region with low phase response, the yellow arrows show those of high phase response, while the red arrow shows the boundaries between the two areas where there appears to be a gradient response. After imaging was complete, H&S staining was performed to yield Fig. 9.3d and e, which correspond with the marked cyan frame in Fig. 9.3a. The zoomed-in view of Fig. 9.3e is the boxed region in 3d. The regions of low phase response correspond to tissues with higher Young's moduli or stiffer regions, according to Eqs. 9.3–9.5. On the histology images, these areas have abnormalities or lesions. The regions of high phase response are softer, associated with healthy tissue. The soft boundaries of the lesions are marked by a gradient of phase responses as expected. This study was the first to show the feasibility of using the ARF-OCE system on vascular tissues.

The relative elasticity measurements provided by the above method is helpful in distinguishing diseased lesions from healthy tissue within a single data set at one time. However, due to the different imaging conditions and external influences, it is difficult to make comparisons between different samples or at various time points. For example, the focal region of the transducer can shift at the sub-millimeter level, altering the ultrasonic force, thus affecting the phase response of the tissue, which can be incorrectly interpreted as a change in stiffness.

To determine the absolute stiffness of the tissues, we have developed a resonance ARF-OCE method. To start, the system can be represented by a simplified mechanical model consisting of a single spring and damper, also known as the Voigt Body Model, which is written as the following differential equation (Liang et al. 2008):

$$F(t) = m\ddot{x}(t) + \gamma\dot{x}(t) + kx(t) \tag{9.9}$$

The sinusoidal force applied to the sample is denoted by $F(t)$, while m is the mass of the object, $x(t)$ is the displacement, γ is the viscosity coefficient, and k is the spring constant. After solving for the displacement in the nonhomogeneous differential equation, and applying Hooke's law to the expression for Young's modulus, the elasticity can be represented by Eq. 9.10:

$$E = \frac{kL}{S} = \frac{(\mu^2 + \lambda^2)mL}{S} \tag{9.10}$$

The thickness of the sample is denoted by L and the contact area is S. The oscillation frequency and the damping coefficient are represented by $\mu = \frac{\sqrt{4mk-\gamma^2}}{2m}$ and $\lambda = \frac{-\gamma}{2m}$, respectively. Based on Eq. 9.10, the Young's modulus is proportional to the square of the oscillation frequency and the damping coefficient, while the mass, thickness, and area of the sample are known and remain relatively constant. The viscosity coefficient of a viscoelastic sample is often insignificant compared to the excitation force of ARF-OCE (Liang et al. 2008; Han et al. 2015). Therefore,

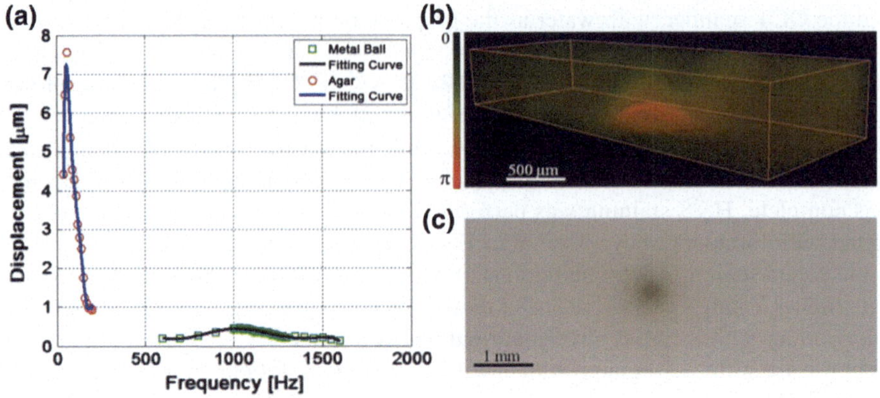

Fig. 9.4 Frequency response of phantom with metal ball inclusion. **a** Frequency response of both phantom and inclusion. **b** 3D OCE. **c** Sample photograph (Qi et al. 2013)

Young's modulus is primarily dependent on the oscillation frequency or the resonance frequency. Based on the principles of physics, the response of the sample will be highest when it is vibrating at its natural resonance frequency. This means that the tissue response will peak to signify the resonance frequency, and the absolute Young's modulus can be calculated.

To demonstrate the dependency of the elasticity on the resonance frequency, it is necessary to sweep across many excitation frequencies to determine the resonance peak of a material with a known stiffness. This has been done in phantoms with consistent mass and geometry but differing stiffness, and the squared relationship was verified (Qi et al. 2013). In Fig. 9.4, an agar phantom was constructed, and a small metal ball was embedded inside. Using the same system as shown in Fig. 9.3, the frequency response of the sample at different frequencies ranging from 50 to 1600 Hz were recorded and plotted in Fig. 9.4a. It is evident that the resonance frequency peaks of the agar and the metal inclusion differ greatly with the agar at 60 Hz and the metal at 1080 Hz. Figure 9.4b shows this large difference in the 3D OCE image, while 4c shows a photograph of the sample.

As expected, the resonance frequency changes with the stiffness of the phantom sample. To verify the same principle for vascular tissues, the same experiments were performed on human cadaver coronary arteries. Figure 9.5a shows the morphological OCT image, from which it is difficult to differentiate the diseased regions from healthy ones. Next, a modulation frequency of 500 Hz was used to excite the tissue, resulting in the OCE response shown in Fig. 9.5b. The modulation frequency was increased to 800 Hz, and the OCE response is illustrated in Fig. 9.5c. After imaging, histological analyses were performed, resulting in the images shown in Fig. 9.5d and e, which is a close-up view of the imaging area. The histology confirmed the presence of necrotic core fibroatheroma, where regions I and III consists of the loose fibrous tissues of the fibrous cap, and region II marks the thicker and denser portions of the fibrous cap. The large necrotic core is underneath the fibrous cap, and so the

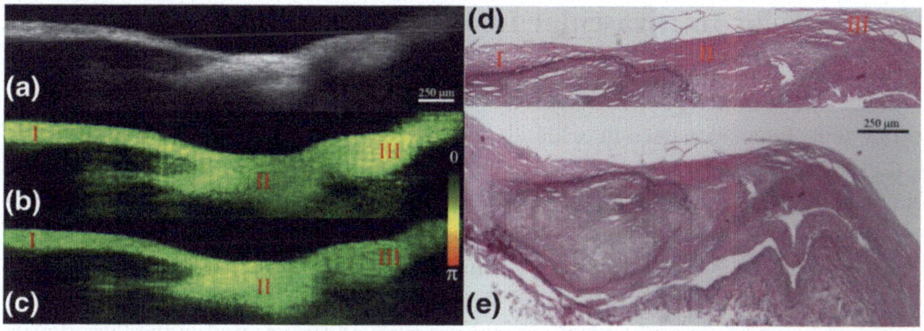

Fig. 9.5 Human cadaver coronary artery imaging. **a** OCT structural image. **b** Resonant OCE image with excitation frequency at 500 Hz. **c** Resonant OCE image with excitation frequency at 800 Hz. **d**, **e** Corresponding histology using H&E staining and zoomed into plaque region, respectively (Qi et al. 2013)

100 μm thick cap in region I is likely to rupture. Regions I and III had a resonance frequency peak at approximately 500 Hz, while region II had the highest response at 800 Hz. The higher frequency response corresponds with a denser and more stable cap tissue, while the lower frequency points to looser and vulnerable plaques. Therefore, the mechanical properties can help determine the stability of lesions and aid in the diagnosis of vulnerable plaques.

A major limitation using this approach is the need for the estimation of the mass and geometry of the sample. In phantom studies, these parameters are known, as in the case of most ex vivo studies, where the values can be measured accurately using mechanical testing and other means. However, in most in vivo animal and clinical studies, these parameters cannot be directly measured. It is necessary to rely on average values or estimations using other imaging modalities, such as ultrasound imaging system to encompass the entire depth of the tissue. Another problem with the imaging system is the opposing direction of the excitation and detection. Again, this is not an issue in ex vivo and phantom imaging, but for in vivo imaging, the set up is not practical as opposite sides of the vascular tissue are often inaccessible for probe placement. In addition, high power is required if the excitation beam has to travel and induce vibrations through the entire depth of the sample. In subsequent studies, this problem was taken care of with a ring transducer placed on the same side as the OCT scanner. There is an aperture in the middle of the transducer to allow the light through, so that the optical and acoustic beams are confocal with one another, leading to a more feasible design as well as lower excitation power (Qi et al. 2014; Qu et al. 2016). The last concern is the miniaturization of the excitation and detection systems to allow for catheter-based intravascular imaging, which will be discussed in the next section.

ARF-OCE for Intravascular Imaging

Catheter-based intravascular ultrasound (IVUS) and intravascular OCT systems and components have been used in research and also developed for commercial purposes in the past years (Cook et al. 2009; Achenbach et al. 2004; Kawasaki et al. 2002; Diaz-Sandoval et al. 2005). In the last 5 years, dual-modal IVUS and OCT systems have been developed to accurately visualize the anatomical structures of plaques as discussed in Chap. 3 (Li et al. 2014a, b; Yin et al. 2011). In its early stages, atherosclerosis will change the composition and the geometry of the vessel walls. These signs of disease can be seen in the mechanical properties of tissues within the three layers of the blood vessels. Since vulnerable plaques are most often characterized with a large lipid pool compressed behind a thin fibrous cap, a high-stress region on the cap indicates a region with a high risk of rupturing. Mechanically, the different compositions of plaques have distinguishable stiffness values. Therefore, intravascular ARF-OCE has been studied to measure these properties.

In order to miniaturize the system described in Fig. 9.3, a small ring transducer along with a corresponding OCT catheter must be built. In Fig. 9.6a and b, a front-facing intravascular ARF-OCE probe is shown, with an outer diameter of 3.5 mm (Qu et al. 2017). An 8.8 MHz miniature ring transducer was used, with a center aperture of approximately 1 mm. The transducer is focused at a depth of 5.5 mm. The optical components are designed into a fiber-based probe, with a 0.7 mm diameter gradient index (GRIN) lens for focusing the light. The fiber and lens are housed in a 0.8 mm diameter polyimide tube for protection, which is inserted into the aperture of the ring transducer. The acoustic and optical beams must focus on precisely the same location for maximum and most efficient excitation and detection. A torque coil has been implemented to protect the optical fiber, which will be essential for clinical translation and lays a foundation for the next generation side-scanning rotational probe. This ARF-OCE probe is, to the best of our knowledge, the first intravascular device of its kind.

After system characterization and phantom studies were performed, a 70 V excitation voltage was determined to be optimal for vascular tissue excitation. A single ARF pulse with duration of 1 mm was applied in order to minimize the excitation power on the tissue, while still maximizing the tissue response to the single pulse. A segment of the human cadaver carotid artery was opened for imaging of the vessel lumen. The artery segment was submerged in a small water bath, with the tip of the probe submerged just above. In addition, a mechanical stage was implemented to scan the probe at 6 μm increments for front-face imaging. The OCT image is shown in Fig. 9.6c, where it is difficult to see any abnormalities in the structure. Figure 9.6d shows the OCE image with quantified small interval displacement values. It is evident that the elasticity is heterogeneous throughout the sample, with stiffer tissue sandwiched in between soft tissues. In addition, it seems that the structures on the left side is stiffer than those on the right. To verify these findings, histology using H&E staining was performed in Fig. 9.6e. It is evident that there are layers of abnormal tissue, likely corresponding to fibrous plaques as shown in the red box, as well as on the left portion of the sample image. Based on calibration results, the diseased

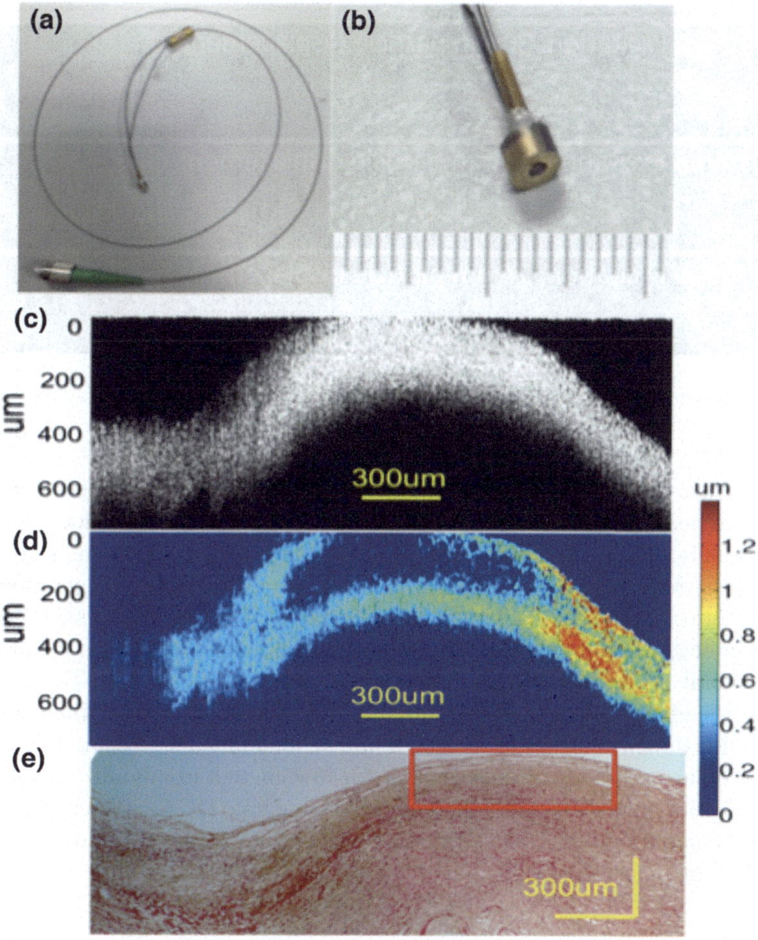

Fig. 9.6 Intravascular single pulse ARF-OCE imaging. **a** Front-facing probe design including ultrasound ring transducer for excitation and 890 nm fiber-based OCT for detection. **b** Enlarged probe tip with 3.5 mm transducer. **c** OCT of human cadaver coronary artery. **d** Corresponding displacement OCE map. **e** Corresponding histology segment showing plaque region (Qu et al. 2017)

region is approximately 30 kPa, while the healthy region on the right side is about 6 kPa.

Finally, the feasibility of using continuous pulse excitation was tested using a 2.5 mm front-facing catheter on a phantom as shown in Fig. 9.7. We demonstrate that even with a small probe, we can generate sufficient force to induce consistent sample vibrations and acquire the phase and displacement data simultaneously. This demonstrates the capability of the ARF-OCE method for real-time imaging with continuous probe rotation and pullback.

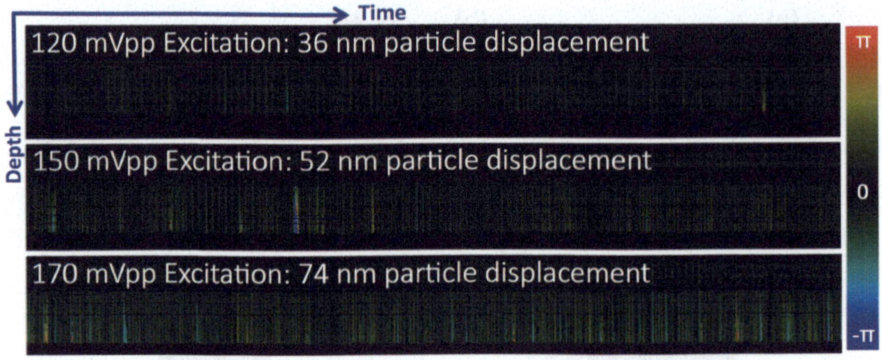

Fig. 9.7 Continuous pulse ARF-OCE imaging of phantom at different excitation voltage

The intravascular ARF-OCE technology has been successfully distinguish diseased from healthy tissues, especially when the OCT image alone does not show this accurately. The current goal is to implement probe rotation during acquisition, as well as to miniaturize the design further. Phase stability and phase wrapping are obstacles in rotational pullback with Doppler OCT. With respect to size, both the transducer design and the OCT portion need to be reduced to fit within the diameter of the femoral artery, similar to the existing OCT and IVUS catheters. The design of the outer sheath and guide wire must also be considered. The main challenge here is to reduce the dimensions of the catheter while maintaining a high enough excitation power to induce vibrations in the sample, but low enough to minimize health risks associated with ARF exposure. Finally, one can integrate OCT catheter with a pressure sensor and use intrinsic blood pressure changes or extrinsic flushing agent as an external force to perform OCE.

Summary

OCE is a valuable tool that has gained momentum in recent years for the characterization and quantification of mechanical properties of tissues. Various methods of excitation and detection have been used, including air puff and ARF for excitation, and shear wave and compressional wave for detection and quantification (Khalil et al. 2005; Wang et al. 2006, 2007; Kennedy et al. 2015; Liang et al. 2010; Manapuram et al. 2012; Rogowska et al. 2004; van Soest et al. 2007; Wang and Larin 2015; Qi et al. 2012, 2013, 2014; Qu et al. 2016, 2017; Fujimoto 2001). All of these methods aim to quantify the elasticity of tissues using the Young's modulus. It has been demonstrated that ARF-OCE has great potential to characterize the mechanical elasticity of vascular lesions in the early diagnosis of atherosclerosis. The surveillance of vulnerable plaques will provide a critically important tool for monitoring disease progression and providing timely intervention in high-risk patients.

References

Achenbach S, Moselewski F, Ropers D, Ferencik M, Hoffmann U, MacNeill B, Pohle K et al (2004) Detection of calcified and noncalcified coronary atherosclerotic plaque by contrast-enhanced, submillimeter multidetector spiral computed tomography a segment-based comparison with intravascular ultrasound. Circulation 109(1):14–17

Ahmad A, Huang P-C, Sobh NA, Pande P, Kim J, Boppart SA (2015) Mechanical contrast in spectroscopic magnetomotive optical coherence elastography. Phys Med Biol 60(17):6655

Amirbekian V, Lipinski MJ, Briley-Saebo KC, Amirbekian S, Aguinaldo JGS, Weinreb DB, Vucic E et al (2007) Detecting and assessing macrophages in vivo to evaluate atherosclerosis noninvasively using molecular MRI. Proc Nat Acad Sci 104(3):961–966

Baldewsing RA, Schaar JA, de Korte CL, Mastik F, Serruys PW, van der Steen AF (2004a) Intravascular ultrasound elastography: a clinician's tool for assessing vulnerability and material composition of plaques. Stud Health Technol Inform 113:75–96

Baldewsing RA, de Korte CL, Schaar JA, Mastik F, van der Steen AFW (2004b) Finite element modeling and intravascular ultrasound elastography of vulnerable plaques: parameter variation. Ultrasonics 42(1):723–729

Baldewsing RA, de Korte CL, Schaar JA, Mastik F, van der Steen AFW (2004c) A finite element model for performing intravascular ultrasound elastography of human atherosclerotic coronary arteries. Ultrasound Med Biol 30(6):803–813

Baldewsing RA, Mastik F, Schaar JA, Serruys PW, van der Steen AFW (2005) Robustness of reconstructing the Young's modulus distribution of vulnerable atherosclerotic plaques using a parametric plaque model. Ultrasound Med Biol 31(12):1631–1645

Baldewsing R, Schaar J, Mastik F, van der Steen A (2006) Atherosclerosis, large arteries and cardiovascular risk, vol 44. Karger Publishers, pp 35–61

Catheline S, Thomas J-L, Wu F, Fink MA (1999) Diffraction field of a low frequency vibrator in soft tissues using transient elastography. IEEE Trans Ultrason Ferroelectr Freq Control 46(4):1013–1019

Catheline S, Gennisson J-L, Delon G, Fink M, Sinkus R, Abouelkaram S, Culioli J (2004) Measurement of viscoelastic properties of homogeneous soft solid using transient elastography: an inverse problem approach. J Acoust Soc Am 116(6):3734–3741

Chai C-K, Speelman L, Oomens CWJ, Baaijens FPT (2014) Compressive mechanical properties of atherosclerotic plaques—indentation test to characterise the local anisotropic behaviour. J Biomech 47(4):784–792

Chen S, Fatemi M, Greenleaf JF (2004) Quantifying elasticity and viscosity from measurement of shear wave speed dispersion. J Acoust Soc Am 115(6):2781–2785

Chen Z, Milner TE, Dave D, Nelson JS (1997a) Optical Doppler tomographic image of fluid flow velocity in highly scattering media. Opt Lett 22(1):64–66

Chen Z, Milner TE, Srinivas S, Malekafzali A, Wang X, Van Gemert MJC, Nelson JS (1997b) Imaging in vivo blood flow velocity using optical Doppler tomography. Opt Lett 22(14):1119–1121

Cheruvu PK, Finn AV, Gardner C, Caplan J, Goldstein J, Stone GW, Virmani R, Muller JE (2007) Frequency and distribution of thin-cap fibroatheroma and ruptured plaques in human coronary arteries: a pathologic study. J Am Coll Cardiol 50(10):940–949

Cook S, Ladich E, Nakazawa G, Eshtehardi P, Neidhart M, Vogel R, Togni M et al (2009) Correlation of intravascular ultrasound findings with histopathological analysis of thrombus aspirates in patients with very late drug-eluting stent thrombosis. Circulation 120(5):391–399

Dariush M, Benjamin EJ, Go AS, Arnett DK, Blaha MJ, Cushman M, Das SR et al (2016) Executive summary: heart disease and stroke statistics—2016 update: a report from the American Heart Association. Circulation 133(4):447

de Korte CL, van der Steen AFW, Céspedes EI, Pasterkamp G (1998) Intravascular ultrasound elastography in human arteries: initial experience in vitro. Ultrasound Med Biol 24(3):401–408

de Korte CL, van der Steen AFW, Céspedes EI, Pasterkamp G, Carlier SG, Mastik F, Schoneveld AH, Serruys PW, Bom N (2000) Characterization of plaque components and vulnerability with intravascular ultrasound elastography. Phys Med Biol 45(6):1465
de Korte CL, Sierevogel MJ, Mastik F, Strijder C, Schaar JA, Velema E, Pasterkamp G, Serruys PW, van der Steen AFW (2002) Identification of atherosclerotic plaque components with intravascular ultrasound elastography in vivo A Yucatan pig study. Circulation 105(14):1627–1630
Diaz-Sandoval LJ, Bouma BE, Tearney GJ, Jang I-K (2005) Optical coherence tomography as a tool for percutaneous coronary interventions. Catheter Cardiovasc Interv 65(4):492–496
Ebenstein DM, Coughlin D, Chapman J, Li C, Pruitt LA (2009) Nanomechanical properties of calcification, fibrous tissue, and hematoma from atherosclerotic plaques. J Biomed Mater Res, Part A 91(4):1028–1037
Evans A, Whelehan P, Thomson K, McLean D, Brauer K, Purdie C, Jordan L, Baker L, Thompson A (2010) Quantitative shear wave ultrasound elastography: initial experience in solid breast masses. Breast Cancer Res 12(6):1
Fujimoto JG (2001) Optical coherence tomography. Comptes Rendus de l'Académie des Sciences-Series IV-Physics 2(8):1099–1111
Gennisson JL, Cornu C, Catheline S, Fink M, Portero P (2005) Human muscle hardness assessment during incremental isometric contraction using transient elastography. J Biomech 38(7):1543–1550
Greenleaf JF, Fatemi M, Insana M (2003) Selected methods for imaging elastic properties of biological tissues. Annu Rev Biomed Eng 5(1):57–78
Han Z, Aglyamov SR, Li J, Singh M, Wang S, Vantipalli S, Chen W, Liu C-h, Twa MD, Larin KV (2015) Quantitative assessment of corneal viscoelasticity using optical coherence elastography and a modified Rayleigh-Lamb equation. J Biomed Opt 20(2):020501
Hansson GK (2005) Inflammation, atherosclerosis, and coronary artery disease. N Engl J Med 352(16):1685–1695
He Y, Qu Y, Zhu J, Zhang Y, Saidi A, Ma T, Zhou Q, Chen Z (2019) Confocal shear wave acoustic radiation force optical coherence elastography for imaging and quantification of the in vivo posterior eye. IEEE J Sel Top Quantum 25(1):7200107
Huang D, Swanson EA, Lin CP, Schuman JS, Stinson WG, Chang W, Hee MR et al (1991) Optical coherence tomography. Science (New York, NY) 254(5035):1178
Inagaki J, Hasegawa H, Kanai H, Ichiki M, Tezuka F (2006) Tissue classification of arterial wall based on elasticity image. Jpn J Appl Phys 45(5S):4732
Kawasaki M, Takatsu H, Noda T, Sano K, Ito Y, Hayakawa K, Tsuchiya K et al (2002) In vivo quantitative tissue characterization of human coronary arterial plaques by use of integrated backscatter intravascular ultrasound and comparison with angioscopic findings. Circulation 105(21):2487–2492
Kennedy BF, Kennedy KM, Oldenburg AL, Adie SG, Boppart SA, Sampson DD (2015) Optical coherence elastography. In: Optical coherence tomography: technology and applications, pp 1007–1054
Khalil AS, Chan RC, Chau AH, Bouma BE, Kaazempur Mofrad MR (2005) Tissue elasticity estimation with optical coherence elastography: toward mechanical characterization of in vivo soft tissue. Ann Biomed Eng 33(11):1631–1639
LaMuraglia GM, Southern JF, Fuster V, Kantor HL (1996) Magnetic resonance images lipid, fibrous, calcified, hemorrhagic, and thrombotic components of human atherosclerosis in vivo. Circulation 94(5):932–938
Li X, Li J, Jing J, Ma T, Liang S, Zhang J, Mohar D et al (2014a) Integrated IVUS-OCT imaging for atherosclerotic plaque characterization. IEEE J Sel Topics Quantum Electron 20(2):196–203
Li J, Li X, Mohar D, Raney A, Jing J, Zhang J, Johnston A et al (2014b) Integrated IVUS-OCT for real-time imaging of coronary atherosclerosis. JACC: Cardiovasc Imaging 7(1):101–103
Liang X, Oldenburg AL, Crecea V, Chaney EJ, Boppart SA (2008) Optical micro-scale mapping of dynamic biomechanical tissue properties. Opt Express 16(15):11052–11065

Liang X, Adie SG, John R, Boppart SA (2010) Dynamic spectral-domain optical coherence elastography for tissue characterization. Opt Express 18(13):14183–14190
Little WC, Constantinescu M, Applegate RJ, Kutcher MA, Burrows MT, Kahl FR, Santamore WP (1988) Can coronary angiography predict the site of a subsequent myocardial infarction in patients with mild-to-moderate coronary artery disease? Circulation 78(5):1157–1166
Loree HM, Tobias BJ, Gibson LJ, Kamm RD, Small DM, Lee RT (1994) Mechanical properties of model atherosclerotic lesion lipid pools. Arterioscler Thromb Vasc Biol 14(2):230–234
Manapuram RK, Aglyamov SR, Monediado FM, Mashiatulla M, Li J, Emelianov SY, Larin KV (2012) In vivo estimation of elastic wave parameters using phase-stabilized swept source optical coherence elastography. J Biomed Opt 17(10):1005011–1005013
Manduca A, Oliphant TE, Dresner MA, Mahowald JL, Kruse SA, Amromin E, Felmlee JP, Greenleaf JF, Ehman RL (2001) Magnetic resonance elastography: non-invasive mapping of tissue elasticity. Med Image Anal 5(4):237–254
Mozaffarian D, Benjamin EJ, Go AS, Arnett DK, Blaha MJ, Cushman M, Das SR et al (2016) Heart disease and stroke statistics—2016 update. Circulation (2005)
Nguyen T-M, Song S, Arnal B, Wong EY, Huang Z, Wang RK, O'Donnell M (2014) Shear wave pulse compression for dynamic elastography using phase-sensitive optical coherence tomography. J Biomed Opt 19(1):016013
Nightingale K, McAleavey S, Trahey G (2003) Shear-wave generation using acoustic radiation force: in vivo and ex vivo results. Ultrasound Med Biol 29(12):1715–1723
O'Donnell M, Skovoroda AR, Shapo BM, Emelianov SY (1994) Internal displacement and strain imaging using ultrasonic speckle tracking. IEEE Trans Ultrason Ferroelectr Freq Control 41(3):314–325
Ophir J, Cespedes I, Ponnekanti H, Yazdi Y, Li X (1991) Elastography: a quantitative method for imaging the elasticity of biological tissues. Ultrason Imaging 13(2):111–134
Prati F, Arbustini E, Labellarte A, Dal Bello B, Sommariva L, Mallus MT, Pagano A, Boccanelli A (2001) Correlation between high frequency intravascular ultrasound and histomorphology in human coronary arteries. Heart 85(5):567–570
Qi W, Chen R, Chou L, Liu G, Zhang J, Zhou Q, Chen Z (2012) Phase-resolved acoustic radiation force optical coherence elastography. J Biomed Opt 17(11):110505
Qi W, Li R, Ma T, Li J, Kirk Shung K, Zhou Q, Chen Z (2013) Resonant acoustic radiation force optical coherence elastography. Appl Phys Lett 103(10):103704
Qi W, Li R, Ma T, Kirk Shung K, Zhou Q, Chen Z (2014) Confocal acoustic radiation force optical coherence elastography using a ring ultrasonic transducer. Appl Phys Lett 104(12):123702
Qu Y, Ma T, He Y, Zhu J, Dai C, Yu M, Huang S et al (2016) Acoustic radiation force optical coherence elastography of corneal tissue. IEEE J Sel Topics Quantum Electron 22(3):1–7
Qu Y, Ma T, He Y, Yu M, Zhu J, Miao Y, Dai C et al (2017) Miniature probe for mapping mechanical properties of vascular lesions using acoustic radiation force optical coherence elastography. Sci Rep 7(1):4731
Qu Y, He Y, Zhang Y, Ma T, Zhu J, Miao Y, Dai C, Zhou Y, Xin Y, Silverman RH, Humayun M, Zhou Q, Chen Z (2018) In-vivo elasticity mapping of retinal layers using synchronized acoustic radiation force optical coherence elastography. Invest Ophthalmol Vis Sci 59(1):455–461
Razani M, Mariampillai A, Sun C, Luk TWH, Yang VXD, Kolios MC (2012) Feasibility of optical coherence elastography measurements of shear wave propagation in homogeneous tissue equivalent phantoms. Biomed Opt Express 3(5):972–980
Rogowska J, Patel NA, Fujimoto JG, Brezinski ME (2004) Optical coherence tomographic elastography technique for measuring deformation and strain of atherosclerotic tissues. Heart 90(5):556–562
Ross R (1999) Atherosclerosis—an inflammatory disease. N Engl J Med 340(2):115–126
Schaar JA, de Korte CL, Mastik F, Strijder C, Pasterkamp G, Boersma E, Serruys PW, van der Steen AFW (2003) Characterizing vulnerable plaque features with intravascular elastography. Circulation 108(21):2636–2641

Sherman CT, Litvack F, Grundfest W, Lee M, Hickey A, Chaux A, Kass R et al (1986) Coronary angioscopy in patients with unstable angina pectoris. New England J Med 315(15):913–919

Sinkus R, Tanter M, Catheline S, Lorenzen J, Kuhl C, Sondermann E, Fink M (2005a) Imaging anisotropic and viscous properties of breast tissue by magnetic resonance-elastography. Magn Reson Med 53(2):372–387

Sinkus R, Tanter M, Xydeas T, Catheline S, Bercoff J, Fink M (2005b) Viscoelastic shear properties of in vivo breast lesions measured by MR elastography. Magn Reson Imaging 23(2):159–165

Sun C, Standish B, Yang VXD (2011) Optical coherence elastography: current status and future applications. J Biomed Opt 16(4):043001–043001

Takano M, Mizuno K, Okamatsu K, Yokoyama S, Ohba T, Sakai S (2001) Mechanical and structural characteristics of vulnerable plaques: analysis by coronary angioscopy and intravascular ultrasound. J Am Coll Cardiol 38(1):99–104

van Soest G, Mastik F, de Jong N, van der Steen AFW (2007) Robust intravascular optical coherence elastography by line correlations. Phys Med Biol 52(9):2445

Virmani R, Burke AP, Kolodgie FD, Farb A (2003) Pathology of the thin-cap fibroatheroma. J Intervent Cardiol 16(3):267–272

Walsh MT, Cunnane EM, Mulvihill JJ, Akyildiz AC, Gijsen FJH, Holzapfel GA (2014) Uniaxial tensile testing approaches for characterisation of atherosclerotic plaques. J Biomech 47(4):793–804

Wang S, Larin KV (2014) Shear wave imaging optical coherence tomography (SWI-OCT) for ocular tissue biomechanics. Opt Lett 39(1):41–44

Wang S, Larin KV (2015) Optical coherence elastography for tissue characterization: a review. J Biophotonics 8(4):279–302

Wang RK, Ma Z, Kirkpatrick SJ (2006) Tissue doppler optical coherence elastography for real time strain rate and strain mapping of soft tissue. Appl Phys Lett 89(14):144103

Wang RK, Kirkpatrick S, Hinds M (2007) Phase-sensitive optical coherence elastography for mapping tissue microstrains in real time. Appl Phys Lett 90(16):164105

Waxman S, Ishibashi F, Muller JE (2006) Detection and treatment of vulnerable plaques and vulnerable patients novel approaches to prevention of coronary events. Circulation 114(22):2390–2411

Xu X, Zhu J, Chen Z (2016) Dynamic and quantitative assessment of blood coagulation using optical coherence elastography. Sci Rep 6

Yamakoshi Y, Sato J, Sato T (1990) Ultrasonic imaging of internal vibration of soft tissue under forced vibration. IEEE Trans Ultrason Ferroelectr Freq Control 37(2):45–53

Yin J, Li X, Jing J, Li J, Mukai D, Mahon S, Edris A et al (2011) Novel combined miniature optical coherence tomography ultrasound probe for in vivo intravascular imaging. J Biomed Opt 16(6):060505–060505

Zhang J, Rao B, Yu L, Chen Z (2009) High-dynamic-range quantitative phase imaging with spectral domain phase microscopy. Opt Lett 34(21):3442–3444

Zhao Y, Chen Z, Saxer C, Xiang S, de Boer JF, Nelson JS (2000a) Phase resolved optical coherence tomography and optical Doppler tomography for imaging blood flow in human skin with fast scanning speed and high velocity sensitivity. Opt Lett 25(2):114–116

Zhao Y, Chen Z, Saxer C, Xiang S, de Boer JF, Nelson JS (2000b) Doppler standard deviation imaging for clinical monitoring of in vivo human skin blood flow. Opt Lett 25(18):1358–1360

Zhu J, Qu Y, Ma T, Li R, Du Y, Huang S, Kirk Shung K, Zhou Q, Chen Z (2015) Imaging and characterizing shear wave and shear modulus under orthogonal acoustic radiation force excitation using OCT Doppler variance method. Opt Lett 40(9):2099–2102

Chapter 10
Therapeutic IVUS and Contrast Imaging

John A. Hossack

Scope

IVUS is the most commonly encountered version of catheter-based ultrasound. The scope of this chapter will address intra-vessel ultrasound imaging. We omit the use of catheter-based ultrasound that is inserted via the peripheral vascular system but is used primarily for imaging inside of organs—typically within heart chambers. This latter field is usually referred to as intracardiac echocardiography (ICE) and involves devices >6F. Readers seeking material in addition to that in this chapter are referred to a number of relevant review articles addressing IVUS contrast imaging and IVUS oriented therapy/drug delivery (Dixon et al. 2015b; Ruiz et al. 2012).

Introduction and Evolution of IVUS Technology

Imaging technology and applications of IVUS are described in detail in other chapters in this volume. However, even within the more limited context of the therapeutic use of IVUS and contrast agent imaging, it is worth briefly reviewing the early evolution of IVUS. Unless stated otherwise, these devices acquire vessel cross-sectional views. As a practical matter, this is the view of most importance in a clinical setting because it provides detailed information relating to vessel wall cross section and is a familiar perspective for physicians who review cross-sectional histological sections acquired *postmortem*. These cross-sectional views are also particularly consequential because the longstanding default imaging modality for cardiovascular vessel imaging involves X-ray fluoroscopy (real-time X-ray) in combination with an X-ray contrast agent. In particular, it should be emphasized that fluoroscopy

J. A. Hossack (✉)
Department of Biomedical Engineering, University of Virginia, Charlottesville, VA 22908, USA
e-mail: hossack@virginia.edu

Q. Zhou and Z. Chen (eds.), *Multimodality Imaging*,
https://doi.org/10.1007/978-981-10-6307-7_10

provides a projection *view* of the vessel lumen. Because X-ray is a projection view, it integrates the signal through the projection direction with the result that, depending on the orientation of any asymmetric narrowing of a vessel lumen, the narrowing may or may not even be detectable in fluoroscopy. Usually, but not always, a narrowing is evident with cautious evaluation from multiple orientations. Secondly, fluoroscopy produces an image showing the extent of the lumen. It provides no information about the vessel cross section—i.e., anything beyond the vessel wall. It should also be emphasized that the degree of vessel narrowing is not well correlated with predicting which lesion may later result in myocardial infarction. Myocardial infarction risk is linked to the degree to which a plaque may be defined as "vulnerable" and this is best predicted by a comprehensive assessment of plaque morphology and composition (Richardson et al. 1989). Fortunately, ultrasound, and some other minimally invasive imaging technologies, can assess plaque morphology and composition. No single modality addresses cardiovascular vessel wall imaging perfectly. This has resulted in a rich field in the development of multimodality devices. Some of these are reviewed in other chapters and in the literature (Ma et al. 2016). IVUS can be applied to a range of vessels in the cardiovascular system. These include coronary, carotid, and peripheral arteries. However, the coronary application is generally considered both the most technically challenging and the most important from a patient mortality and morbidity perspective. Unless specifically stated otherwise, references to technology in this chapter refer to coronary oriented devices. A typical technical objective cross-sectional dimension for this application is approximately 3F (1 mm diameter). Prototyping typically occurs using larger devices, but ~3F is a typical clinical translation goal.

In human coronary IVUS imaging, the required imaging depth is typically around 1 mm and no more than 4 mm. This imaging depth is appropriately addressed using ultrasound frequencies in the 20–80 MHz range (Li et al. 2011; Ma et al. 2015; van der Steen et al. 2006). Consequently, spatial resolution (as low as 70 μm axially (Li et al. 2011; Ma et al. 2015) enables assessment of critically important, fine scale, anatomical structures such as a fibrous cap on an atheromatous plaque (Hiro et al. 2001). The intravascular placement of the transducer also eliminates several technical challenges frequently encountered in ultrasound imaging. There is no shadowing due to bone or gas. There is no significant phase aberration due to differential sound speed (e.g., in fat). Because the transducer is located relative to the vessel itself, the beating heart causes the catheter to move with the vessel. There will be some residual offset during the cardiac cycle, but it is far less than in the case of transcutaneous ultrasound registered to a static point on the abdomen. The quality of the imaging environment also makes advanced signal analysis, with the objective of using spectral qualities of echo signal signature to assess atherosclerotic plaque composition, more tractable (Schartl et al. 2001; Rodriguez-Granillo et al. 2006). Being placed inside the lumen, it is possible to step the transducer in axial direction increments and acquire a 3D volume with relative ease (Slager et al. 2000). Currently, intravascular ultrasound is most frequently used as an imaging method to assess lesions that cannot be definitively categorized using angiography. For example, left main coronary artery lesions are frequently challenging, and IVUS has been shown to be useful in

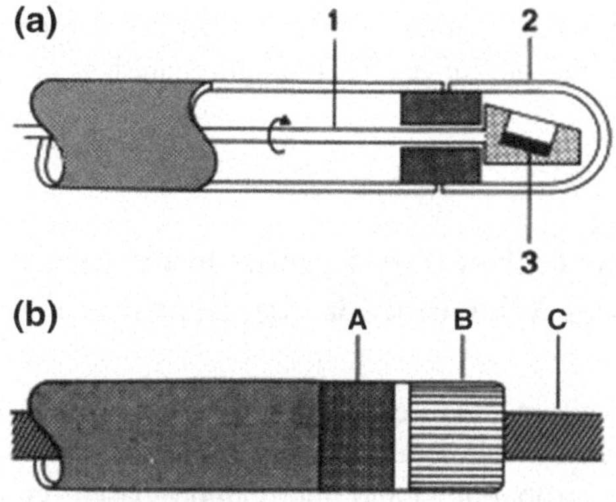

Fig. 10.1 Schematic illustration of IVUS catheter configurations. **a** Mechanical rotating single-element catheter tip consisting of rotating shaft (1); transparent dome (2); and transducer element (3). **b** Electronically switched phased array catheter tip consisting of integrated circuitry for reduction of the number of wires (A); multi-element transducer (B); and guide wire (C). *Source* Bom et al. (1998)

their assessment (Ragosta 2015). IVUS also has a well-established role in planning percutaneous coronary interventions (PCI) and for guiding and verifying coronary stent placement (Claessen et al. 2011). These, and other, applications are discussed in greater detail in other chapters in this book.

Early prototype IVUS was mainly evaluated using animal models or in an ex vivo human setting. Early IVUS catheters, designed for acquiring vessel cross-sectional views, generally fall into one of the two classifications: (1) mechanically scanned single element transducers, and (2) circumferential phased transducer array. These are illustrated schematically in Fig. 10.1.

There are two variants of the scanned single element transducer catheter. In the first variant, an ultrasound element is mounted to a rotating drive cable operating internally to the IVUS catheter (Fig. 10.1a). A variant of this design involves using a static axial oriented transducer and a rotating acoustic mirror that projects the ultrasound in pulse-echo mode around the catheter circumference. In the second classification, an ultrasound array is used (Fig. 10.1b). The elements are spaced circumferentially around the catheter as if a microscale phased array was "wrapped" around the catheter. The first array-based design was described by Bom in 1972 (Bom et al. 1972). Despite the technical challenges implicit in designing and fabricating a phased array IVUS, significant progress was made in this technology and in the late 1980s and the early 1990s resulted in the endosonics (Rancho Cordova, CA, USA) human coronary compatible commercial product. A clinical compatible mechanically scanned single element IVUS catheter evolved approximately in parallel with the phased array devices. Working with founder/inventor Dr. Paul Yock, cardiovascular imaging systems (Sunnyvale, CA, USA) developed a clinical IVUS. Derivatives of both these products are still in widespread clinical use. Their continued success owes much to the widely appreciated value of vessel cross-sectional views in daily clinical practical practice. More advanced designs, such as required for forward-looking imaging, or real-time 3D imaging, are less critical to daily clinical needs.

IVUS Transducer Technology Evolution in the Context of Therapeutic and Microbubble Applications

The technological evolution and direction of IVUS designs are discussed in the context of this chapter's theme—therapeutic applications (primarily, but not exclusively in conjunction with microbubbles) and microbubble-based (contrast) imaging. For therapeutic applications, higher power and frequency agility are desirable. Specifically, although a lower frequency and high power are desirable, fine-resolution imaging performance is also a necessity. As a practical matter, offsetting transducers (therapeutic transducer axially offset from an imaging transducer) are not an acceptable design compromise. Because the transducers are adjacent to the tissue target, offset transducers cannot operate without a near field "blind zone"—i.e., where the ultrasound beams of each of the offset transducers do not overlap. Translating the device back and forth to make up for this blind zone is not viable in a clinical setting and involves increased complexity due to the continuous motion implicit when operating inside a beating heart. In the case of contrast imaging, frequency agility and sensitivity are necessary. It may be desirable to use a harmonic mode of agent detection and even if that is not the case, it is frequently desirable to use a relatively low frequency for contrast agent imaging to achieve higher sensitivity. Meanwhile, the highest possible frequency and bandwidth will provide the best anatomic B-mode image for placing a contrast-specific signal in anatomic context. Thus, each of the therapeutic application and the contrast imaging application places an emphasis on highly versatile transducer design.

Single Element IVUS

The scanned single element device can generally be fabricated more simply with a larger aperture and higher frequency. The relative simplicity of the scanned single element device generally makes it more suitable for laboratory prototyping. In fact, one practice involves partially dismantling a commercial IVUS and replacing with the single transducer element with one adapted for a new application—e.g., higher power and lower frequency (Kilroy et al. 2014a). Higher power and lower frequency are, of course, typically preferred in therapeutic applications that are central to this chapter. In a typical commercial IVUS imaging catheter, the center frequency is 40 MHz, and the aperture is circular and occupies approximately 80% of the diameter of the device. Thus, the mechanically scanned device can achieve very fine anatomic resolution by virtue of the high center frequency and relatively large aperture. However, it should be noted that currently available mechanically scanned IVUS possesses a fixed focus and therefore achieves degraded imaging resolution away from the focal depth. Electronic focusing on a range of depths may be achieved using an annular array. In 2012, Lui et al. (Liu et al. 2012) reported a 1.5 mm diameter annular array. Consequently, it is conceivable that a ~3F annular array IVUS will be developed in the near future. It is worth noting that the small channel count (<10 channels) and very short transit times imply that a synthetic aperture approach may be used to achieve dynamic focusing in both transmit and receive mode when using an annular array. Mechanically scanned single element devices were also plagued by an artifact resulting from non-uniformly rotating drive cables causing reconstruction errors encountered when the device was tightly curved or subjected to strong pulsatile flow effects. This artifact, referred to as non-uniform rotational distortion (NURD) (Mintz et al. 2001), is now less commonly encountered due to improvements in device drive cable technology.

Progressive refinements in design and material choices have resulted in improvements in mechanically scanned single element performance. Piezoelectric polymers (polyvinylidene difluoride—PVDF) were initially favored due to their amenability to high-frequency fabrication—i.e., ease of realization 40 MHz operation. However, PVDF has been largely replaced by higher efficiency, higher power, piezoelectric ceramics (mainly lead zirconate titanates—PZTs) that yield higher signal-to-noise ratio (SNR) and resulting imaging penetration depth. A number of challenges are involved in designing an IVUS catheter for delivering localized therapeutic intervention/drug delivery. An IVUS catheter, designed for imaging, possesses high frequency, and therefore small, transducer element(s). The small volume of active transducer material limits energy conversion (in both transmit and receive modes) and can create an electrical matching challenge—primarily in the case of phased array devices. Some piezoelectric material choices that provide the highest imaging signal to noise are less robust and have lower upper bounds on peak and continuous power handling. For example, in the case of widely used lead zirconate titanates (PZT), there is a choice to be made between a "soft" ceramic possessing a high dielectric constant (implying lower electrical impedance) and high electromechanical coupling coefficient (resulting in high signal bandwidth) and "hard" ceramics with lower dielectric constant and lower coupling coefficient but higher peak and continuous power handling capability. The upper power limit is determined by consideration of both the maximum voltage that may result in depoling of a synthetic piezoelectric ceramic and the maximum mechanical stress conditions tolerated by any particular design choice. More recently, ceramics have been replaced by single crystal piezoelectric materials that yield yet higher levels of performance enabling the preferred combination of very high center frequency (up to 80 MHz), high imaging bandwidth (65%), and high sensitivity (Li et al. 2011; Yoon et al. 2015). Each of these successive innovations typically results in a more challenging fabrication process that is rewarded with improved imaging performance. (The single crystal materials require alignment with crystal direction and require a more demanding approach to machining than is the case for ceramics.) At each stage, device design is guided by the use of comprehensive transducer computer models that account for electrical matching considerations, backing design, and matching layer design (Li et al. 2011).

Phased Array IVUS

The phased array IVUS approach has a very different set of qualities. Implicitly, there is no possibility of NURD and no moving parts. Obviously, device complexity is far greater resulting in considerably more expensive prototype development. There is no practical possibility of swapping out the transducer elements for a new application as is the case for a scanned single element device. In terms of imaging resolution, it has proven practical to place 64 elements (~20 MHz center frequency) at approximately half wavelength spacing around a 3.5 F catheter. It should also be noted that the available aperture for imaging is also less than for the mechanically scanned single element in which the element occupies most of the cross-sectional dimension. For these reasons (smaller aperture and lower frequency), the imaging resolution is typically measurably inferior to that of the mechanically scanned single element device. Because phased array IVUS frequently operate using in a synthetic aperture beamforming approach (O'Donnell and Thomas 1992; O'Donnell et al. 1997), in which a transmit aperture is synthesized in a serial manner over multiple discrete firings, it is possible to achieve dynamic focusing, in both transmit and receive mode, providing for more uniform resolution performance through the field of view up to the diffraction limit. The phased array approach confers a number of inherent advantages that may mitigate the apparent loss of raw anatomic resolution. Being a solid-state array, it is possible to enable Doppler processing (O'Donnell and Thomas 1992), although, as a practical matter in a catheter–vessel configuration, blood flow is primarily in a direction orthogonal to the ultrasound beam axis, and therefore, blood flow is better detected using speckle decorrelation. Speckle decorrelation-based blood flow is now widely employed and clinically available (Lee et al. 2012). A solid-state array means that successive acoustic line firings can be placed along arbitrarily selected trajectories—as opposed to being dictated by the rotational speed of a drive cable present in the case of mechanically scanned transducers. This transmit beam design versatility enables multi-pulse techniques of widespread use in contrast agent imaging—such as pulse inversion.

Capacitive Micromachined Transducers (cMUT)

In parallel with innovations in piezoelectric materials, the technology of capacitive micromachined ultrasound transducers (cMUTs) has evolved rapidly. The very small single cell dimensions (typically $\leq$ 50 m across), combined with a very thin active membrane, yield high-frequency and high-bandwidth capabilities that are highly desirable for contrast agent imaging. cMUTs can also be configured for high-power operation. To date, high-power (therapeutic level) cMUTs have been limited to intracardiac, not intravascular, applications (Wong et al. 2010; Stephens et al. 2012). However, there is no fundamental reason why IVUS cMUTs cannot be designed with sufficient transmit power levels. The substantial costs of cMUT prototype design and fabrication have proven to present a challenge and only a relatively few research groups possess the expertise, infrastructure, and equipment to make prototype cMUT IVUS. Although cMUT technology is not readily suited to early-stage prototyping because of high capital, maintenance, and processing costs, it does possess attributes making it compelling for future commercial usage. Specifically, cMUT technology, being based on integrated circuit (IC) design and fabrication processes, is best suited to commercial applications involving high product volumes and small devices (i.e., many devices per wafer). Thus, the IVUS field, involving high volumes of small, single-use, complex device designs, more than any other application area in medical ultrasound, is suited to cMUT technology from a commercial perspective. Additionally, once the complexities of cMUT design and processing are solved, cMUT technology possesses a range of unique attributes. cMUT designs are based on microlithographical mask processes, and therefore, arbitrary transducer element and array designs are possible. Because the elements can be formed immediately over the associated transmit and receive electronics, complex element interconnect challenges can be more readily solved (Gurun et al. 2014). Additionally, varying levels of signal processing can be performed on silicon immediately below the elements—transmit–receive switch, pre-amplification, analog to digital conversion, and beamforming/multiplexing (Gurun et al. 2014). The versatility in design offered by cMUTs has resulted in a wide range of designs being reported. These include designs for forward-looking IVUS (FL-IVUS) (Gurun et al. 2014) and forward-looking 3D acquisition (Choe et al. 2012). However, it is worth observing that while forward-looking and 3D IVUS have been described, there have been relatively few articles on circumferential array for 2D cross-sectional imaging. In 2008, Xuefeng et al. (2008) described progress toward circumferential imaging using cMUTs in conjunction with a discussion of practical challenges (membrane cracking). It is worth also pointing out that while prototyping costs are relatively high when using cMUT technology, considerable advances in cMUT modeling have been made recently (Satir and Degertekin 2015). Consequently, similar to the example when considering modern IC design, one can have sufficient confidence in a comprehensive model-based design to know that a fabricated design will perform in accordance with the model and therefore a model-based "rapid iterative" (sometimes also referred to as "virtual prototyping") approach to new design development is enabled (Satir and Degertekin 2015).

Therapeutic Applications of IVUS

A powerful concept in cardiovascular clinical care, enabled by IVUS, is the potential to make a high sensitivity, high specificity, diagnosis that is immediately followed by an IVUS-based therapeutic intervention during the same catheterization procedure. In some instances, it may be possible to use the same multi-functional IVUS device for each of diagnosis and therapy. Alternatively, dedicated imaging and therapy IVUS devices are used. However, it is desirable that the therapeutic device possess, at least, a basic imaging capability to guide the procedure. The ideal mode of use involves observing the target region of interest prior to intervention, guiding the intervention in real time, and having some degree of validation/verification that the therapy has addressed the targeted completely with zero, or minimal, off-target impact. Ultrasound, practically uniquely, is well positioned to address this model of usage. For example, combining a catheter-based therapy delivery system with a fixed reference, external, imaging system (such as X-ray fluoroscopy or MRI) would be unwieldy and unlikely to meet the simultaneous imaging and therapy goal. X-ray fluoroscopy visualizes only the lumen, and MRI lacks the required spatial resolution and may be incompatible with a catheter depending on the choice of materials used (i.e., danger of RF-induced heating, etc.)

Some IVUS transducers are operated off-resonance at a frequency lower than the device's natural resonant frequency or a special low frequency, high power, the design is used (Phillips et al. 2010). In therapeutic applications, IVUS origin ultrasound energy can be used in isolation or in combination with microbubbles. For example, using 2 MHz ultrasound, it has been shown that IVUS originated ultrasound can improve gene transfection (Amabile et al. 2001). Reduced neointimal hyperplasia has been also reported following 0.7–1 MHz insonation with an IVUS catheter (Fitzgerald et al. 2001; Regar et al. 2003). It should be noted that these results described above were obtained without a drug.

Using a 2 MHz IVUS catheter, investigators have also studied whether catheter origin ultrasound enhances the efficacy of the thrombolytic recombinant tissue plasminogen activator (r-tPA). When compared with standard therapy (catheter-directed thrombolysis CDT), the results did not indicate a statistically significant clinical outcome improvement (Baker et al. 2012; Engelberger et al. 2015). In these latter studies, a fully commercialized and regulatory approved therapeutic IVUS (EKOS corp.) was employed. This 2 MHz catheter has no imaging capability and is used without microbubbles. However, the vast majority of recent work, described below, indicates that the use of ultrasound in combination with drug and microbubbles has far superior efficacy. (It is probable that some of the studies involving the use of ultrasound alone, or in combination with an approved drug, were motivated less by a pursuit of optimal therapy and more by the perception that there exists a simplified (no new drug) regulatory approval pathway.)

Therapeutic Applications of IVUS: Microbubble-Enhanced Drug and Gene Delivery

The use of microbubbles, in combination with applied ultrasound, to improve cellular uptake of drugs and genetic material occurs through a process most often referred to as "sonoporation" (van Wamel et al. 2006; Ferrara et al. 2007). In sonoporation, pores are transiently formed in the cell membrane as a result of adjacent microbubble oscillation, cavitation, or jetting resulting from ultrasound stimulation (van Wamel et al. 2006). The key to successful sonoporation is to generate a pore that is sufficiently "open" and long-lived that drug or gene can pass through the membrane but also is small enough that the pore is short-lived and the cell can recover. Pores that are too large will, unsurprisingly, result in cell death. This topic is covered in impressive microscopy-based detail in an article from Hu et al. (2013) in relation to relatively large MRC-5 fetal fibroblast cells. Extensive previous studies have indicated that high acoustic pressures (>1 MPa), peak negative pressure [PNP], low frequencies (~1 MHz), and long pulses (>4 cycles) are effective when using the most commonly encountered microbubbles (i.e., those in 1–4 μm diameter range) (Karshafian et al. 2009). Obviously, these conditions are not easily achievable using an unmodified clinical imaging IVUS catheter. It therefore becomes necessary to use custom transmitter electronics and/or IVUS transducer elements designed for therapeutic applications. As an example, in a study of IVUS-based gene delivery, a clinical 40 MHz IVUS was operated using a custom transmitter system to produce 200 kPa PNP pulses at 1.5 MHz (Phillips et al. 2012). A similar clinical IVUS catheter was also used for in vivo gene delivery in a swine model (Phillips et al. 2010). The swine left anterior descending (LAD) coronary artery received balloon angioplasty and cytomegalovirus–red fluorescent protein (CMV-RFP) plasmid loaded microbubbles were administered via a port on the catheter. In this example, insonation used a 5 MHz pulses at 2 MPa PNP, and CMV-RFP transfection was observed along a segment of the vessel wall associated with the direction of the acoustic beam (Fig. 10.2). It is preferable to use an IVUS transducer design optimized for low MHz operation rather than using a clinical imaging device operating off-resonance. This reduces the required transmit voltage and consequent risk of transducer damage. Unfortunately, the need to also fit within a 1 mm diameter (3F) catheter for coronary applications presents fundamental challenges when using common piezoelectric ceramics or crystals. The speed of sound in these materials is sufficiently high that cross-sectional dimensions consistent with low MHz resonance do not fit within target 3F IVUS dimension. One solution to this design challenge involves designing the transducer to use a long-dimension (i.e., axial direction) width-mode resonance. The desired thickness direction output is created via mechanical and electromechanical cross-coupling. Using this approach enabled a design yielding 1 MPa PNP at a frequency of 1.75 MHz (Kilroy et al. 2014b). In an in vitro experiment, this catheter was used in combination with microbubbles to deliver a model drug (DiI) to smooth muscle cells (SMCs) with minimal cell death. A similar catheter was used in a porcine coronary model and yielded in vivo delivery of the rapamycin (an anti-proliferative

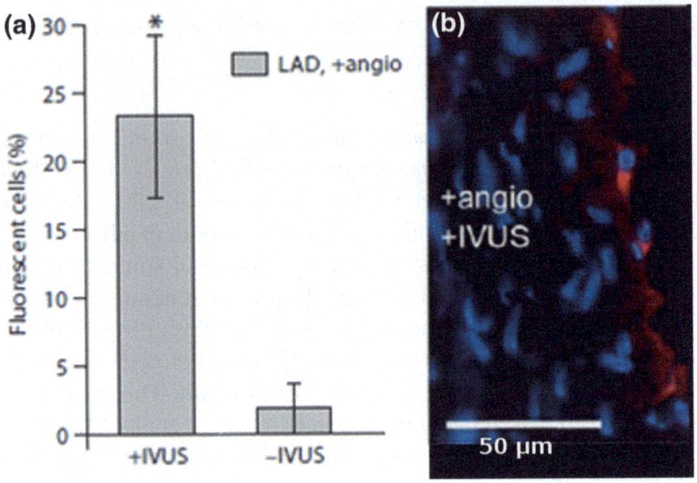

Fig. 10.2 IVUS insonation of gene-loaded microbubbles enhanced gene delivery post balloon angioplasty. **a** The insonated left anterior descending (LAD) coronary artery showed a 6.5 fold increase in cells expressing the cytomegalovirus-red fluorescent protein plasmid over the non-insonated control. **b** Fluorescent image of a section of the LAD wall that received both angioplasty and IVUS. Blue indicates cell nuclei; red indicates cells successfully transfected with the plasmid. *Source* Phillips et al. (2010)

drug used in some drug-eluting stents) using rapamycin-loaded microbubbles (Kilroy et al. 2015). Arteries receiving rapamycin-loaded microbubbles and 5 MHz ultrasound experienced a 50% reduction in neointimal formation compared to arteries treated with rapamycin-loaded microbubbles and no ultrasound (Fig. 10.3) (Kilroy et al. 2015). In conclusion, these studies show the potential of using modified IVUS catheters to achieve localized drug delivery in the context of coronary artery disease.

Primary acoustic radiation force (ARF), whereby particles in the field are subjected to a force causing them to move along the axis of ultrasound beam propagation can be used in the context of IVUS to push microbubbles toward the vessel wall. It is worth recalling that if ARF is applied transcutaneously, it will cause microbubbles in one portion of a vessel oriented perpendicular to the acoustic beam to move toward the distal vessel wall while on the opposite (proximal) side of the vessel, the microbubbles will be caused to move away from the wall. Thus, the IVUS approach, operating from the inside facing outwards works particularly well in this context. ARF has been used to displace microbubbles in IVUS for both molecular imaging (Rychak et al. 2007) and drug delivery (Phillips et al. 2011). ARF is optimized when using low frequency and high duty factor pulses to maximize microbubble imposed velocity (Dayton et al. 2002). An IVUS catheter designed specifically with ARF in mind has been designed to emit 3.5 MHz ultrasound (Kilroy et al. 2012). In order to maximize the "residence time," during which microbubbles are within the field of the ARF transducer, the transducer was elongated (3.35 mm) in the axial (blood

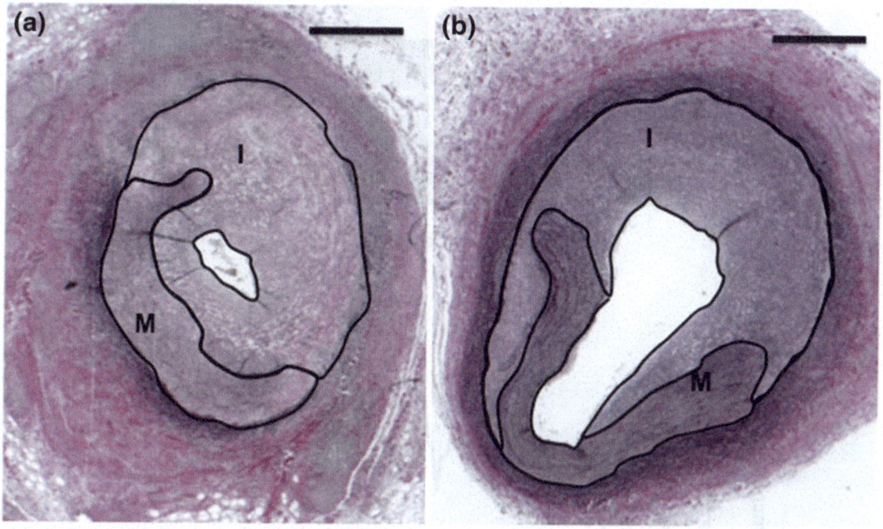

Fig. 10.3 Representative histology images of porcine coronary arteries treated with rapamycin microbubbles and IVUS. **a** Artery section treated with rapamycin microbubbles only. **b** Artery section treated with rapamycin microbubbles and therapeutic intravascular ultrasound. "I" and "M" denote vessel intima and media, respectively. Scale bars are 500 μm. *Source* Kilroy et al. (2015)

flow) direction. Clearly, for maximum ARF effect, the transducer should be as long in the axial direction as possible consistent also with the desire to minimize the rigid length near the device tip. A rigid tip creates challenges when negotiating tortuous vessels. Using a similar ARF optimized IVUS transducer, enhanced delivery of a model drug was demonstrated in ex vivo and in vivo arteries (Kilroy et al. 2014a). The long duration, moderate pressure, and ARF pulses were used to affect model drug delivery from the microbubbles accumulated on the vessel. It was observed that drug delivery was restricted to within the −6 dB beamwidth (Fig. 10.4). This demonstrates the potential for ARF to assist in providing a spatially controlled drug delivery effect to a vessel wall. Using both axial and rotational translations, it follows that it is possible, in principle at least, to "paint" the vessel wall according to a therapy plan individualized to any unique asymmetric lesion.

Microbubbles in IVUS Imaging

Shell-stabilized gas-filled microbubbles are widely used as the basis of a contrast agent in ultrasound imaging. Comprehensive reviews of microbubble design and clinical applications are available in the literature (Cosgrove and Lassau 2010; Unnikrishnan and Klibanov 2012). As a blood flow tracer, in their simplest usage, microbubbles provide a bright echo signal that delineates the

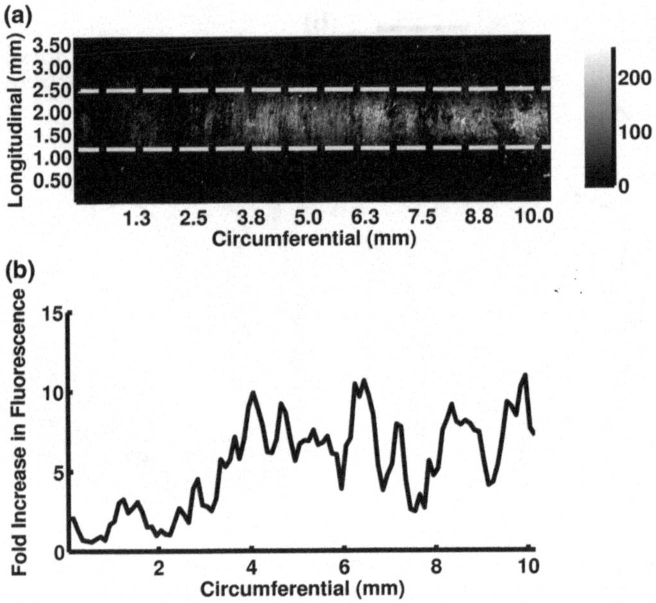

Fig. 10.4 **a** Microscope image of DiI delivery along the vessel wall. Gray-scale bar measures the pixel intensity in arbitrary units of fluorescence. The dotted yellow lines denote the −6-dB beam width of the ultrasound transducer (1.3 mm). **b** Fluorescence intensity increase over background in the artery plotted along the circumference in an ex vivo artery where delivery was performed in phosphate-buffered saline solution. Plot axes were truncated to show that the fluorescence intensity increase occurs along the entire vessel wall. A maximum fluorescence intensity increase of 10× was measured in this artery with a mean of 5.3 ± 2.9 fold. This result was produced with a sonoporation pulse with a peak negative pressure of 2 MPa and pulse repetition frequency of 1 kHz. The region from 0 to 2 mm has a lower fluorescence intensity increase because the catheter was not centered and may have blocked the flow of microbubbles to this region. *Source* Kilroy et al. (2014a)

extent of the vasculature and the blood filled chambers of the heart. In this context, microbubbles have been used for assessing myocardial perfusion defects following myocardial infarction (Wei et al. 1998). They have also been used to assess enhanced perfusion and angiogenesis in cancer applications (Hohmann et al. 2003; Zhao et al. 2010). In additional to having a role in assessing anatomy (via enhanced of vessel boundaries) and function (i.e., blood flow/perfusion), microbubbles are now finding applications using ligand-based molecular targeting to vascular endothelial growth factor receptor-2 (VEGFR-2) in cancer settings (Pochon et al. 2010). Microbubbles have received regulatory approval in most advanced countries for a range of cardiovascular, abdominal imaging, and molecular imaging applications. In particular, in the USA, microbubbles have been approved for better delineating the chamber boundaries in the heart and, in April 2016, were FDA approved for the characterization of liver lesions (FDA 2016). Initial clinical trials have been completed on VEGFR-2 targeted microbubbles in the USA and Europe and it seems probable that molecular targeted microbubbles will achieve regulatory approval for clinical usage in coming years.

A somewhat similar challenge exists for contrast agents as for IVUS from a clinical translation perspective: Both IVUS and contrast agents take a previously non-invasive imaging modality and make it minimally invasive. This inevitably adds cost, very moderate level of pain and a small risk (primarily of infection) associated with the required needle-based veinous access. As in the case of IVUS versus transcutaneous ultrasound, the penalty involved in migrating into an invasive modality is rewarded with expanded research and clinical information. Nevertheless, the slow rate of adoption, at least in the USA, has been disappointing to those performing research in the field. However, IVUS benefits via the fact that it is minimally invasive from the outset so the addition of microbubbles creates only marginal increases to cost and risk.

The high echo signal obtained from microbubbles is a consequence of their very low density and high compressibility relative to adjacent tissues and structures. However, because other targets in an ultrasound field can also give rise to a strong echo signal (e.g., vessel/blood interface, etc.), it is highly desirable to exploit the unique echo signal "signature" of a microbubble. The high compressibility of a microbubble gives rise to a nonlinear vibrational response to impinging compressional ultrasound wave energy. As a consequence, harmonics of the transmitted waveform are present in the echo signal. Most often the first harmonic of the fundamental (transmitted) signal is used in signal isolation methods. It should be noted that in the literature this harmonic signal is usually referred to as the "second harmonic" to mean the first harmonic of the transmitted signal—i.e., centered at $2f_0$ where f_0 is the fundamental, transmitted, ultrasound center frequency. However, higher harmonics and subharmonics are sometimes employed. Given the very high frequencies typically used in IVUS, it can be observed that achieving transducer and system bandwidth to encompass both transmitted and nonlinear harmonic signal presents an obvious technical challenge. A number of signal isolation methods have been developed to isolate the nonlinear microbubble signal and thereby isolate it from adjacent tissue signal. The most common approaches involve multiple pulse excitation along a common acoustic beamline (Shen et al. 2005). These methods include "amplitude modulation" (AM) (Brock-Fisher et al. 1996), "pulse inversion" (PI) (Chapman and Lazenby 1997), and "contrast pulse sequences" (CPS) (Phillips 2001; Phillips and Gardner 2004). These methods all involve the use of the observation that serial signals acquired of an identical, stationary, target behave as a linear system when all components of the pulse-echo system are linear. More simply stated, in a linear system, the echo signal is a scaled replica according to the amplitude/phase of the excitation (transmitted) signal. Thus, these acquired linear signals can be scaled and added, or subtracted, in order that linear origin signals cancel. In the case of amplitude modulation, a scaled replica of an initial pulse is used (e.g., "1," "½"). The two received signals have a compensatory scaling applied to the, for example, smaller of the two signals such that when subtract, perfect cancelation occurs (i.e., in this case just described, the second pulse "½" is scaled by a factor of 2 resulting in a "1"). In the case of pulse inversion, the two transmitted signals are inverted replicas of each other (i.e., "1," "−1"). Upon summing the received signals, the linear origin signals cancel. "Contrast Pulse Sequences" involves a combination of both AM and PI (e.g., ½, −1, ½). In this CPS example, the three received signals can be summed, without scaling, to achieve cancelation of the linear origin signal. Notice that in all of these three methods, any nonlinear origin signal will not perfectly cancel. It should also be noticed that these nonlinear signal separation methods are sensitive, to varying degrees, to both subharmonics and superharmonics—not just the second harmonic. It is also possible to isolate nonlinear signals using conventional frequency filtering. Additionally, cancelation techniques are frequently combined with frequency filtering (Shen et al. 2005). It may be observed that these multi-pulse methods, especially the longer pulse methods (e.g., CPS), are susceptible to imperfect fundamental signal cancelation whenever there is tissue motion between successive pulses (Shen et al. 2005). In the context of IVUS, this is a concern even in the presence of the very short inter-pulse intervals implicit in very short pulse-echo transit durations.

IVUS Specific Applications of Microbubbles for Imaging

Microbubbles have long been recognized in IVUS imaging as making a contribution toward better delineating the boundary between the lumen and the wall of a blood vessel (Vavuranakis et al. 2005; Masuda et al. 2001). The increased contrast obtained when using microbubbles affords improved detection of in-stent neointima and reduces observer variability when assessing stenosis and neointima area (Masuda et al. 2001). It is now established that plaque microvascularity and plaque vulnerability are closely correlated (Naghavi and Falk 2010; Howard et al. 2015). Consequently, the use of microbubbles with IVUS has been used to assess the *vasa vasorum* that develops in coronary atherosclerotic lesions (Vavuranakis et al. 2005, 2008; Goertz et al. 2007a; Carlier et al. 2005) [As an aside, although IVUS is necessary for assessing the *vasa vasorum* in the coronary arteries, a transcutaneous ultrasound approach for assessing the *vasa vasorum* in carotid arteries is feasible (Song and Zhang 2015)]. Microbubbles, even when using a non-optimized IVUS device, yield enough vascular contrast enhancement to enable assessment of vascular density (Carlier et al. 2005). It is possible to measure intra-plaque perfusion density by contrasting the gray-scale image intensity acquired before, during, and after microbubble infusion (Vavuranakis et al. 2008). Using this approach, it has been observed that increased microbubble image signal between the intima-media boundary and the adventitia, following infusion, was associated with increased plaque vascularity (Fig. 10.5). Thus, it is believed that a combination of IVUS plaque morphological assessment (e.g., identifying cap thickness, calcification, lipid core) and *vasa vasorum* density imaging using microbubble-enhanced IVUS provides a powerful instrument and method for identification of plaques susceptible to rupture—i.e., "vulnerable plaques" (Carlier et al. 2005).

Efforts have been made to improve the sensitivity and specificity of IVUS instrumentation for microbubble detection. In this context, the very high frequencies com-

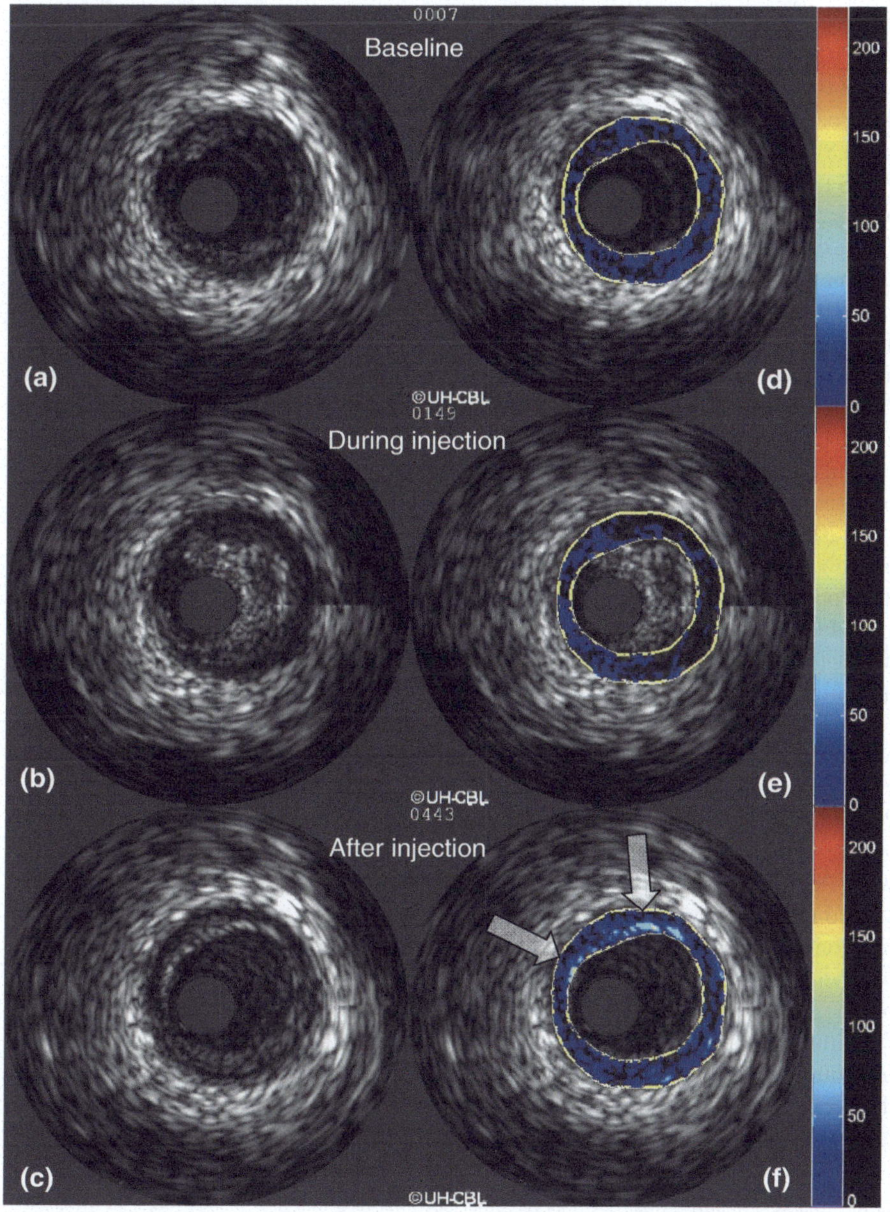

Fig. 10.5 Depiction of qualitative representation of enhancement. Unprocessed images are displayed **a** before, **b** during, and **c** after injection of microbubbles. Corresponding processed images are displayed in **d**–**f**. Enhancement is graded from minimal (blue) to maximal (red). Values are a percentage of the maximum grey level intensity difference (255). Arrows indicate points of intense, stable enhancement at the media–adventitia border. Diffuse points of enhancement are present nearby. *Source* Vavuranakis et al. (2008)

monly used in IVUS present significant technical challenges. Super harmonics occur at very high frequencies (e.g., 40–80 MHz) and require very high signal bandwidth. It needs to be emphasized that this infers very high signal bandwidth at every "link" in the processing chain—transmitter electronics, transducer, transducer interconnecting cable, receiver electronics, and digitization. The mismatch between commonly used IVUS frequencies and the natural resonant frequencies of commonly used microbubbles (<5 MHz typically), has motivated the consideration of subharmonic detection. These challenges have motivated development of dual-frequency IVUS transducers and methods to suppress echoes originating from tissue and preserve the nonlinear signal originating from microbubbles (Goertz et al. 2007b). IVUS imaging using subharmonic and second-harmonic signals have been evaluated in vessel phantoms and have yielded up to 15 dB of microbubble-specific contrast enhancement (Goertz et al. 2006a, 2007b). These approaches have also been in tested a fully developed rabbit aortic plaque (in vivo) (Goertz et al. 2006a, b). Harmonic imaging provided a successful means of imaging the structure of the *vasa vasorum* with an ~8 dB increase in sensitivity relative to standard, fundamental frequency-based, IVUS imaging (Goertz et al. 2006a, b) (Fig. 10.6). These early successes provided evidence that specialized microbubble imaging is feasible at IVUS imaging frequencies and motivated further development of newer techniques that improve microbubble-specific detection in the vasculature. These new methods include radial modulation (Yu et al. 2014), superharmonic (Ma et al. 2014), ultraharmonic imaging (Maresca et al. 2014), and chirp-based imaging (Shekhar and Doyley 2013; Shekhar et al. 2016). For example, Maresca et al. recently demonstrated the use of conventional band-limited IVUS transducers with chirp transmit sequences for ultraharmonic contrast-enhanced imaging of microvasculature not readily visualized by conventional IVUS imaging (Maresca et al. 2013, 2014) (Fig. 10.7). The data provided by contrast-enhanced IVUS can also be used to extend the quality and accuracy of existing methods for characterizing the vessel wall. For example, Huntzicker et al. (Huntzicker et al. 2016) have recently demonstrated that in model-based IVUS elastography, the additional information gained in relation to the geometry of detected microvessels can be used to improve the accuracy of assessments of stress distribution in the vicinity of the fibrous cap on a vulnerable plaque.

Although there has been considerable success using existing commercial IVUS transducers with conventional levels of signal bandwidth (<80% −6 dB pulse-echo), there exists a high level of interest in creating transducers true multiple frequency capability. Although 50% −6 dB bandwidth (i.e., approximating a two-cycle pulse) is sufficient to achieving λ axial resolution—and lateral resolution is only able to approach λ resolution in near ideal focusing conditions (i.e., f/number ~1, at focus), higher bandwidth has immediate value if attempting to assess sub or super harmonic signals. Additionally, the higher bandwidth will improve any signal analysis dependent on spectral qualities of the received signal. For example, "virtual histology," using radiofrequency analysis (Nair et al. 2002, 2007) has encountered significant limitations in terms of reported accuracy (Thim et al. 2000) and these are probably attributable at least in part to limited signal bandwidth. Higher transducer bandwidth can be achieved using high coupling coefficient single crystal materials, more

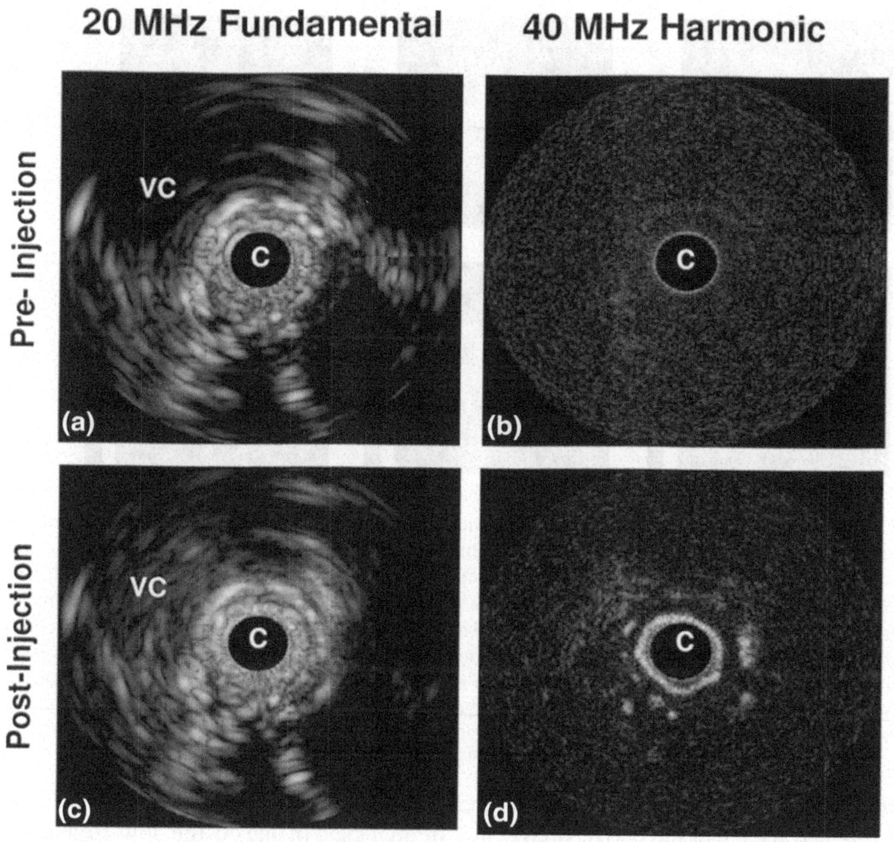

Fig. 10.6 In vivo contrast imaging results in an atherosclerotic rabbit aorta. **a** Fundamental mode before agent injection, where "C" is the catheter and "VC" is the vena cava. **b** Fundamental mode 10 seconds after injection where changes in adventitial enhancement are not evident, except for in a region at 4 o'clock and within the vena cava. **c** Harmonic mode before injection shows the tissue signals to be largely suppressed. **d** At 10 seconds after injection, the harmonic mode shows significant adventitial enhancement, consistent with the detection of adventitial microvessels. Scale of images is 12 mm across. The dynamic range of the fundamental and harmonic images are 40 and 25 dB, respectively. *Source* Goertz et al. (2006b)

complex matching layer schemes or cMUT technology. However, in the IVUS field, the small dimensions (physical fabrication difficulties, electrical matching challenges, cable loss effects, etc.) present more extreme practical challenges to the transducer designer. It also worth reiterating that due to the importance of tissue/contrast signal arising immediately adjacent to the transducer, using an offset transmitter/receiver approach is not a viable option. In one approach, multiple passive layers are used or provide 20 and 40 MHz operation from a single element transducer (Vos et al. 2005). Stacked, high- and low-frequency element-based, transducer designs have been proposed. Generally, the two components of the stack will

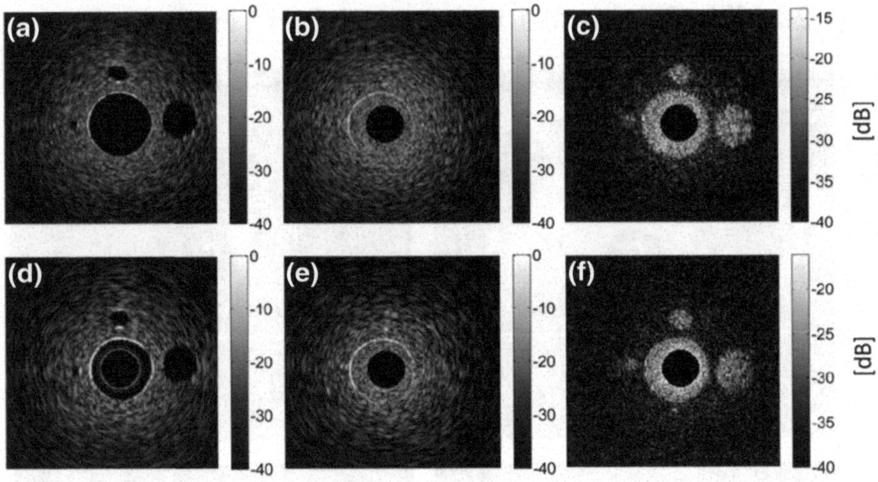

Fig. 10.7 Images acquired in the tissue-mimicking phantom using the chirp reversal sequence and the pulse inversion sequence. **a** IVUS chirp image without contrast. **b** IVUS chirp image with contrast. **c** IVUS chirp reversal contrast image. **d** 22 MHz conventional IVUS image without contrast. **e** 22 MHz conventional IVUS image with contrast. **f** Ultraharmonic IVUS contrast image. Intensity scales in contrast modes are displayed relative to their respective B-mode intensity. In both the chirp reversal (c) and ultraharmonic (f) contrast modes, small model vessels (0.5 mm vessel to left) can be identified above background. In the ultraharmonic example, (f), a 0.2 mm vessel to bottom is readily visible. *Source* Maresca et al. (2013)

mechanically cross couple making for a challenge in obtaining a "clean" spectral performance at each of the two desired center frequencies. In one design, low-frequency (35 MHz) and high-frequency (90–150 MHz) transducer elements were stacked to provide for conventional resolution and imaging depth performance at 35 MHz and very fine, but shallow, imaging performance at the higher frequency (Ma et al. 2015). In a second stacked design, a low-frequency transducer element (6.5 MHz) was placed under a high-frequency transducer element (30 MHz) (Fig. 10.8) (Ma et al. 2014). The lower-frequency transducer insonated microbubbles near their resonance while emitted superharmonics were detected using the high-frequency element. This "transmit low, receive high" (TLRH) technique permits much high-resolution imaging of microbubbles than is possible when limited to imaging near the microbubble resonance frequency (Hu et al. 2010). Using this approach, an in vitro microbubble SNR of 12 dB has been demonstrated by Goertz et al. (2006a). More recently, the stacked element design has been extended to use lateral-mode transmission to yield yet higher versatility and performance (Wang et al. 2016). The transmit low, receive high imaging approach has been extended recently by Lindsey et al. (2016) to include adaptive (minimum variance) beamforming to achieve further refinements to imaging resolution. In vitro, combining minimum variance and phase coherence factor processing improved the spatial resolution by 41.7% while improvements of

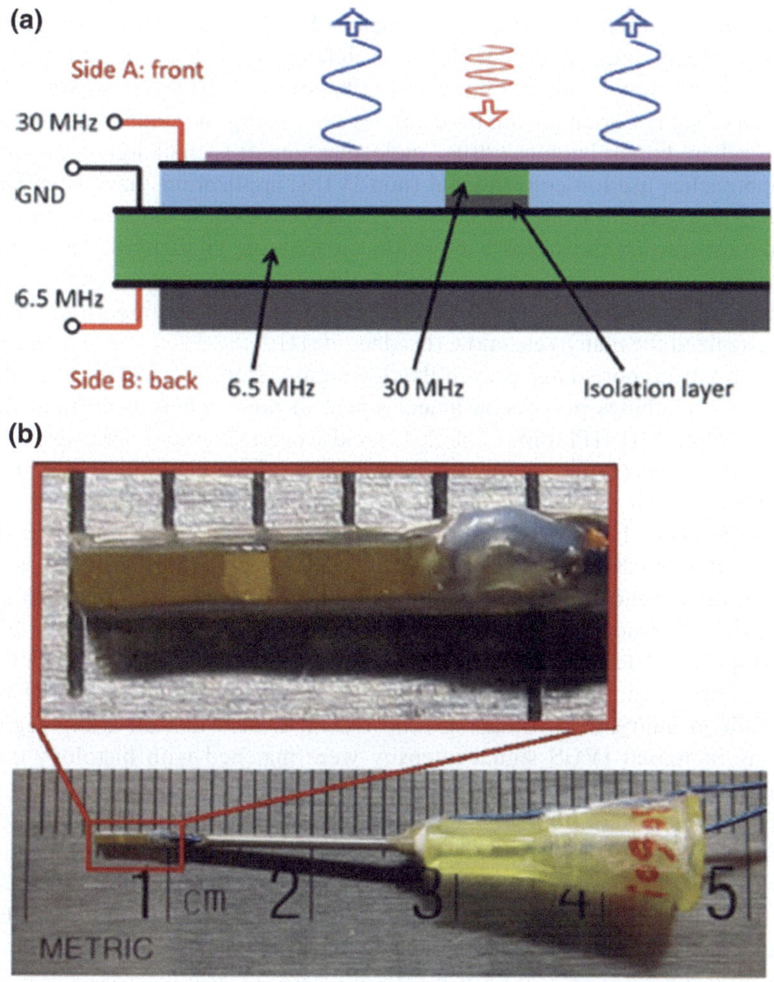

Fig. 10.8 **a** Design of the dual-frequency IVUS transducer. **b** Prototype transducer housed on the tip of a 20-gauge hypodermic needle. *Source* Ma et al. (2014)

2.2 dB in contrast to tissue ratio and 37.2% in spatial resolution were achieved in vivo (Lindsey et al. 2016).

Molecular Imaging in the IVUS Field

Ultrasound molecular imaging is based on the detection of microbubbles that have adhered to a vessel wall resulting from a binding event between molecule specific ligand on the microbubble surface and a matched molecular receptor expressed on the vessel surface. Single microbubble origin signal to noise does not present a major challenge—certainly for the in vitro case (Klibanov et al. 2004). However, reliable separation of signals arising from tissue, freely circulating (non-adhered) microbubbles and adhered microbubbles ("true" molecular signal) is challenging. The underlying approaches used in conventional (non-IVUS) applications have been largely mapped over to use in IVUS. The same targeting and signal separation methods can be used—except that the signal bandwidths create more challenges. The potential of ultrasound molecular imaging to elucidate the molecular status of cardiovascular (and other) disease and thus enable the earlier detection, or staging, of disease has been recognized for many years now (Lindner 2004).

Early in vitro reports have proven that the signals from molecular targeted, and adhered, microbubbles possess sufficient signal to noise when using both fundamental mode 40 MHz (Phillips et al. 2012) and when using nonlinear signal-based sensing approaches (Goertz et al. 2007a). An early (1999) in vivo IVUS molecular imaging experiment involved echogenic liposomes (ELIP) targeted to intercellular adhesion molecule-1 (ICAM-1), a Yucatan mini-swine model with atherosclerotic carotid arteries and 20 MHz IVUS (Demos et al. 1999). In 2000, Lanza et al. studied perfluorocarbon nanoemulsions for detection of tissue factor expression in swine carotid arteries. A more extensive study by Hamilton et al. (2004), also using ELIPs, reported specific detection of ICAM-1, vascular cell adhesion molecule-1 (VCAM-1), fibrin, fibrinogen, and tissue factor in swine carotid arteries subsequent to angioplasty balloon injury. IVUS imaging (20 MHz) indicated that all artery segments producing increased IVUS signal intensity were matched with histology consistent with increased expression of the target molecule. More recently, Phillips et al. (2012) presented in vitro 40 MHz IVUS imaging of VCAM-1 targeted microbubbles (Fig. 10.9). Together, these in vivo studies suggest IVUS-based identification of the molecular-specific information may enable early characterization of endothelial injury following angioplasty balloon injury and identify atheroma components that may guide the eventual choice of therapeutic intervention.

Integrated IVUS and Microbubble Generation Devices

Usually, when IVUS and microbubbles are combined, the microbubbles are introduced via a separate intravenous injection or from a lumen and port on the catheter device itself. Of course, using the device to dispense microbubbles directly increases the local concentration of microbubbles allowing for greater potential for efficacy with minimal systemic-related risk. Recently, it has been proposed that microbubbles actually be formed in the catheter device itself (Dhanaliwala et al. 2013). Microbubbles generated using a flow focusing microfluidics device have been tested in contrast-enhanced imaging (Hettiarachchi et al. 2007; Kaya et al. 2010; Dhanaliwala et al. 2013), molecular imaging (Talu et al. 2007; Seo et al. 2010) and therapeutic delivery

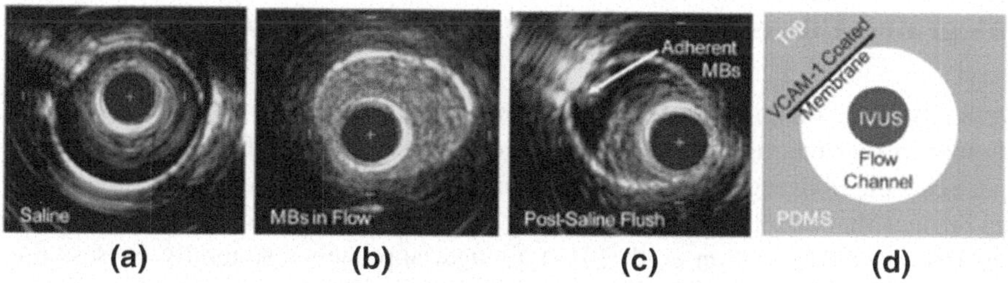

Fig. 10.9 **a-b** IVUS imaging within PDMS flow chambers. VCAM-1 coated membranes are located on the upper left corner and simulate inflamed cells. VCAM-1-targeted microbubbles were sent through the flow chambers at 5 mL/min. Adherent microbubbles were observed following a flush with saline. Image intensity was quantified with ImageJ. Tick marks are spaced in increments of 1 mm. (h) Schematic of flow channel with VCAM-1 coated membrane in the upper left corner. *Source* Phillips et al. (2012)

applications (Hettiarachchi et al. 2009; Dixon et al. 2013; Chen et al. 2014). Microfluidics devices produce microbubbles which have a number of unique characteristics. The foremost positive characteristic is that, when operated in a stable regime, they are capable of producing essentially monodisperse (single size) microbubbles. This characteristic has driven the majority of interest in these devices. Unfortunately, microfluidics devices have limited production rates—typically possessing an upper limit in the hundreds of thousands of microbubbles per second range. Since a human intravenous dose may be of the order of several hundred million microbubbles, this presents a practical production challenge. Additionally, and particularly unfortunate when coupled with the previous limitation, the microbubbles have relative low stability (typically tens of minutes). Finally, at least until the present, it has been feasible to produce microbubbles that are generally on the upper bound of what might be tolerated for intravenous use—i.e., it is challenging to produce microbubbles smaller than approximately 6–8 μm in diameter. However, the limited stability, low production rate and large diameters are of lesser importance when the device is placed on a catheter tip and the microbubbles are intended to be used in conjunction with applied ultrasound within seconds of dispensation. It can also be argued that relatively large microbubbles can also be better tolerated if the rate of disintegration off-target is sufficiently high (e.g., disintegrated within 30 s). Taken together, the characteristics of relatively large microbubbles and rapid off-target disintegration, makes the approach particularly well suited to sonothrombolysis where an aggressive bioeffect mediated by the combination of ultrasound, large microbubbles and thrombolytic drug (typically tissue plasminogen activator), combined with minimal off-target effects, is desired (Dixon et al. 2016). A 1.5 mm diameter prototype microfluidic catheter device has been developed and tested in vitro for the same application (Dixon et al. 2015c).

Integrated Ultrasound and Optical Imaging Devices

Optically based imaging methods, including: optical coherence tomography (OCT), photoacoustic imaging (PAI), near-infrared fluorescence (NIRF), and near-infrared spectroscopic imaging have all been explored in the context of catheter-based imaging devices that also include IVUS (Jaffer et al. 2008; Karpiouk et al. 2010; Dixon and Hossack 2013; Abran et al. 2014). Optical approaches generally possess limited imaging penetration but provide useful information related to atherosclerotic plaque composition using the different optical attenuation, and related absorption, coefficients (Brugaletta et al. 2011; Jansen et al. 2011; Wang et al. 2011). The finer spatial resolution of optical methods, combined with the ability to identify a soft lipid core based on optical absorption spectra provides information frequently unavailable from IVUS. It is also possible to combine these approaches using multimodal ultrasound/optical contrast agents (Wilson et al. 2012; Dove et al. 2013; Hannah et al. 2014; Dixon et al. 2015a; Castelino et al. 2016) to yield a versatile multimodality device yielding the potential for intravascular molecular imaging and therapy using the combination of ultrasound and optical energy. Optical methods can be used to measure fibrous cap thickness (i.e., anatomic imaging) (Jang et al. 2002), evaluate plaque inflammation (i.e., function) (Jaffer et al. 2008) or characterize lipid content (i.e., tissue characterization) (Jansen et al. 2013). Therapeutic IVUS and microbubbles can then be used to affect therapy/drug delivery. Additionally, if an optically labeled drug or microbubble is used, optical methods may be used to verify site-specific drug delivery. Thus, the long-term goal of being able to diagnose, treat, and verify treatment within a single procedure, and possibly using even a single multimodal device, may be realized. Although a fully integrated device is yet to be tested in vivo, various combination devices have been described in the literature (Karpiouk et al. 2010; Jansen et al. 2011; Dixon and Hossack 2013; Abran et al. 2014).

Conclusions

IVUS provides a highly versatile platform technology that can be tailored to meet both diagnostic and therapeutic clinical needs. The technical specifications underlying the needs for both imaging and therapy have motivated a number of recent innovations in IVUS and this trend will continue. Mechanically scanned single elements continue to dominate in early-stage prototyping due to the reduced challenges in design and fabrication when compared to array-based systems. Transducer material choice and designs (e.g., stacked layers) continue to evolve progressing from synthetic piezoelectric ceramics to single crystal materials (possessing higher SNR and bandwidth). Whereas piezoelectric materials are amenable to "one-off" prototyping, cMUT technology is particularly well suited to the long-term commercial and clinical product. Most recently, and probably with the greatest long-term potential, silicon-based cMUT technology offers immense versatility in terms of frequency, array configuration and dual imaging/therapy capability. Solid-state circumferential arrays continue to be favored for some applications. As devices migrate from prototype stage to high volume clinical usage, one might expect progressively greater use to be made of cMUT and phased array approaches.

Microbubbles combined with IVUS energy provide for molecular imaging or localized drug delivery. Relatively little work has progressed as far as human clinical trial but new applications enabled by new versatile transducer designs and novel microbubble formulations have been demonstrated in both in vitro and in vivo animal models. Molecular targeted microbubbles, paired with IVUS, enable molecular imaging and drug delivery that address applications unique to the intravascular setting. Combining these anatomic, functional and molecular imaging capabilities, a drug/gene delivery capability and considering the use of optical detection component, brings us close to a device which, within a single catheterization procedure, makes it possible to assess disease (using anatomic, functional or molecular characteristics), choose and effect an optimal therapy and verify successful therapy. In particular, the ability to orient an ultrasound beam to address specific subregions across the wall of vessel encompasses a degree of versatility probably unmatched by any conceivable competing technology. Thus, it may be argued, especially in view of the slow uptake of IVUS technology in some parts of the world (notably in the USA), that only a very small fraction of the total contribution offered by IVUS technology has yet been realized.

Acknowledgements The author is grateful for valuable contributions made by current and former members of his research group (Adam J. Dixon, Ph.D., Joseph P. Kilroy, Ph.D., Ali H. Dhanaliwala, Ph.D., Shiying Wang, Ph.D., Linsey C. Phillips, Ph.D., and Johnny L Chen, B.S) and current and former University of Virginia colleagues: M. Ragosta, M.D., Alexander L. Klibanov, Ph.D., and Brian R. Wamhoff, Ph.D. The research program has been supported by NIH, University of Virginia Coulter Partnership, and the Virginia Commonwealth Health Research Board. Opinions expressed are those of the author and do not necessarily represent official views of sources of funding.

References

Abran M, Cloutier G, Cardinal M-HR, Chayer B, Tardif J-C, Lesage F (2014) Development of a photoacoustic, ultrasound and fluorescence imaging catheter for the study of atherosclerotic plaque. IEEE Trans Biomed Circ Syst 8(5):696–703. https://doi.org/10.1109/TBCAS.2014.2360560

Amabile PG, Waugh JM, Lewis TN, Elkins CJ, Janas W, Dake MD (2001) High-efficiency endovascular gene delivery via therapeutic ultrasound. J Am Coll Cardiol 37(7):1975–1980

Baker R, Samuels S, Benenati JF, Powell A, Uthoff H (2012) Ultrasound-accelerated versus standard catheter-directed thrombolysis—a comparative study in patients with iliofemoral deep vein thrombosis. J Vasc Intervent Radiol 23(11):1460–1466. https://doi.org/10.1016/j.jvir.2012.08.008

Bom N, Lancee CT, van Egmond FC (1972) An ultrasonic intracardiac scanner. Ultrasonics 10(2):72–77

Bom N, Li W, van der Steen AFW, Lancée CT, Céspedes EI, Slager CJ, de Korte CL (1998) New developments in intravascular ultrasound imaging. Eur J Ultrasound 7(1):9–14

Brock-Fisher GA, Poland MD, Rafter P (1996) Means for increasing sensitivity in non-linear ultrasound imaging systems. United States Patent 5(577):505
Brugaletta S, Garcia-Garcia HM, Serruys PW, de Boer S, Ligthart J, Gomez-Lara J, Witberg K, Diletti R, Wykrzykowska J, van Geuns R-J, Schultz C, Regar E, Duckers HJ, van Mieghem N, de Jaegere P, Madden SP, Muller JE, van der Steen AFW, van der Giessen WJ, Boersma E (2011) NIRS and IVUS for characterization of atherosclerosis in patients undergoing coronary angiography. JACC Cardiovasc Imag 4(6):647–655. https://doi.org/10.1016/j.jcmg.2011.03.013
Carlier S, Kakadiaris IA, Dib N, Vavuranakis M, O'Malley SM, Gul K, Hartley CJ, Metcalfe R, Mehran R, Stefanadis C, Falk E, Stone G, Leon M, Naghavi M (2005) Vasa vasorum imaging: A new window to the clinical detection of vulnerable atherosclerotic plaques. Curr Atheroscler Rep 7(2):164–169. https://doi.org/10.1007/s11883-005-0040-2
Castelino RF, Hynes M, Munding CE, Telenkov S, Foster FS (2016) Combined frequency domain photoacoustic and ultrasound imaging for intravascular applications. Biomed Opt Express 7(11):4441–4449. https://doi.org/10.1364/boe.7.004441
Chapman C, Lazenby J (1997) US patent 5,632,277 ultrasound imaging system employing phase inversion subtraction to enhance the image
Chen JL, Dhanaliwala AH, Dixon AJ, Klibanov AL, Hossack JA (2014) Synthesis and characterization of transiently stable albumin-coated microbubbles via a flow-focusing microfluidic device. Ultrasound Med Biol 40(2):400–409. https://doi.org/10.1016/j.ultrasmedbio.2013.09.024
Choe JW, Oralkan Ö, Nikoozadeh A, Gencel M, Stephens DN, O'Donnell M, Sahn DJ, Khuri-Yakub BT (2012) Volumetric real-time imaging using a CMUT ring array. IEEE Trans Ultrason Ferroelectr Freq Control 59(6):1201–1211. https://doi.org/10.1109/TUFFC.2012.2310
Claessen BE, Mehran R, Mintz GS, Weisz G, Leon MB, Dogan O, de Ribamar Costa J, Stone GW, Apostolidou I, Morales A, Chantziara V, Syros G, Sanidas E, Xu K, Tijssen JGP, Henriques JPS, Piek JJ, Moses JW, Maehara A, Dangas GD (2011) Impact of intravascular ultrasound imaging on early and late clinical outcomes following percutaneous coronary intervention with drug-eluting stents. JACC Cardiovasc Interv 4(9):974–981. https://doi.org/10.1016/j.jcin.2011.07.005
Cosgrove D, Lassau N (2010) Imaging of perfusion using ultrasound. Eur J Nucl Med Mol I 37(1):65–85. https://doi.org/10.1007/s00259-010-1537-7
Dayton PA, Allen JS, Ferrara KW (2002) The magnitude of radiation force on ultrasound contrast agents. J Acoust Soc Am 112(5 Pt 1):2183–2192
Demos SM, Alkan-Onyuksel H, Kane BJ, Ramani K, Nagaraj A, Greene R, Klegerman M, McPherson DD (1999) In vivo targeting of acoustically reflective liposomes for intravascular and transvascular ultrasonic enhancement. J Am Coll Cardiol 33(3):867–875
Dhanaliwala AH, Chen JL, Wang S, Hossack JA (2013) Liquid flooded flow-focusing microfluidic device for in situ generation of monodisperse microbubbles. Microfluid Nanofluid 14(3–4):457–467. https://doi.org/10.1007/s10404-012-1064-x
Dixon AJ, Jun L, Rickel JMR, Shin B, Zuo Z, Hossack JA Sonothrombolysis efficacy of microbubbles produced by a microfluidic device in a rat ischemic stroke model. In: 2016 IEEE international ultrasonics symposium (IUS), 18–21 Sept 2016, pp 1–4. https://doi.org/10.1109/ultsym.2016.7728879
Dixon AJ, Hossack JA (2013) Intravascular near-infrared fluorescence catheter with ultrasound guidance and blood attenuation correction. J Biomed Optics 18(5):056009. https://doi.org/10.1117/1.JBO.18.5.056009
Dixon AJ, Dhanaliwala AH, Chen JL, Hossack JA (2013) Enhanced intracellular delivery of a model drug using microbubbles produced by a microfluidic device. Ultrasound Med Biol 39(7):1267–1276
Dixon AJ, Hu S, Klibanov AL, Hossack JA (2015a) Oscillatory dynamics and in vivo photoacoustic imaging performance of plasmonic nanoparticle-coated microbubbles. Small In Press
Dixon AJ, Kilroy JP, Dhanaliwala AH, Chen JL, Phillips LC, Ragosta M, Klibanov AL, Wamhoff BR, Hossack JA (2015b) Microbubble-mediated intravascular ultrasound imaging and drug delivery. IEEE Trans Ultrason Ferroelectr Freq Control 62(9):1674–1685. https://doi.org/10.1109/tuffc.2015.007143

Dixon AJ, Kilroy JP, Klibanov AL, Hossack JA (2015) Microbubbles produced by a catheter-based flow-focusing microfluidic device for sonothrombolysis. In: European symposium on ultrasound contrast imaging, 2015/01/23/ 2015c. Rotterdam, The Netherlands

Dove JD, Murray TW, Borden MA (2013) Enhanced photoacoustic response with plasmonic nanoparticle-templated microbubbles. Soft Matter 9(32):7743. https://doi.org/10.1039/c3sm51690c

Engelberger RP, Spirk D, Willenberg T, Alatri A, Do D-D, Baumgartner I, Kucher N (2015) Ultrasound-assisted versus conventional catheter-directed thrombolysis for acute iliofemoral deep vein thrombosis. Circ: Cardiovasc Intervent 8(1). https://doi.org/10.1161/circinterventions.114.002027

FDA (2016) http://www.accessdata.fda.gov/drugsatfda_docs/label/2016/203684s001lbl.pdf

Ferrara KW, Pollard R, Borden M (2007) Ultrasound microbubble contrast agents: fundamentals and application to gene and drug delivery. Annu Rev Biomed Eng 9:415–447. https://doi.org/10.1146/annurev.bioeng.8.061505.095852

Fitzgerald PJ, Takagi A, Moore MP, Hayase M, Kolodgie FD, Corl D, Nassi M, Virmani R, Yock PG (2001) Intravascular sonotherapy decreases neointimal hyperplasia after stent implantation in swine. Circulation 103(14):1828–1831. https://doi.org/10.1161/01.cir.103.14.1828

Goertz DE, Frijlink ME, de Jong N, van der Steen AF (2006a) Nonlinear intravascular ultrasound contrast imaging. Ultrasound Med Biol 32(4):491–502. https://doi.org/10.1016/j.ultrasmedbio.2006.01.001

Goertz DE, Frijlink ME, Tempel D, van Damme LCA, Krams R, Schaar JA, Ten Cate FJ, Serruys PW, de Jong N, van der Steen AFW (2006b) Contrast harmonic intravascular ultrasound: a feasibility study for vasa vasorum imaging. Invest Radiol 41(8):631–638. https://doi.org/10.1097/01.rli.0000229773.11715.da

Goertz DE, Frijlink ME, Krams R, de Jong N, van der Steen AFW (2007a) Vasa vasorum and molecular imaging of atherosclerotic plaques using nonlinear contrast intravascular ultrasound. Netherlands Heart Journal 15(2):77–80

Goertz DE, Frijlink ME, Tempel D, Bhagwandas V, Gisolf A, Krams R, de Jong N, van der Steen AFW (2007b) Subharmonic contrast intravascular ultrasound for vasa vasorum imaging. Ultrasound Med Biol 33(12):1859–1872. https://doi.org/10.1016/j.ultrasmedbio.2007.05.023

Gurun G, Tekes C, Zahorian J, Xu T, Satir S, Karaman M, Hasler J, Degertekin FL (2014) Single-chip CMUT-on-CMOS front-end system for real-time volumetric IVUS and ICE imaging. IEEE Trans Ultrason Ferroelectr Freq Control 61(2):239–250. https://doi.org/10.1109/TUFFC.2014.6722610

Hamilton AJ, Huang S-L, Warnick D, Rabbat M, Kane B, Nagaraj A, Klegerman M, McPherson DD (2004) Intravascular ultrasound molecular imaging of atheroma components in vivo. J Am Coll Cardiol 43(3):453–460. https://doi.org/10.1016/j.jacc.2003.07.048

Hannah A, Luke G, Wilson K, Homan K, Emelianov S (2014) Indocyanine green-loaded photoacoustic nanodroplets: dual contrast nanoconstructs for enhanced photoacoustic and ultrasound imaging. ACS Nano 8(1):250–259. https://doi.org/10.1021/nn403527r

Hettiarachchi K, Talu E, Longo ML, Dayton PA, Lee AP (2007) On-chip generation of microbubbles as a practical technology for manufacturing contrast agents for ultrasonic imaging. Lab Chip 7(4):463. https://doi.org/10.1039/b701481n

Hettiarachchi K, Zhang S, Feingold S, Lee AP, Dayton PA (2009) Controllable microfluidic synthesis of multiphase drug-carrying lipospheres for site-targeted therapy. Biotechnol Prog 25(4):938–945. https://doi.org/10.1002/btpr.214

Hiro T, Fujii T, Yasumoto K, Murata T, Murashige A, Matsuzaki M (2001) Detection of fibrous cap in atherosclerotic plaque by intravascular ultrasound by use of color mapping of angle-dependent echo-intensity variation. Circulation 103(9):1206–1211

Hohmann J, Albrecht T, Hoffmann CW, Wolf KJ (2003) Ultrasonographic detection of focal liver lesions: increased sensitivity and specificity with microbubble contrast agents. Eur J Radiol 46(2):147–159

Howard DPJ, van Lammeren GW, Rothwell PM, Redgrave JN, Moll FL, de Vries J-PPM, de Kleijn DPV, den Ruijter HM, de Borst GJ, Pasterkamp G (2015) Symptomatic carotid atherosclerotic disease: correlations between plaque composition and ipsilateral stroke risk. Stroke; A J Cereb Circ 46(1):182–189. https://doi.org/10.1161/STROKEAHA.114.007221

Hu X, Zheng H, Kruse DE, Sutcliffe P, Stephens DN, Ferrara KW (2010) A sensitive TLRH targeted imaging technique for ultrasonic molecular imaging. IEEE Trans Ultrason Ferroelectr Freq Control 57(2):305–316. https://doi.org/10.1109/TUFFC.2010.1411

Hu Y, Wan JM, Yu AC (2013) Membrane perforation and recovery dynamics in microbubble-mediated sonoporation. Ultrasound Med Biol 39(12):2393–2405. https://doi.org/10.1016/j.ultrasmedbio.2013.08.003

Huntzicker S, Shekhar H, Doyley MM (2016) Contrast-enhanced quantitative intravascular elastography: The impact of microvasculature on model-based elastography. Ultrasound Med Biol 42(5):1167–1181. https://doi.org/10.1016/j.ultrasmedbio.2015.12.024

Jaffer FA, Vinegoni C, John MC, Aikawa E, Gold HK, Finn AV, Ntziachristos V, Libby P, Weissleder R (2008) Real-time catheter molecular sensing of inflammation in proteolytically active atherosclerosis. Circulation 118(18):1802–1809. https://doi.org/10.1161/CIRCULATIONAHA.108.785881

Jang I-K, Bouma BE, Kang D-H, Park S-J, Park S-W, Seung K-B, Choi K-B, Shishkov M, Schlendorf K, Pomerantsev E, Houser SL, Aretz HT, Tearney GJ (2002) Visualization of coronary atherosclerotic plaques in patients using optical coherence tomography: comparison with intravascular ultrasound. J Am Coll Cardiol 39(4):604–609

Jansen K, van der Steen AFW, van Beusekom HMM, Oosterhuis JW, van Soest G (2011) Intravascular photoacoustic imaging of human coronary atherosclerosis. Opt Lett 36(5):597–599

Jansen K, Wu M, van der Steen AFW, van Soest G (2013) Lipid detection in atherosclerotic human coronaries by spectroscopic intravascular photoacoustic imaging. Opt Express 21(18):21472–21484

Karpiouk AB, Wang B, Emelianov SY (2010) Development of a catheter for combined intravascular ultrasound and photoacoustic imaging. Rev Sci Instrum 81(1):014901. https://doi.org/10.1063/1.3274197

Karshafian R, Bevan PD, Williams R, Samac S, Burns PN (2009) Sonoporation by ultrasound-activated microbubble contrast agents: effect of acoustic exposure parameters on cell membrane permeability and cell viability. Ultrasound Med Biol 35(5):847–860. https://doi.org/S0301-5629(08)00503-6 [pii], 10.1016/j.ultrasmedbio.2008.10.013

Kaya M, Feingold S, Hettiarachchi K, Lee AP, Dayton PA (2010) Acoustic responses of monodisperse lipid encapsulated microbubble contrast agents produced by flow focusing. Bubble Sci Eng Technol 2(2):33–40. https://doi.org/10.1179/175889610X12779105661532

Kilroy JP, Klibanov AL, Wamhoff BR, Hossack JA (2012) Intravascular ultrasound catheter to enhance microbubble-based drug delivery via acoustic radiation force. IEEE Trans Ultrason Ferroelectr Freq Control 59(10):2156–2166. https://doi.org/10.1109/TUFFC.2012.2442

Kilroy JP, Klibanov AL, Wamhoff BR, Bowles DK, Hossack JA (2014a) Localized in vivo model drug delivery with intravascular ultrasound and microbubbles. Ultrasound Med Biol 40(10):2458–2467. https://doi.org/10.1016/j.ultrasmedbio.2014.04.007

Kilroy JP, Patil AV, Rychak JJ, Hossack JA (2014b) An IVUS transducer for microbubble therapies. IEEE Trans Ultrason Ferroelectr Freq Control 61(3):441–449. https://doi.org/10.1109/TUFFC.2014.2929

Kilroy JP, Dhanaliwala AH, Klibanov A, Bowles DK, Wamhoff BR, Hossack JA (2015) Reducing neointima formation in a swine model with IVUS and sirolimus microbubbles. Ann Biomed Eng. In Press

Klibanov AL, Rasche PT, Hughes MS, Wojdyla JK, Galen KP, Wible JH Jr, Brandenburger GH (2004) Detection of individual microbubbles of ultrasound contrast agents: imaging of free-floating and targeted bubbles. Invest Radiol 39(3):187–195

Lanza GM, Abendschein DR, Hall CS, Marsh JN, Scott MJ, Scherrer DE, Wickline SA (2000) Molecular imaging of stretch-induced tissue factor expression in carotid arteries with intravascular ultrasound. Invest Radiol 35(4):227–234

Lee JH, Ahn SG, Yoon J (2012) Endovascular stent grafting via the left radial artery for a spontaneous isolated dissecting rupture of the superior mesenteric artery. Korean Circ J 42(2):140–141. https://doi.org/10.4070/kcj.2012.42.2.140

Li X, Wu W, Chung Y, Shih WY, Shih W-H, Zhou Q, Shung KK (2011) 80-MHz intravascular ultrasound transducer using PMN-PT free-standing film. IEEE Trans Ultrason Ferroelectr Freq Control 58(11):2281–2288. https://doi.org/10.1109/TUFFC.2011.2085

Lindner JR (2004) Microbubbles in medical imaging: current applications and future directions. Nat Rev Drug Discovery 3(6):527–533. https://doi.org/10.1038/nrd1417

Lindsey BD, Martin KH, Jiang X, Dayton PA (2016) Adaptive windowing in contrast-enhanced intravascular ultrasound imaging. Ultrasonics 70:123–135. https://doi.org/10.1016/j.ultras.2016.04.022

Liu C, Djuth F, Li X, Chen R, Zhou Q, Shung KK (2012) Micromachined high frequency PMN-PT/epoxy 1-3 composite ultrasonic annular array. Ultrasonics 52(4):497–502. https://doi.org/10.1016/j.ultras.2011.11.001

Ma J, Martin KH, Dayton PA, Jiang X (2014) A preliminary engineering design of intravascular dual-frequency transducers for contrast-enhanced acoustic angiography and molecular imaging. IEEE Trans Ultrason Ferroelectr Freq Control 61(5):870–880. https://doi.org/10.1109/TUFFC.2014.6805699

Ma T, Yu M, Chen Z, Fei C, Shung K, Zhou Q (2015) Multi-frequency intravascular ultrasound (IVUS) imaging. IEEE Trans Ultrason Ferroelectr Freq Control 62(1):97–107. https://doi.org/10.1109/TUFFC.2014.006679

Ma T, Zhou B, Hsiai TK, Shung KK (2016) A review of intravascular ultrasound-based multimodal intravascular imaging: the synergistic approach to characterizing vulnerable plaques. Ultrason Imag 38(5):314–331. https://doi.org/10.1177/0161734615604829

Maresca D, Renaud G, van Soest G, Li X, Zhou Q, Shung KK, de Jong N, van der Steen AFW (2013) Contrast-enhanced intravascular ultrasound pulse sequences for bandwidth-limited transducers. Ultrasound Med Biol 39(4):706–713. https://doi.org/10.1016/j.ultrasmedbio.2012.10.020

Maresca D, Skachkov I, Renaud G, Jansen K, van Soest G, de Jong N, van der Steen AFW (2014) Imaging microvasculature with contrast-enhanced ultraharmonic ultrasound. Ultrasound Med Biol 40(6):1318–1328. https://doi.org/10.1016/j.ultrasmedbio.2013.12.029

Masuda J, Terashima M, Yokoyama M (2001) Improved reproducibility of intravascular ultrasound assessment of coronary in-stent neointima with the use of an echogenic contrast agent. Jpn Circ J 65(7):632–636. https://doi.org/10.1253/jcj.65.632

Mintz GS, Nissen SE, Anderson WD, Bailey SR, Erbel R, Fitzgerald PJ, Pinto FJ, Rosenfield K, Siegel RJ, Tuzcu EM, Yock PG, O'Rourke RA, Abrams J, Bates ER, Brodie BR, Douglas PS, Gregoratos G, Hlatky MA, Hochman JS, Kaul S, Tracy CM, Waters DD, Winters JWL (2001) American College of Cardiology clinical expert consensus document on standards for acquisition, measurement and reporting of intravascular ultrasound studies (ivus)33A report of the american college of cardiology task force on clinical expert consensus documents developed in collaboration with the european society of cardiology endorsed by the society of cardiac angiography and interventions. J Am Coll Cardiol 37(5):1478–1492. https://doi.org/10.1016/S0735-1097(01)01175-5

Naghavi M, Falk E (2010) Asymptomatic atherosclerosis: from vulnerable plaque to vulnerable patient, vol 2. Humana Press, Totowa, NJ

Nair A, Kuban BD, Tuzcu EM, Schoenhagen P, Nissen SE, Vince DG (2002) Coronary plaque classification with intravascular ultrasound radiofrequency data analysis. Circulation 106(17):2200–2206

Nair A, Margolis MP, Kuban BD, Vince DG (2007) Automated coronary plaque characterisation with intravascular ultrasound backscatter: ex vivo validation. EuroIntervention 3(1):113–120

O'Donnell M, Thomas LJ (1992) Efficient synthetic aperture imaging from a circular aperture with possible application to catheter-based imaging. IEEE Trans Ultrason Ferroelectr Freq Control 39(3):366–380. https://doi.org/10.1109/58.143171

O'Donnell M, Eberle MJ, Stephens DN, Litzza JL, San Vicente K, Shapo BM (1997) Synthetic phased arrays for intraluminal imaging of coronary arteries. IEEE Trans Ultrason Ferroelectr Freq Control 44(3):714–721. https://doi.org/10.1109/58.658335

Phillips P (2001) Contrast pulse sequences (CPS): imaging non-linear microbubbles. Proc 2001 IEEE Ultrason Symp 2:1739–1745

Phillips P, Gardner E (2004) Contrast-agent detection and quantification. Eur Radiol 14(Suppl 8):P4–10

Phillips LC, Klibanov AL, Bowles DK, Ragosta M, Hossack JA, Wamhoff BR (2010) Focused in vivo delivery of plasmid DNA to the porcine vascular wall via intravascular ultrasound destruction of microbubbles. J Vasc Res 47(3):270–274. https://doi.org/10.1159/000258905

Phillips LC, Dhanaliwala AH, Klibanov AL, Hossack JA, Wamhoff BR (2011) Focused ultrasound-mediated drug delivery from microbubbles reduces drug dose necessary for therapeutic effect on neointima formation–brief report. Arterioscler Thromb Vasc Biol 31(12):2853–2855. https://doi.org/10.1161/ATVBAHA.111.238170

Phillips LC, Klibanov AL, Wamhoff BR, Hossack JA (2012) Intravascular ultrasound detection and delivery of molecularly targeted microbubbles for gene delivery. IEEE Trans Ultrason Ferroelectr Freq Control 59(7):1596–1601. https://doi.org/10.1109/TUFFC.2012.2359

Pochon S, Tardy I, Bussat P, Bettinger T, Brochot J, von Wronski M, Passantino L, Schneider M (2010) BR55: a lipopeptide-based VEGFR2-targeted ultrasound contrast agent for molecular imaging of angiogenesis. Invest Radiol 45(2):89–95. https://doi.org/10.1097/RLI.0b013e3181c5927c

Ragosta M (2015) Left main coronary artery disease: importance, diagnosis, assessment, and management. Curr Prob Cardiology 40(3):93–126. https://doi.org/10.1016/j.cpcardiol.2014.11.003

Regar E, Thury A, van der Giessen WJ, Sianos G, Vos J, Smits PC, Carlier SG, de Feyter P, Foley DP, Serruys PW (2003) Sonotherapy, antirestenotic therapeutic ultrasound in coronary arteries: The first clinical experience. Catheter Cardiovasc Interv 60(1):9–17. https://doi.org/10.1002/ccd.10617

Richardson PD, Davies MJ, Born GV (1989) Influence of plaque configuration and stress distribution on fissuring of coronary atherosclerotic plaques. Lancet 2(8669):941–944

Rodriguez-Granillo GA, McFadden EP, Valgimigli M, van Mieghem CAG, Regar E, de Feyter PJ, Serruys PW (2006) Coronary plaque composition of nonculprit lesions, assessed by in vivo intracoronary ultrasound radio frequency data analysis, is related to clinical presentation. Am Heart J 151(5):1020–1024. https://doi.org/10.1016/j.ahj.2005.06.040

Ruiz EMG, Papaioannou TG, Vavuranakis M, Stefanadis C, Naghavi M, Kakadiaris IA (2012) Analysis of contrast-enhanced intravascular ultrasound images for the assessment of coronary plaque neoangiogenesis: another step closer to the identification of the vulnerable plaque. Curr Pharm Des 18(15):2207–2213

Rychak JJ, Klibanov AL, Ley KF, Hossack JA (2007) Enhanced targeting of ultrasound contrast agents using acoustic radiation force. Ultrasound Med Biol 33(7):1132–1139. https://doi.org/10.1016/j.ultrasmedbio.2007.01.005

Satir S, Degertekin FL (2015) A nonlinear lumped model for ultrasound systems using CMUT arrays. IEEE Trans Ultrason Ferroelectr Freq Control 62(10):1865–1879. https://doi.org/10.1109/TUFFC.2015.007145

Schartl M, Bocksch W, Koschyk DH, Voelker W, Karsch KR, Kreuzer J, Hausmann D, Beckmann S, Gross M (2001) Use of intravascular ultrasound to compare effects of different strategies of lipid-lowering therapy on plaque volume and composition in patients with coronary artery disease. Circulation 104(4):387–392

Seo M, Gorelikov I, Williams R, Matsuura N (2010) Microfluidic assembly of monodisperse, nanoparticle-incorporated perfluorocarbon microbubbles for medical imaging and therapy. Langmuir 26(17):13855–13860. https://doi.org/10.1021/la102272d

Shekhar H, Doyley MM (2013) The response of phospholipid-encapsulated microbubbles to chirp-coded excitation: Implications for high-frequency nonlinear imaging. J Acoust Soc Am 133(5):3145–3158. https://doi.org/10.1121/1.4798677

Shekhar H, Huntzicker S, Awuor I, Doyley MM (2016) Chirp-coded ultraharmonic imaging with a modified clinical intravascular ultrasound system. Ultrason Imaging 38(6):403–419. https://doi.org/10.1177/0161734615618639

Shen C-C, Chou Y-H, Li P-C (2005) Pulse inversion techniques in ultrasonic nonlinear imaging. J Med Ultrasound 13(1):3–17. https://doi.org/10.1016/S0929-6441(09)60073-4

Slager CJ, Wentzel JJ, Schuurbiers JCH, Oomen JAF, Kloet J, Krams R, von Birgelen C, van der Giessen WJ, Serruys PW, de Feyter PJ (2000) True 3-dimensional reconstruction of coronary arteries in patients by fusion of angiography and IVUS (ANGUS) and its quantitative validation. Circulation 102(5):511–516. https://doi.org/10.1161/01.cir.102.5.511

Song Z-Z, Zhang Y-M (2015) Contrast-enhanced ultrasound imaging of the vasa vasorum of carotid artery plaque. World J Radiol 7(6):131–133. https://doi.org/10.4329/wjr.v7.i6.131

Stephens DN, Truong UT, Nikoozadeh A, Oralkan Ö, Seo CH, Cannata J, Dentinger A, Thomenius K, de la Rama A, Nguyen T, Lin F, Khuri-Yakub P, Mahajan A, Shivkumar K, O'Donnell M, Sahn DJ (2012) First in vivo use of a capacitive micromachined ultrasound transducer array-based imaging and ablation catheter. J Ultrasound Med: Off J Am Inst Ultrasound Med 31(2):247–256

Talu E, Hettiarachchi K, Zhao S, Powell RL, Lee AP, Longo L, Dayton PA (2007) Possible method for improving sensitivity in molecular imaging. Molec Imag 6(6):384–392

Thim T, Hagensen MK, Wallace-Bradley D, Granada JF, Kaluza GL, Drouet L, Paaske WP, Bøtker HE, Falk E (2000) Unreliable assessment of necrotic core by virtual histology intravascular ultrasound in porcine coronary artery disease. Circ: Cardiovasc Imag 5(384):391–516. https://doi.org/10.1161/circimaging.109.919357

Unnikrishnan S, Klibanov AL (2012) Microbubbles as ultrasound contrast agents for molecular imaging: preparation and application. AJR Am J Roentgenol 199(2):292–299. https://doi.org/10.2214/AJR.12.8826

van der Steen AFW, Ra Baldewsing, Levent Degertekin F, Emelianov S, Frijlink ME, Furukawa Y, Goertz D, Karaman M, Khuri-Yakub PT, Kim K, Mastik F, Moriya T, Oralkan O, Saijo Y, Ja Schaar, Serruys PW, Sethuraman S, Tanaka A, Vos HJ, Witte R, O'Donnell M (2006) IVUS beyond the horizon. EuroIntervention 2(1):132–142

van Wamel A, Kooiman K, Harteveld M, Emmer M, ten Cate FJ, Versluis M, de Jong N (2006) Vibrating microbubbles poking individual cells: drug transfer into cells via sonoporation. J Control Release: Off J Control Release Soc 112(2):149–155. https://doi.org/10.1016/j.jconrel.2006.02.007

Vavuranakis M, Kakadiaris IA, O'Malley SM, Stefanadis C, Vaina S, Drakopoulou M, Mitropoulos I, Carlier S, Naghavi M (2005) Detection of luminal-intimal border and coronary wall enhancement in intravascular ultrasound imaging after injection of microbubbles and simultaneous sonication with transthoracic echocardiography. Circulation 112(1):e1–e2. https://doi.org/10.1161/circulationaha.104.479915

Vavuranakis M, Kakadiaris IA, O'Malley SM, Papaioannou TG, Sanidas EA, Naghavi M, Carlier S, Tousoulis D, Stefanadis C (2008) A new method for assessment of plaque vulnerability based on vasa vasorum imaging, by using contrast-enhanced intravascular ultrasound and differential image analysis. Int J Cardiol 130(1):23–29. https://doi.org/10.1016/j.ijcard.2007.07.170

Vos HJ, Frijlink ME, Droog E, Goertz DE, Blacquière G, Gisolf A, de Jong N, van der Steen AFW (2005) Transducer for harmonic intravascular ultrasound imaging. IEEE Trans Ultrason Ferroelectr Freq Control 52(12):2418–2422

Wang H-W, Chai N, Wang P, Hu S, Dou W, Umulis D, Wang LV, Sturek M, Lucht R, Cheng J-X (2011) Label-free bond-selective imaging by listening to vibrationally excited molecules. Phys Rev Lett 106(23):238106. https://doi.org/10.1103/PhysRevLett.106.238106

Wang Z, Martin KH, Huang W, Dayton PA, Jiang X (2016) Contrast enhanced superharmonic imaging for acoustic angiography using reduced form-factor lateral mode transmitters for intravascular

and intracavity applications. IEEE Trans Ultrason Ferroelectr Freq Control. https://doi.org/10.1109/tuffc.2016.2619687
Wei K, Jayaweera AR, Firoozan S, Linka A, Skyba DM, Kaul S (1998) Quantification of myocardial blood flow with ultrasound-induced destruction of microbubbles administered as a constant venous infusion. Circulation 97(5):473–483
Wilson K, Homan K, Emelianov S (2012) Biomedical photoacoustics beyond thermal expansion using triggered nanodroplet vaporization for contrast-enhanced imaging. Nat Commun 3:618. https://doi.org/10.1038/ncomms1627
Wong SH, Kupnik M, Watkins RD, Butts-Pauly K, Khuri-Yakub BT (2010) Capacitive micromachined ultrasonic transducers for therapeutic ultrasound applications. IEEE Trans Bio-Med Eng 57(1):114–123. https://doi.org/10.1109/TBME.2009.2026909
Xuefeng Z, Lin D, Oralkan O, Khuri-Yakub B (2008) Fabrication of flexible transducer arrays with through-wafer electrical interconnects based on trench refilling with PDMS. J Microelectromech Syst 17(2):446–452
Yoon S, Williams J, Kang BJ, Yoon C, Cabrera-Munoz N, Jeong JS, Lee SG, Shung KK, Kim HH (2015) Angled-focused 45 MHz PMN-PT single element transducer for intravascular ultrasound imaging. Sens Actuators, A 228:16–22. https://doi.org/10.1016/j.sna.2015.02.037
Yu FTH, Villanueva FS, Chen X (2014) Radial modulation contrast imaging using a 20-MHz single-element intravascular ultrasound catheter. IEEE Trans Ultrason Ferroelectr Freq Control 61(5):779–791. https://doi.org/10.1109/TUFFC.2014.6805692
Zhao H, Xu R, Ouyang Q, Chen L, Dong B, Huihua Y (2010) Contrast-enhanced ultrasound is helpful in the differentiation of malignant and benign breast lesions. Eur J Radiol 73(2):288–293. https://doi.org/10.1016/j.ejrad.2009.05.043

Chapter 11
High-Resolution Ultrasound Imaging System

Weibao Qiu and Hairong Zheng

Introduction

Intravascular ultrasound (IVUS) usually uses ultrasound with center frequency of 20–60 MHz, which is noted down as high frequency ultrasound (Sun et al. 2008a; Silverman et al. 2007), ultrasound biomicroscopy (Vogt and Ermert 2007; Foster et al. 2000) or micro-ultrasound (Stuart Foster et al. 2011; Foster et al. 2009) in ultrasound society. Since the center frequency of ultrasound is higher than traditional ultrasound (2–15 MHz) for high-resolution imaging, the system design is different from the traditional scheme. Separate commercial modules with high performance are usually chosen for research application (Sun et al. 2008b) regardless of their high cost and large size. It would provide a platform for concept verification due to their fast implementations. But for industry, the system would usually consider the cost and integration. Therefore, the design of the electric system should be easy to manufacture and mass produce.

The most widely available systems are using a single element transducer based catheter (Römer et al. 2000; Qiu et al. 2012); therefore, this chapter would only cover the system aspect for rotational IVUS. Array-based system requires multiplexer IC in the catheter which is not included in this chapter. This chapter introduces some design considerations of a high frequency ultrasound system specifically for IVUS applications. Four parts are included in this chapter: pulse generation, echo receiver, basic imaging algorithm, and some advanced IVUS technology developed in recent years.

W. Qiu (✉) · H. Zheng
Paul C. Lauterbur Research Center for Biomedical Imaging, Institute of Biomedical and Health Engineering, Shenzhen Institutes of Advanced Technology, Chinese Academy of Sciences, Shenzhen 518055, China
e-mail: wb.qiu@siat.ac.cn

H. Zheng
e-mail: hr.zheng@siat.ac.cn

Q. Zhou and Z. Chen (eds.), *Multimodality Imaging*,
https://doi.org/10.1007/978-981-10-6307-7_11

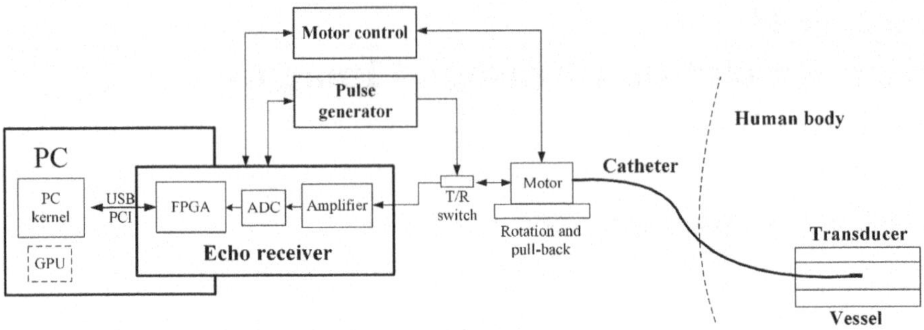

Fig. 11.1 The block diagram of an imaging system for intravascular ultrasound

A typical high frequency ultrasound imaging system for IVUS is shown in Fig. 11.1 (Qiu et al. 2012). A miniaturized ultrasound transducer with a surface size of about 0.6 * 0.6 mm^2 is placed at distal end of the catheter to go inside the vessel (Li et al. 2014; Ma et al. 2015). It is able to transmit and receive an ultrasound signal to acquire a cross-sectional view of the coronary artery. A motor control unit is employed to rotate and pull back the catheter for the imaging (Ma et al. 2015). The catheter, motor control unit, and the electronic system are usually separate parts to compose a whole IVUS system. In terms of an electronic system, a pulse generation and echo receiver are needed for ultrasound imaging. A high-voltage pulse is used to excite the transducer at desired frequency and spectrum to optimize the performance of the transducer (Qiu et al. 2012). There are several pulse types for different imaging strategy, which will be present in the second section of this chapter. As for the echo receiver, signal amplifier, band-specific filter, analog to digital converter (ADC), and processing unit are required (Qiu et al. 2013). Field programmable gate array (FPGA) based digital processor is frequently employed to process the ultrasound echo signal for flexibility and programmability (Qiu et al. 2012). After being processed by the local field processor, the data are transferred to the computer for further investigations and displayed. The scheme of high-speed data transfers could be USB 3.0 and PCIE interface in current transmission technology (Qiu et al. 2012). Graphical user interface (GUI) software is employed to receive and display the IVUS images for real-time visualization. A graphic processing unit (GPU) could be used for the data processing, depending on the algorithms employed. The details of the echo receiver will be presented in the third section. There are several ways to deal with the IVUS imaging, and some basic imaging processes will be introduced in the fourth section. In the end, some advanced technologies from the technical point of view will be presented.

Pulse Generation

In this section, basic requirements of the pulse generation for the excitation of IVUS transducer are introduced. Optimized excitation would increase the efficacy of the excitation. The transducer is a piezoelectric material which requires a short electrical pulse to generate acoustic waveform. The spectrum of the electrical pulse should cover the inherent resonant frequency of the transducer. Several waveform schemes which can be used for IVUS have been listed in Table 11.1. The properties of the electrical pulses are listed to elaborate the performance of each one.

The unipolar pulse is the easiest one to be generated. It is a high-voltage ultra-short unipolar spike, usually a negative unipolar pulse. It has been extensively used due to relatively easy accomplishment and the wide bandwidth for a variety of frequencies (Qiu et al. 2012). It is able to cover from sub-MHz to 100 MHz frequency range with very high peak amplitude (more than 200 V), which has been applied for B-mode medical imaging and non-destructive testing (NDT). The shorter the pulse, the wider the frequency range that can be covered. 10 ns negative pulse could cover to as high as 100 MHz (−3 dB). There are several systems available for the unipolar pulse generation (e.g., Panametrics 5900 pulser/receiver, Olympus NDT Inc., Waltham, MA, USA; UT340, Utex scientific instruments Inc., Mississauga, Ontario, Canada). However, the drawback of this kind of pulse is also obvious that some of the peak energy is concentrated outside of the bandwidth of the transducer, especially at relatively low frequency range, which would slow down the center frequency of the pulse. Unwanted signals and heat may be generated. The efficacy is relatively low in for a unipolar pulse, especially for high-resolution ultrasound imaging.

Bipolar high-voltage pulse is the most widely used waveform in medical ultrasound equipment including the IVUS imaging system (Brown and Lockwood 2002; Xu et al. 2007). The bandwidth of a bipolar pulse can be adjusted to match the bandwidth of the transducer for high energy transmission. A high signal-to-noise ratio (SNR) can then be achieved. For this reason, the center frequency of the bipolar pulse should be tunable for the transducer in the catheter with different center frequencies. There are several implemented methods for the bipolar pulse. Metal-oxide-semiconductor field effect transistor (MOSFET) is usually used for ultra-short high-voltage pulse (Brown and Lockwood 2002; Xu et al. 2007). Paired MOSFET (e.g., TC6320, Supertex Inc., Sunnyvale, CA) with both N- and P-channel is able to provide bipolar pulse. Some parameters should be aware in IVUS system is the timing related value and current related value. Since the center frequency is usually in the range of 20–50 MHz, which is high frequency range. The spectrum of the pulse should be also in this range, which would require the pulse to have a fast transaction time. Rise time, turn-on delay time, fall time, and turn-off delay time are the important parameters for the good waveform. The on-state resistance should be as low as possible to low down the power cost when the pulse is excited. Larger resistance would decrease the current of the pulse and consequently increase the power consumption in the device and low down the driving performance of the pulse. The number of the pulse cycle is also a parameter could be used to adjust the image. Longer pulse

Table 11.1 Electrical pulses for the excitation of the transducer

Pulse type	Waveform	Typical cycle	Imaging mode	Bandwidth range	Typical implementation
Negative pulse (Sun et al. 2008)	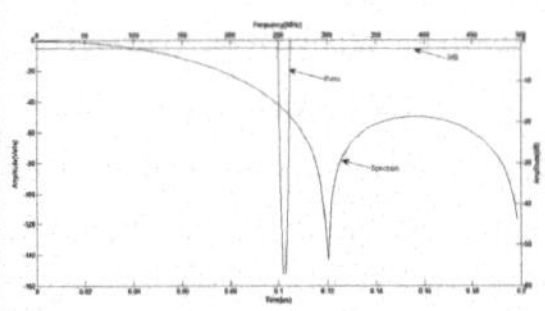	1	B-mode	>100 MHz (pulse length < 10 ns)	N-MOSFET
Quasi-sine wave pulse (Qiu et al. 2012; Xu et al. 2007)	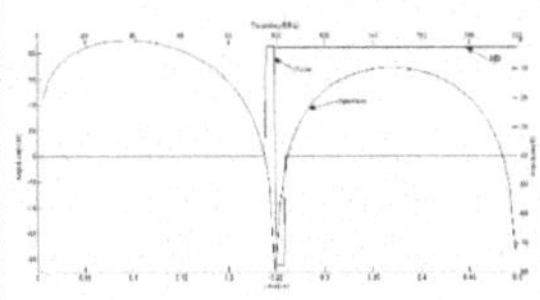	1–2	B-mode	~30 MHz	Paired MOSFET and transformer
	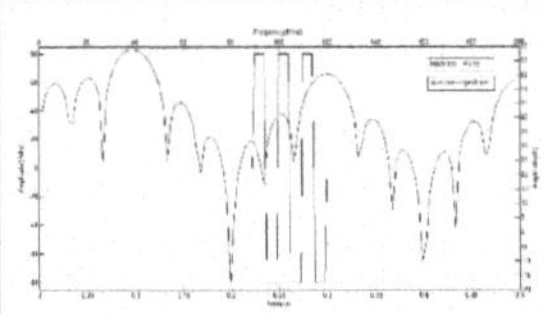	3–7	Flow-mode	~15 MHz	Paired MOSFET and transformer
Sine wave pulse (Qiu et al. 2012)	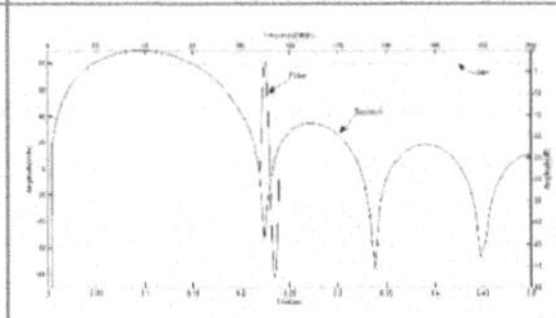	1–2	B-mode	~30 MHz	Power amplifier
	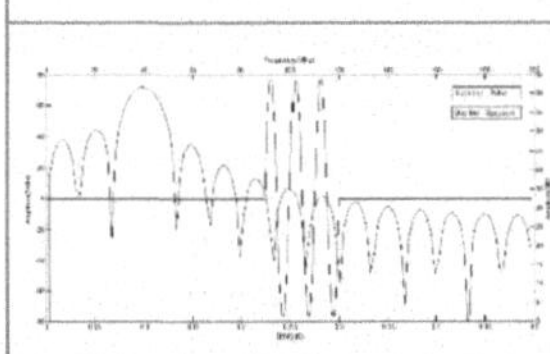	3–7	Flow-mode	~15 MHz	Power amplifier
Arbitrary wave (Qiu et al. 2012)	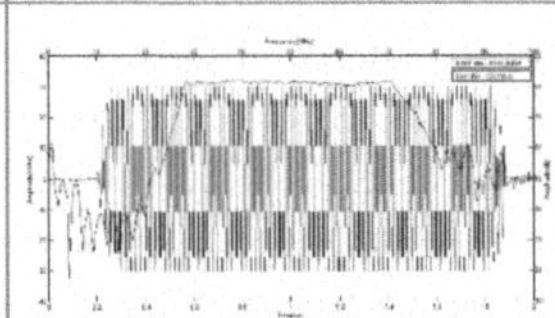	~5 μs	Coded excitation mode	~50 MHz	Power amplifier

could generate more power to penetrate deeper into the tissue. However, the spatial resolution will be sacrificed. It could be used for Doppler measurement (Xu et al. 2007), differentiating the flow from the tissue (Zhou et al. 2007). Although it has been widely applied for a general imaging system, it really depends on the system scheme for flow measurement in the IVUS application. It is a trade-off for the whole system. The bipolar pulse scheme provides higher efficiency than the unipolar pulse (Qiu et al. 2012). However, the amplitude of the bipolar pulse is limited to 150 *V*pp, since

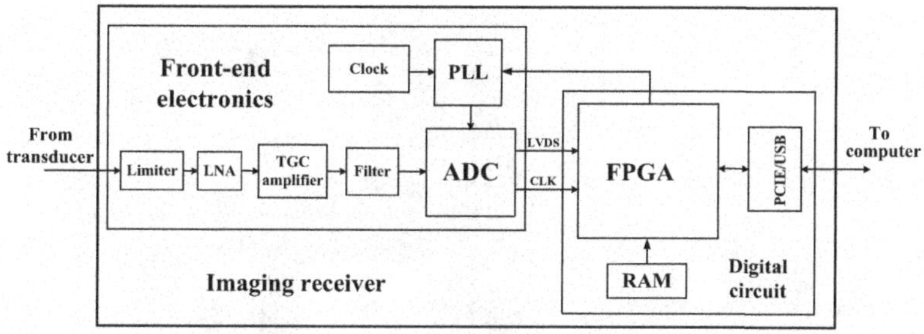

Fig. 11.2 Schematic of the receiver for IVUS imaging

high power performance P-channel MOSFET is usually not as good as N-channel MOSFET.

Although the MOSFET is still popular in an ultrasound system, the shape is not a sinusoidal wave due to its inherent features. We called quasi-sine wave for this method. The spectrum would be better with pure sinusoidal wave for the excitation, and special power amplifier could be designed for this purpose. Single cycle and multiple cycles of sinusoidal waveform could be implemented in this scheme (Qiu et al. 2012, 2013, 2017). The power amplifier for this application should be a linear power amplifier to maintain wide spectrum. There are two things to be aware of when designing this type of power amplifier. Firstly, this power amplifier should have a fast start feature, which makes sure the first pulse could be generated quickly. In addition, the power amplifier should be switched off after the pulse generation. Since the linear power amplifier does have static current to maintain the working state, consequently, there is some noise also generated by the power amplifier. The output of the amplifier is usually connected to the receiver part directly, which would definitely transfer some noises to the receiver. So that, the power amplifier should have a switch-off circuitry inside for a high-resolution imaging. For the more advanced imaging technique, a linear power amplifier can also show its beauty of nature. Arbitrary waveform can be generated for coded excitation to improve the penetration depth of the imaging, and the imaging resolution can be maintained in a good effort (Qiu et al. 2017). Chirp waveform is the typical one for this application.

Receiver Circuitry

The ultrasound echo signal can be produced when ultrasound propagates in the targeted tissue. The echo signal would be collected by the transducer to be converted to electrical signal by piezoelectric effect. A receiver circuit should be designed to receive and process the echo signal, in which Fig. 11.2 shows a typical schematic of such a receiver circuit (Qiu et al. 2012).

Fig. 11.3 Photograph of the integrated PCBs for IVUS imaging

The system could be implemented by separate modules or an integrated PCB. The echo signal received from the catheter is sent to the electronics for data acquisition and image processing. After passing the protection circuitry, signal amplification, and band-pass filtering, the echo signal is converted to a digital signal by an ADC. An FPGA is usually employed as a core processor of the imaging platform, which processes the digitized signal for real-time images. The algorithms implemented in the FPGA can be programmed for different setting. An external random access memory (RAM) device with large capacity and high-speed interface is employed for FPGA to support sophisticated algorithm such as digital scan converter (DSC). High-speed PCIE or USB interface could be used for real-time imaging and fast radio frequency (RF) data transfer. The following sections explain the detailed design for IVUS imaging. Table 11.2 shows the typical electronics requirement for an IVUS system. Figure 11.3 shows some PCBs specifically designed for IVUS data acquisition and processing.

Table 11.2 Electronics requirement for an IVUS system

Items	Performance
Frequency range	10–80 MHz
High-voltage tunable pulse	>100 *V*pp
Adjustable gain	>45 dB
Gain fluctuation	<±1.5 dB
ADC	>11 bits, 150 MSPS
Dynamic range	>45 dB
Data transferring speed	100 MByte/s

Termination and Protection

The transmit path and receive path are connected together for the imaging application in the circuit since there is only one transducer employed in the IVUS. There are two points that need to be paid attention to for this part: electrical matching and high-voltage protection (Choi et al. 2011, 2014a, b). The impedance of the transducer is usually designed to 50 Ω, and therefore, the receiver should be also 50 Ω for the matched transmission. It is better to tune the circuit to 50 Ω to avoid the reflection over ultrasound cable. The reflection may occur if the circuit is mismatched electrically, which will cause a ring down signal. The imaging resolution will be degraded by the elongated ringing signal. Once the impedance of the receiver is matched to the cable impedance, the ringing effect will be minimized.

The protection circuits, also named as limiter, are used to block the high-voltage waveform from the pulse generator (Choi et al. 2014a, b). The high-voltage excitation pulse will get into the receive path which may cause the saturation of ADC. It needs time to be recovered to the normal state from the saturation state. The imaging performance is decreased during this time period. Moreover, the high-voltage waveform may damage the receiver since the amplitude of the waveform will hit the root of the maximal signal of the circuit. There are several methods for the protection circuit including passive circuit and active circuit, which are shown in Fig. 11.4. There is no need for external power for passive circuit, whereas external power or control is needed for active circuit. Resistor diode circuit is the easiest way for the protection; however, the bandwidth is poor and some energy is absorbed during the transmission period. Figure 11.4b shows another passive protection circuit with the use of depletion MOSFETs (Choi et al. 2014b). Figure 11.4c shows an active protection circuit with biased diode bridge.

Low Noise Amplifier and TGC

The quality of IVUS images is significantly influenced by the front-end analog electronics, including the low noise amplifier and time gain compensation (TGC) amplifier. Poor electrical components and circuitry design can increase the noise level and corrupt the weak echo signals. This is particularly important for the IVUS imaging, where the echo signal is extremely weak. The lowest amplitude of the signal is lower than 50 μV, which is not able to be acquired directly by the digital converter. About 45 dB gain (about 180 times amplitude increase) could be applied to the small signal for the data acquisition (Qiu et al. 2017). The noise figure, which is one important parameter for the low noise amplifier, should be taken care for the amplification strategy. The dynamic range of the IVUS image is influenced by the performance of the amplifier. Another rule is to use smaller number of amplifiers because each amplifier can definitely introduce noise. It should be noted that, once the echo signal is distorted in the front-end section, it is very difficult to be recovered by post-processing even

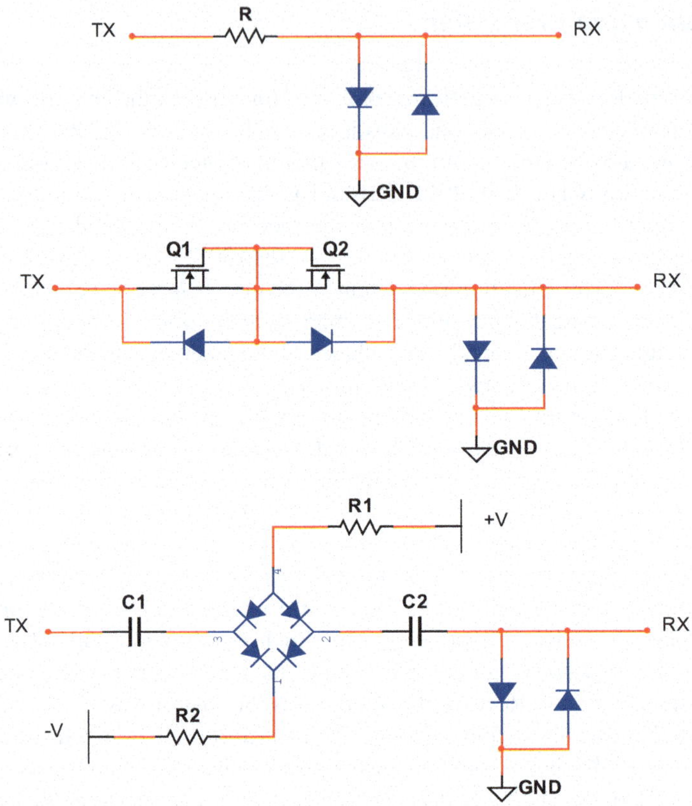

Fig. 11.4 Different protection schemes for IVUS imaging

Table 11.3 Selection of the amplifier for IVUS imaging

Vendor	Brand	Gain (dB)	Bandwidth (MHz)	Noise figure	Cost (USD)
Miteq	1114	32	500	1.1	>200[a]
Tyco electronics	SMA231	26	250	1.7	183.18[b]
ADI	AD8331	43.5	120	4.15	11.45[a]
Texas instruments	THS4509	20	1900	17.1	10.46[a]

[a]Digikey price (acquired from www.digikey.com), [b]Mouser price (acquired from www.mouser.com)

with sophisticated algorithms. Therefore, more attention should be paid to diminish the noise level and to achieve a high sensitivity (Table 11.3).

Analog to Digital Converter

Digital echo signal is needed for the imaging of intravascular vessels. After the amplification of the echo signal, the signal will be digitized by the ADC. The sampling rate of the ADC must be at least double than the maximum frequency of the ultrasonic echo signal to avoid signal distortion based on the Nyquist sampling theorem. The center frequency of ultrasound for single element IVUS imaging is usually around 40 MHz. Assuming the −6 dB bandwidth is about 60%, where the upper frequency range is higher than 52 MHz. Therefore, the sampling frequency should be higher than 104 MHz. Usually, the sampling rate should be higher than 150 MHz for high frequency ultrasound to acquire wider spectrum data. In the early stage, quadrature sampling techniques were employed to minimize the requirement of sampling rate by using two separate ADCs as half of the above requirements (Foster et al. 2002). But the circuitry control is complicated, which may affect the performance of the digitized process. An excitation delay method was proposed recently. By delaying half or a quarter of full cycle of excitation pulse would enable a low sampling frequency (Qiu et al. 2018). However, it may not be suitable for IVUS (mechanical rotation mode) application since the excitation is needed multiple times. Table 11.4 shows the high-speed ADCs which could be used for IVUS imaging.

There are actually several choices for the high-speed ADCs. Attention should be given to the sampling rate, effective number of bits (ENOB), and signal-to-noise ratio (SNR). Main vendors for ADC are Texas Instruments Inc. from Dallas, TX, and Analog Devices, Canton, MA. In addition, a low-jitter clock generator is important for accurate clock source of high precision data acquisition.

A low pass filter or band-pass filter should be employed before the ADC to remove noise before the digitized procedure. Some comments (RLP83+, Mini-Circuits, Brooklyn, NY) can be used for anti-aliasing filter. The insertion loss of the low pass filter should be minimized as low as possible, e.g., less than 1 dB.

Data Process Unit

When the digitization process is complete, the digital signal is transmitted to the FPGA by high-speed interfaces such as low-voltage differential signaling (LVDS)

Table 11.4 High-speed ADCs for IVUS imaging

Vendor	Brand	Speed (MSPS)	Bits	SNR (dBFS)	Cost (USD)[a]
TI	ADS4129	250	12	70.4	56.21
ADI	AD9230	250	12	65.5	98.53
Linear	LTC2152	250	12	68.5	60.17

[a]Digikey price

bus or CMOS bus. There are several data process units in local field such as FPGA, digital signal processor (DSP), ARM, etc. FPGA is a field process unit, which is the most important digital processor unit in medical ultrasound equipment (Qiu et al. 2012). There are plenty of I/O interfaces in a FPGA chip, which enables the chip can be connected to multiple ADCs with a same timing sequence. FPGA is an integrated component which contains a hierarchy of reconfigurable interconnections designed to be programmed with a different code. Reconfigurable properties and high-speed implementation make FPGA very suitable for the design of flexible and programmable system for IVUS application. In ultrasound imaging system, FPGA is employed to support high-speed ADC interface, signal processing, and data transfer to the computer. Table 11.5 shows the commercially available FPGA for IVUS application. There are two vendors named Altera and Xilinx, who are major suppliers for the FPGA components. There are plenty of choices for the selection of FPGA. Several criterions can be used for the selection of an FPGA: IO number, logic element resources, speed level, and special requirement such as LVDS interface, PLL, or DSP block, etc. The designer should consider carefully and simulate the algorithms for the image processing. Complex image processing algorithms definitely require large logic elements and DSP blocks (including multipliers).

Computer Connection

After image processing in the FPGA, the data need to be transferred to a computer for post-processing and display. Different connection methods could be used for IVUS imaging. Table 11.6 shows the features of different connection methods. USB 3.0 now becomes a convenient method for the data transfer. It is able to support more than 300 MB/s data throughput for the imaging applications. PCIE is a stronger transfer strategy for the imaging application. The highest speed is about 15.8 GB/s, which enable a super-speed data transfer. However, the driver firmware is more difficult for both the PC side and FPGA side for the PCIE interface.

Table 11.5 FPGA components for high-resolution ultrasound imaging

Brand	Components	LE (K)	Memory (Mb)	Pin	Cost (USD)[a]
Altera	5CGXFC7D7F31C8N	149.5	7.5	480	249.95
	10AX057H4F34E3SG	570	40.1	492	1795
Xilinx	XC7S50-2FGGA484C	52.2	2.6	250	63.77
	XC7K410T-1FFG900C	406.7	27.9	500	1496

[a]Digikey price

Table 11.6 Connection scheme between FPGA and computer

Interface	Data throughput		Features
USB3.0	>300 MB/s		USB devices can be connected and disconnected at any time
PCIE3.0	X1	984.6 MB/s	Physical PCI Express links may contain from one to 32 lanes
	X16	15.8 GB/s	

Data Process

Previous sections introduce key parts in an imaging circuit for the IVUS application. They are crucial for data acquisition to ensure a low noise level RF data. When the data are transferred to the data process unit, such as FPGA, digital signal processing is required to get an ultrasound image. There are several methods to deal with the algorithms. Usually, the processing algorithms can be divided into two steps: The first step is achieved by local field processing unit, and the second step is programmed in a computer. Generally, the signal processing algorithms before the image do not require large size data storage, which is suitable to be achieved on a FPGA. In contrast, post-image processing algorithms require more data storage, which are suitable to be achieved in a computer. The strategy of the implementation of the imaging algorithms is determined by the designer. Here in this section, the algorithms which are suitable for the implementation in FPGA will be introduced.

As the core processor, the implemented algorithms in FPGA greatly influence the performance of the imaging system. Figure 11.5 shows the structure of the implemented algorithms for real-time grayscale IVUS imaging (Qiu et al. 2012). Digital band-pass filter (BPF), envelope detection, digital scan conversion, and other algorithms can be implemented (Hu et al. 2006; Xu et al. 2008; Zhang et al. 2010). Hardware interface in the FPGA enables the connection of high-speed ADC. The data were filtered first by BPF to remove the noise outside the spectrum. The envelope was then extracted by a detector and then followed by a digital scan converter (DSC) and logarithmic compression. The DSC was employed to convert the polar data to Cartesian coordinates. Finally, image data will be transferred to the computer by high-speed PCIE or USB interface. Other algorithms including digital time gain compensation (DTGC) and data smoothing may also be applicable in the FPGA. DTGC provides that the data are digitally amplified by configurable coefficients to compensate for ultrasound attenuation in the tissue. The following sections give more details on the individual algorithms implemented in the FPGA.

Digital Filter

The echo signal received by the FPGA may contain some noises from electronics, cable, etc., which degrades the signal-to-noise ratio (SNR) of the images. Although an

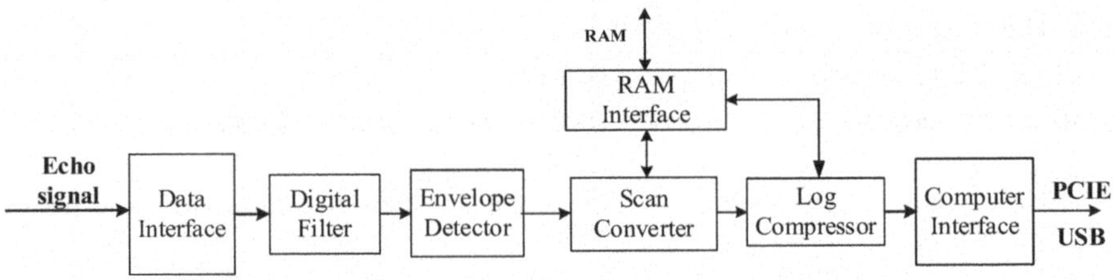

Fig. 11.5 The algorithms implemented in FPGA for real-time IVUS imaging

analog filter has been applied before the ADC, there are still some noises left that get into the digital circuits. A digital filter in the FPGA would eliminate the undesirable noises and improve the SNR further. The easiest way in FPGA for the filter is FIR filter (Qiu et al. 2012). Since FPGA is competent for multiplication and addition, it is very suitable for the FIR calculation. Moreover, the FIR filter coefficients are symmetric; therefore, the number of multipliers can be reduced by half, which saves resources in the FPGA. Some advanced functions such as DSP block in the FPGA can be used to achieve fast processing. Although high order FIR provides better noise suppression, it results in elongated ripples which degraded the imaging resolution. Therefore, it is a trade-off for the selection of SNR improvement and unwanted ripples. Figure 11.6 shows a representative signal processed by a digital 31-tap FIR band-pass filter. More than 40 dB additional noise suppression can be attained with this setting. FPGA also supports high-speed process, which the FIR filter can run in a clock higher than 240 MHz. The coefficients of the digital filter are flexible and reconfigurable during the imaging process.

Envelope Detection

Grayscale B-mode imaging is usually used for the IVUS imaging to represent the backscattering of the ultrasound from the vessel. It requires a positive data for the display. Usually, the envelope is extracted from the filtered ultrasound data. Since ultrasound echo signals are broadband in frequency spectrum, the Hilbert transform could be used to extract the envelope. An architectural example of the envelope detection method is shown in Fig. 11.7. The in-phase (I) and quadrature (Q) signal could be obtained by the Hilbert transform. The modulus of I/Q signals is calculated for the acquisition of envelope by removing the carrier frequency. To fit for FPGA architecture, cordic algorithm could be employed for this action by an iterative process using a series of adders and shifters (Qiu et al. 2012). Simulation with cordic algorithm in the FPGA shows that it can run at a 250 MHz clock, which supports a high throughput process for modulus calculation.

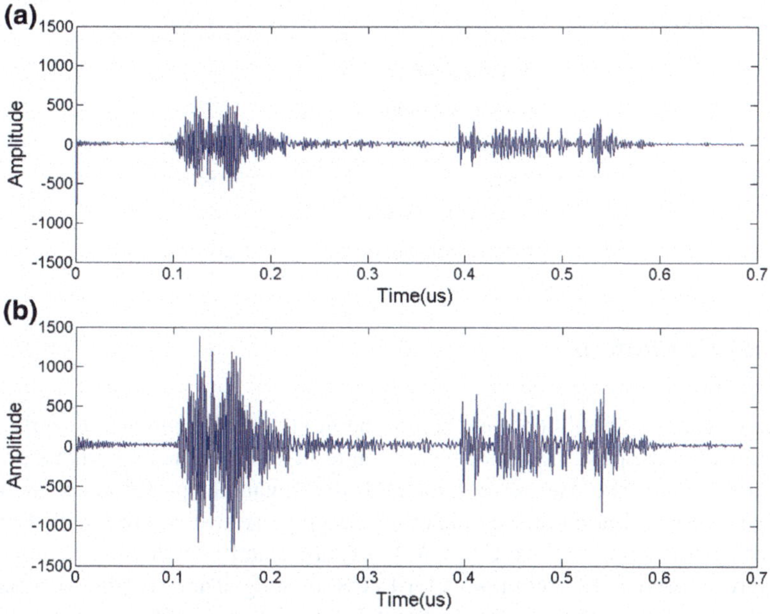

Fig. 11.6 **a** Unprocessed echo waveform and **b** processed waveform with a 31-tap digital filter in the FPGA

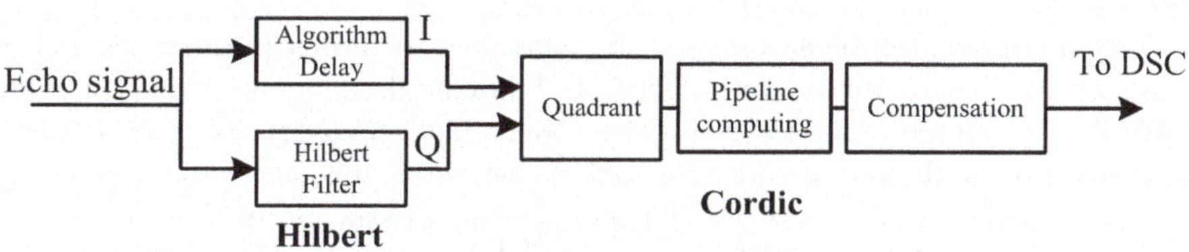

Fig. 11.7 Hilbert transform and cordic algorithm based envelope detector

Digital Scan Convertor

A single element transducer is usually employed for IVUS imaging, which requires a mechanical rotation of the transducer to acquire a sectional image. Therefore, the ultrasound signals are obtained at different angles (~500 angles), which restores as a polar coordinate. On the other hand, the screen for the imaging is presented with Cartesian coordinates. So that, a conversion named as digital scan convertor (DSC) is required to transform the polar coordinate's data to Cartesian coordinate's data (Chang et al. 2008; Levesque and Sawan 2009). Figure 11.8 demonstrates the principle of DSC. C-R represents the Cartesian coordinate and *P* represents the Polar coordinate. The value in pixel *S* can be calculated by relevant P_{xi} and P_{yi}. Linear interpolation can be achieved in the FPGA (Qiu et al. 2012).

Table 11.7 Evaluation items and reference value

Articles	Referred value
Axial resolution	~50 μm
Lateral resolution	~150 μm
Dynamic range	>50 dB
CNR	>5
Imaging frame rate	>20
Penetration depth	>5 mm

Imaging Evaluation

Imaging evaluation should be done before the in vivo study, which is useful for the improvement of the imaging performance. Table 11.7 shows the imaging parameters and referred value. There are several methods to evaluate the performance including wire phantom imaging and tissue phantom imaging. Tungsten wires with 12.5 μm diameter (California Fine Wire Co., CA, USA) are usually employed to evaluate the imaging resolution at different points for IVUS imaging. They are placed at specific positions, showing the imaging resolution at those points. Quantitative measurements of the axial and lateral resolution can be achieved by measuring the full width at half maximum (FWHM) of the wire targets (Brown and Lockwood 2005; Ketterling et al. 2006).

A tissue mimicking phantom could be fabricated to further evaluate the IVUS image quality. There are several methods to fabricate the phantom (Madsen et al. 2010). It may contain a mixture of deionized water, high-grade agarose, preservative, propylene glycol, filtered bovine milk, and glass-bead to generate tissue mimicking attenuation and backscattering. This phantom provides tissue mimicking attenuation of about −32 dB/cm for 40 MHz ultrasound and backscattering to test the imaging

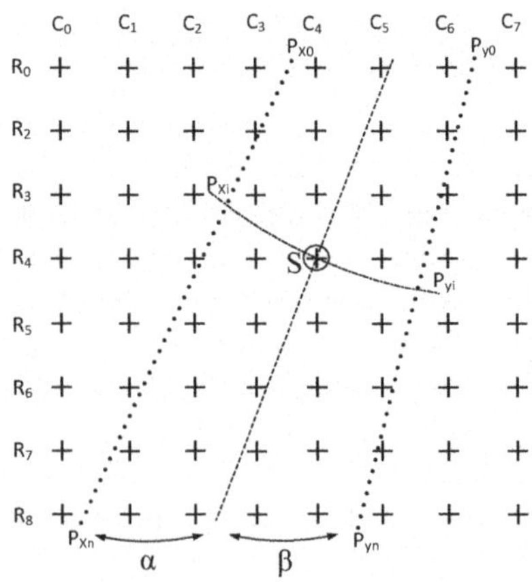

Fig. 11.8 Block diagram of digital scan converter implemented in FPGA with linear interpolation

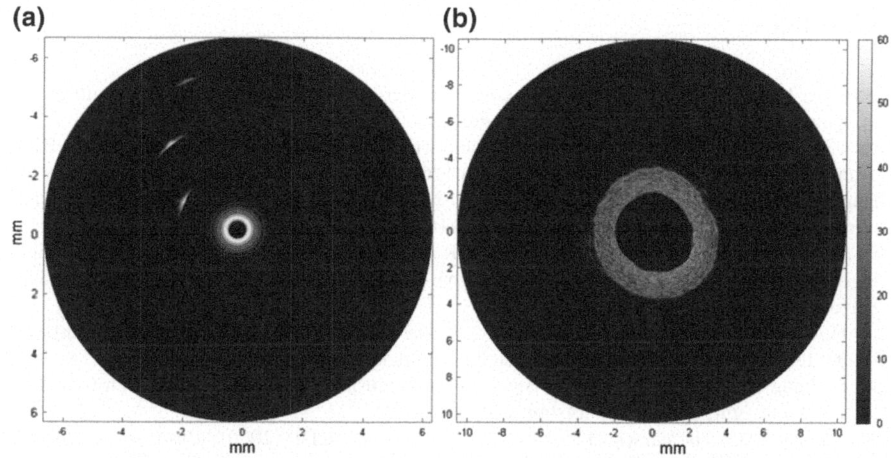

Fig. 11.9 Phantom evaluation of the ultrasound system **a** Tungsten wire phantom image, **b** Image of tissue mimicking phantom

resolution and penetration depth. Some anechoic holes are made inside the phantom to evaluate the value of contrast-to-noise ratio (CNR), which offers an indirect characterization of spatial resolution in all directions simultaneously (Mamou et al. 2009; Lediju et al. 2011). The CNR was calculated as:

$$\mathrm{CNR} = \frac{|\mathrm{mean}_t - \mathrm{mean}_n|}{\sqrt{\mathrm{sta}_t^2 + \mathrm{sta}_n^2}}$$

where mean_t and mean_n represent the mean backscatter amplitude of the phantom and the anechoic hole. sta_t and sta_n represent the corresponding standard deviations.

Figure 11.9a shows the wire phantom image. The axial and lateral resolutions in the third wire are 57.8 and 181.6 μm, respectively. The dynamic range of the images is set to 50 dB. Figure 11.9b presents the tissue phantom image. High-resolution ultrasound can provide a certain imaging depth.

Summary

This chapter provides a basic introduction to the IVUS system. Some essential parts including pulse generation, echo receiver, and image processing algorithms are introduced. In addition, the evaluation method is also introduced in the last session. The contents in this chapter can be useful for designing a system for IVUS application.

References

Brown JA, Lockwood GR (2002) A low-cost, high-performance pulse generator for ultrasound imaging. IEEE Trans Ultrason Ferroelectr Freq Control 49(6):848–851

Brown JA, Lockwood GR (2005) A digital beamformer for high-frequency annular arrays. IEEE Trans Ultrason Ferroelectr Freq Control 52(8):1262–1269

Chang JH, Yen JT, Kirk Shung K (2008) High-speed digital scan converter for high-frequency ultrasound sector scanners. Ultrasonics 48(5):444–452

Choi H, Li X, Lau S-T, Hu C, Zhou Q, Shung KK (2011) Development of integrated preamplifier for high-frequency ultrasonic transducers and low-power handheld receiver. IEEE Trans Ultrason Ferroelect Freq Control 58(12):2646–2658

Choi H, Kim MG, Cummins TM, Hwang JY, Shung KK (2014a) Power MOSFET-diode-based limiter for high-frequency ultrasound systems. Ultrason Imaging 36(4):317–330

Choi H, Yang HC, Shung KK (2014) Bipolar-power-transistor-based limiter for high frequency ultrasound imaging systems. Ultrasonics 54(3):754–758

Foster FS, Pavlin CJ, Harasiewicz KA, Christopher DA, Turnbull DH (2000) Advances in ultrasound biomicroscopy. Ultrasound Med Biol 26(1):1–27

Foster FS, Zhang MY, Zhou YQ, Liu G, Mehi J, Cherin E, Harasiewicz KA, Starkoski BG, Zan L, Knapik DA, Adamson SL (2002) A new ultrasound instrument for In vivo microimaging of mice. Ultrasound Med Biol 28(9):1165–1172

Foster FS, Mehi J, Lukacs M, Hirson D, White C, Chaggares C, Needles A (2009) A new 15–50 MHz array-based micro-ultrasound scanner for preclinical imaging. Ultrasound Med Biol 35(10):1700–1708

Hu C, Xu X, Cannata JM, Yen JT, Shung KK (2006) Development of a real-time, high-frequency ultrasound digital beamformer for high-frequency linear array transducers. IEEE Trans Ultrason Ferroelectr Freq Control 53(2):317–323

Ketterling JA, Ramachandran S, Aristizabal O (2006) Operational verification of a 40-MHz annular array transducer. IEEE Trans Ultrason Ferroelectr Freq Control 53(3):623–630

Lediju MA, Trahey GE, Byram BC, Dahl JJ (2011) Short-lag spatial coherence of backscattered echoes: imaging characteristics. IEEE Trans Ultrason Ferroelectr Freq Control 58(7):1377–1388

Levesque P, Sawan M (2009) Real-time hand-held ultrasound medical-imaging device based on a new digital quadrature demodulation processor. IEEE Trans Ultrason Ferroelectr Freq Control 56(8):1654–1665

Li X, Ma T, Tian J, Han P, Zhou Q, Shung KK (2014) Micromachined PIN-PMN-PT crystal composite transducer for high-frequency intravascular ultrasound (IVUS) imaging. IEEE Trans Ultrason Ferroelectr Freq Control 61(7):1171–1178

Ma J, Martin KH, Li Y, Dayton PA, Shung KK, Zhou Q, Jiang X (2015a) Design factors of intravascular dual frequency transducers for super-harmonic contrast imaging and acoustic angiography. Phys Med Biol 60(9):3441–3457

Ma T, Yu M, Li J, Munding CE, Chen Z, Fei C, Shung KK, Zhou Q (2015b) Multi-frequency intravascular ultrasound (IVUS) imaging. IEEE Trans Ultrason Ferroelectr Freq Control 62(3):97–107

Madsen EL, Frank GR, McCormick MM, Deaner ME, Stiles TA (2010) Anechoic sphere phantoms for estimating 3-D resolution of very-high-frequency ultrasound scanners. IEEE Trans Ultrason Ferroelectr Freq Control 57(10):2284–2292

Mamou J, Aristizábal O, Silverman RH, Ketterling JA, Turnbull DH (2009) High-frequency chirp ultrasound imaging with an annular array for ophthalmologic and small-animal imaging. Ultrasound Med Biol 35(7):1198–1208

Qiu W, Chen Y, Li X, Yu Y, Cheng WF, Tsang FK, Zhou Q, Shung KK, Dai J, Sun L (2012a) An open system for intravascular ultrasound imaging. IEEE Trans Ultrason Ferroelectr Freq Control 59(10):2201–2209

Qiu W, Yu Y, Tsang FK, Sun L (2012b) A multi-functional, reconfigurable pulse generator for high frequency ultrasound imaging. IEEE Trans Ultrason Ferroelectr Freq Control 59(7):1432–1442

Qiu W, Yu Y, Tsang FK, Sun L (2012c) An FPGA based open platform for ultrasound biomicroscopy. IEEE Trans Ultrason Ferroelectr Freq Control 59(7):1558–1567

Qiu W, Yu Y, Chabok HR, Liu C, Tsang FK, Zhou Q, Shung KK, Zheng H, Sun L (2013a) A flexible annular array imaging platform for micro-ultrasound. IEEE Trans Ultrason Ferroelectr Freq Control 60(1):178–186

Qiu W, Yu Y, Tsang FK, Zheng H, Sun L (2013b) A novel modulated excitation imaging system for micro-ultrasound. IEEE Trans Biomed Eng 60(7):1884–1890

Qiu W, Wang X, Chen Y, Fu Q, Su M, Zhang L, Xia J, Dai J, Zhang Y, Zheng H (2017) Modulated excitation imaging system for intravascular ultrasound. IEEE Trans Biomed Eng 64(8):1935–1942

Qiu W, Xia J, Shi Y, Mu P, Wang X, Gao M, Wang C, Xiao Y, Yang G, Liu J, Sun L, Zheng H (2018) A delayed-excitation data acquisition method for high-frequency ultrasound imaging. IEEE Trans Biomed Eng 65(1):15–20

Römer TJ, Brennan JF III, Puppels GJ, Zwinderman AH, van Duinen SG, van der Laarse A, van der Steen AF, Bom NA, Bruschke AV (2000) Intravascular ultrasound combined with Raman spectroscopy to localize and quantify cholesterol and calcium salts in atherosclerotic coronary arteries. Arterioscler Thromb Vasc Biol 20(2):478–483

Silverman RH, Ketterling JA, Coleman DJ (2007) High-frequency ultrasonic imaging of the anterior segment using an annular array transducer. Ophthalmology 114(4):816–822

Stuart Foster F, Hossack J, Lee Adamson S (2011) Micro-ultrasound for preclinical imaging. Interface Focus 1:576–601

Sun L, Lien C, Xu X, Shung KK (2008a) In vivo cardiac imaging of adult zebrafish using high frequency ultrasound (45–75 MHz). Ultrasound Med Biol 34(1):31–39

Sun L, Xu X, Richard WD, Feng C, Johnson JA, Shung KK (2008b) A high-frame rate duplex ultrasound biomicroscopy for small animal imaging in vivo. IEEE Trans Biomed Eng 55(8):2039–2049

Vogt M, Ermert H (2007) In vivo ultrasound biomicroscopy of skin: spectral system characteristics and inverse filtering optimization. IEEE Trans Ultrason Ferroelectr Freq Control 54(8):1551–1559

Xu X, Yen JT, Shung KK (2007) A low-cost bipolar pulse generator for high frequency ultrasound applications. IEEE Trans Ultrason Ferroelectr Freq Control 54(2):443–447

Xu X, Sun L, Cannata JM, Yen JT, Shung KK (2008) High-frequency ultrasound doppler system for biomedical applications with a 30-MHz linear array. Ultrasound Med Biol 34(4):638–646

Zhang L, Xu X, Hu C, Sun L, Yen JT, Cannata JM, Shung KK (2010) A high-frequency, high frame rate duplex ultrasound linear array imaging system for small animal imaging. IEEE Trans Ultrason Ferroelectr Freq Control 57(7):1548–1557

Zhou Q, Xu X, Gottlieb EJ, Sun L, Cannata JM, Ameri H, Humayun MS, Han P, Shung KK (2007) PMN-PT single crystal, high-frequency ultrasonic needle transducers for pulsed-wave Doppler application. IEEE Trans Ultrason Ferroelectr Freq Control 54(3):668–675

The manufacturer's authorised representative in the EU is Springer Nature Customer Service Centre GmbH, Europaplatz 3, 69115 Heidelberg, Germany. If you have any concerns regarding our products, please contact ProductSafety@springernature.com

Printed and bound by CPI Group (UK) Ltd, Croydon, CR0 4YY
25/09/2026
02222900-0012